Advances in Industrial Control

Other titles published in this Series:

Radiotherapy Treatment Planning
Olivier Haas

Performance Assessment of Control Loops
Biao Huang and Sirish L. Shah

Data-driven Techniques for Fault Detection and Diagnosis in Chemical Processes
Evan L. Russell, Leo H. Chiang and Richard D. Braatz

Non-linear Model-based Process Control
Rashid M. Ansari and Moses O. Tadé

Nonlinear Identification and Control
Guoping Liu

Digital Controller Implementation and Fragility
Robert S.H. Istepanian and James F. Whidborne (Eds.)

Optimisation of Industrial Processes at Supervisory Level
Doris Sáez, Aldo Cipriano and Andrzej W. Ordys

Applied Predictive Control
Huang Sunan, Tan Kok Kiong and Lee Tong Heng

Hard Disk Drive Servo Systems
Ben M. Chen, Tong H. Lee and Venkatakrishnan Venkataramanan

Robust Control of Diesel Ship Propulsion
Nikolaos Xiros

Model-based Fault Diagnosis in Dynamic Systems Using Identification Techniques
Silvio Simani, Cesare Fantuzzi and Ron J. Patton

Strategies for Feedback Linearisation
Freddy Garces, Victor M. Becerra, Chandrasekhar Kambhampati and Kevin Warwick

Robust Autonomous Guidance
Alberto Isidori, Lorenzo Marconi and Andrea Serrani

Dynamic Modelling of Gas Turbines
Gennady G. Kulikov and Haydn A. Thompson (Eds.)

Control of Fuel Cell Power Systems
Jay T. Pukrushpan, Anna G. Stefanopoulou and Huei Peng

Optimal Real-time Control of Sewer Networks
Magdalene Marinaki and Markos Papageorgiou
Publication due December 2004

Adaptive Voltage Control in Power Systems
Giuseppe Fusco and Mario Russo
Publication due August 2005

Jairo Espinosa, Joos Vandewalle and
Vincent Wertz

Fuzzy Logic, Identification and Predictive Control

With 138 Figures

Springer

Jairo Espinosa, Ph.D. Eng.M.Sc.
IPCOS Belgium, Technologielaan 11-0101, B-3001 Heverlee (Leuven), Belgium

Joos Vandewalle, Prof.Dr.Ir.
ESAT, Faculty of Engineering, KU Leuven, Kasteelpark Arenberg 10,
3001 Heverlee (Leuven), Belgium

Vincent Wertz, Prof.Dr.Ir.
CESAME, Université Catholique de Louvain, av. Georges Lemaître 4,
B-1348 Louvain-la-Neuve, Belgium

British Library Cataloguing in Publication Data
Espinosa, Jairo
Fuzzy logic, identification and predictive control. —
(Advances in industrial control)
1. Automatic control 2. Predictive control 3. Fuzzy logic
4. Data mining
I. Title II. Vandewalle, Joos III. Wertz, Vincent
629.8

Library of Congress Cataloging-in-Publication Data
Espinosa, Jairo.
Fuzzy logic, identification and predictive control / Jairo Espinosa, Joos Vandewalle,
Vincent Wertz.
p. cm — (Advances in industrial control)
Includes bibliographical references and index.

1. Predictive control. 2. Fuzzy systems. I. Vandewalle, J. (Joos), 1948– II. Wertz,
Vincent. III. Title. IV. Series.
TJ217.6.E87 2004
629.8—dc22 2004046864

Advances in Industrial Control series ISSN 1430-9491
ISBN 978-1-84996-931-4 e-ISBN 978-1-84628-087-0
Springer Science+Business Media
springeronline.com

Printed in the United States of America

69/3830-543210 Printed on acid-free paper

Advances in Industrial Control

Professor Emeritus O.P. Malik
Department of Electrical and Computer Engineering
University of Calgary
2500, University Drive, NW
Calgary
Alberta
T2N 1N4
Canada

Professor K.-F. Man
Electronic Engineering Department
City University of Hong Kong
Tat Chee Avenue
Kowloon
Hong Kong

Professor G. Olsson
Department of Industrial Electrical Engineering and Automation
Lund Institute of Technology
Box 118
S-221 00 Lund
Sweden

Professor A. Ray
Pennsylvania State University
Department of Mechanical Engineering
0329 Reber Building
University Park
PA 16802
USA

Professor D.E. Seborg
Chemical Engineering
3335 Engineering II
University of California Santa Barbara
Santa Barbara
CA 93106
USA

Doctor I. Yamamoto
Technical Headquarters
Nagasaki Research & Development Center
Mitsubishi Heavy Industries Ltd
5-717-1, Fukahori-Machi
Nagasaki 851-0392
Japan

A mi familia y mis hijos

Series Editors' Foreword

The series *Advances in Industrial Control* aims to report and encourage technology transfer in control engineering. The rapid development of control technology has an impact on all areas of the control discipline. New theory, new controllers, actuators, sensors, new industrial processes, computer methods, new applications, new philosophies . . . , new challenges. Much of this development work resides in industrial reports, feasibility study papers and the reports of advanced collaborative projects. The series offers an opportunity for researchers to present an extended exposition of such new work in all aspects of industrial control for wider and rapid dissemination.

It seems quite surprising to realise that the notion of fuzzy sets and all the related concepts and applications have been developing now for nearly forty years! A theory which was regarded by some as problematic has survived into various successful industrial applications. The theory itself has continued to develop as this excellent and insightful monograph by authors Jairo Espinosa, Joos Vandewalle and Vincert Wertz shows. The presentation is almost equally balanced between fuzzy models and fuzzy control. The already extensive results presented are also further supported by nearly forty pages of proofs, explanations and illustrative examples in the Appendices.

The first part of the monograph is comprised of four chapters on different aspects of and developments in fuzzy modelling. An important point to note here is that the modelling approaches presented often have a wider applications interest than just for the control community.

From experience, when the material to be presented is a little difficult, illustration by typical examples can be very useful. All through the monograph the authors have introduced academic and industrial examples to provide the necessary insight. In the modelling chapters these examples include a heat exchanger system, a reactor process and data from a gas furnace installation.

The second part of the book concentrates on fuzzy control. This part opens with a very interesting discussion of PID-like fuzzy controllers and proceeds to demonstrate that under some reasonably relaxed conditions "any linear controller . . . can be made into an exactly equivalent fuzzy controller" and the proof is presented in one of the appendices for completeness. This type of result characterizes the work presented in these three chapters where useful and insightful presentations are interwoven with interesting and important results. The final chapter of the monograph presents some well targeted discussion of successes and areas for further work within the fuzzy systems and control paradigm. Researchers and postgraduate students will find this chapter full of stimulating challenges and unsolved research issues.

An interesting feature of this monograph is the bold attempt to produce repeatable engineering procedures that make the engineering application of fuzzy methods in identification and control much more accessible and therefore more valuable. We believe this volume to be a fine contribution to the *Advances in Industrial Control* series and believe it will find a wide readership within the control community.

M.J. Grimble and M.A. Johnson
Industrial Control Centre
Glasgow, Scotland, U.K.

Preface

Since the idea of the fuzzy set was proposed in 1965, many developments have occurred in this area. Applications have been made in such diverse areas as medicine, engineering, management, behavioral science, just to mention some. The application of the fuzzy sets involves different technologies, such as fuzzy clustering on image processing, classification, identification and fault detection, fuzzy controllers to map expert knowledge into control systems, fuzzy modeling combining expert knowledge, fuzzy optimization to solve design problems.

Fuzzy systems are used in the area of artificial intelligence as a way to represent knowledge. This representation belongs to the paradigm of behavioral representation in opposition to the structural representation (neural networks). The foundation of this paradigm is that intelligent behavior can be obtained by the use of structures that not necessarily resemble the human brain.

A very interesting characteristic of the fuzzy systems is their capability to handle in the same framework numeric and linguistic information. This characteristic made these systems very useful to handle expert control tasks.

This book is divided in two parts. The first part is devoted to the construction of static and dynamic fuzzy models from numerical information. Such models are important in areas such as data mining and control of dynamical systems. The second part shows how to exploit these models to design control systems. The book is organized into 8 chapters and 5 appendices.

Chapter 1 is entirely dedicated to the problem of function approximation and modeling. The first part of the chapter shows the approximation capabilities of fuzzy systems with triangular, polynomial and Gaussian membership functions. In this part, this book presents an analytical study of the approximation capabilities of the different types of membership functions.

Chapter 2 describes different techniques to construct fuzzy models from input–output data. Gradient descent techniques, clustering and evolutionary techniques are explained in this chapter. Some gradient expressions are derived to illustrate the expressions used to adjust the fuzzy models. This section is complemented with the Appendix B where the main clustering techniques used in modeling are explained. In the second part of the chapter the problems of generalization and consequence estimation are studied. An initialization and training method for the consequences is one of the highlights of this chapter. The method improves the generalization capabilities of the fuzzy models. It is illustrated by means of a graphical example.

Chapter 3 introduces the concept of linguistic integrity and presents an algorithm to build fuzzy models with linguistic meaning. This algorithm is especially important in tasks such as knowledge discovery and data mining. The contributed algorithm is named AFRELI (Autonomous Fuzzy Rule Extractor with Linguistic Integrity). This algorithm is complemented with an algorithm to reduce the complexity of the fuzzy models (FuZion algorithm). Several examples are presented where complex nonlinear functions, chaotic nonlinear systems and industrial processes are modeled using this algorithm.

Chapter 4 is devoted to the problem of nonlinear identification of dynamic systems using fuzzy models. In this chapter the tools developed in previous chapters are used to develop a framework for system identification using fuzzy models. The chapter begins by formulating the problem of system identification using fuzzy models. The chapter includes an analysis of the structure of the fuzzy models, which are more suitable to be applied in system identification. Thereafter the chapter studies the problem of experiment design and proposes types of signals that are considered to deliver "sufficient excitation" to guarantee the reliable construction of the model. The regressors selection is considered a complex problem also analyzed in this book. The main methods are reviewed and the advantages and disadvantages of the methods are analyzed. The possible structures of the models for nonlinear systems are enumerated and a short analysis of the applicability of some of these structures to identification using fuzzy models is included. Parameter calculations for different type of structures are studied under the assumption that gradient descent methods are used in this calculation. The issue of dynamic calculation of the gradients is emphasized, and Appendix C includes the derived expressions of these gradients. The chapter closes with a short discussion of the validation issues. This discussion points out the fact that fuzzy models can be validated using not only quantitative criteria but also qualitative criteria based on the information given by the linguistic rules. The chapter presents an example of identification of a gas furnace process proposed by Box and Jenkins [1]. In this example, most of the elements presented in the chapter are included.

Chapter 5 presents an overview of different techniques that have been designed to construct fuzzy controllers. The chapter starts by making a classification of the different methods where fuzzy sets are applied to control. The chapter explains some of the methods starting with the first early ideas employing pure expert knowledge. Then the discussion focuses on a very practical method, which is the design of PID-like fuzzy controllers. This section is complemented with the theorem presented in Appendix D. This contributed theorem guarantees that any discrete linear controller can be copied exactly by a fuzzy system. This property is exploited to initialize fuzzy controllers such that their initial performance (before tuning) at least equals the performance of a given linear controller. Then, the overview presented in this chapter focus on the adaptive control techniques based on fuzzy models. Inverse learning and direct learning are the two methods studied in this section. The chapter then pays attention to methods based on direct synthesis, and the method of feedback linearization is proposed where the models used to linearize the affine system are fuzzy models. The main drawbacks of this method are analyzed and the sliding mode fuzzy control method is explained as an alternative. Finally, the chapter is completed by a description of the fuzzy gain scheduling method. The advantages of the method, including the existence of methods to directly design stable controllers and test controllers for stability, are studied and discussed. The section is complemented with an industrial example presented in Appendix E where a stable fuzzy scheduling controller is designed for an automotive application a continuous variable transmission (CVT) system.

Chapters 6 and 7 are devoted to the construction of predictive controllers based on fuzzy models. Chapter 6 begins with the simplest idea of unconstrained predictive fuzzy control. The problem is formulated and a method to reduce the problem to a quadratic program is presented. The method includes the formulation of a predictor based on the concept of free and forced response; the estimation of the forced response is improved by a method proposed in this chapter. The chapter includes an application example where the control strategy is applied to a continuous stirred tank reactor (CSTR). In this example, the strategy is shown to perform quite similarly to the most optimal strategy. The second part of Chapter 6 studies the constrained predictive control problem. This part shows three different algorithms that exploit the information provided by the different types of dynamic fuzzy models. The chapter closes with an example where the control methods are applied to a steam generator model of a power plant and a gas-phase polymerization reactor for the production of polyethylene. The strategies are compared with classical linear predictive control strategies, and the improvement in performance can be clearly observed.

Chapter 7 presents a novel extension to the concepts presented in Chapter 6. This chapter, studies the problem of robust nonlinear predictive control based on fuzzy models. The chapter begins with the formulation of the problem. Then the problem is reduced to a robust quadratic program. The robust quadratic program is written as a second-order cone program using a new formulation presented in this book. Advantages of the use of these algorithms are mentioned including the computational complexity. The use of this robust optimization technique guarantees a minimum performance despite the mismatch of the models or the linearization errors introduced by the control algorithm. The chapter is completed with a list of possible ways to describe the uncertainty of the models.

Finally, Chapter 8 condenses the main contributions of the book and proposes new challenges for researchers.

The book also includes five appendices. The first two appendices show the fundamentals of fuzzy set operations and clustering. The last three are extensions to the content of the chapters.

The reader will be guided by summary boxes that contain the main ideas of the different sections, facilitating the comprehension and the goals of each section of the book.

The book is written at a level suitable for use in a graduate course on applications of fuzzy systems in data mining and nonlinear modeling and control. The book discusses novel ideas and provides a new insight into the studied topics. For this reason, the book is a valuable source for researchers in the areas of artificial intelligence, data mining, modeling and control. The realistic examples also provide a good opportunity to people in industry to evaluate these new technologies, which have been applied with success especially in the areas of monitoring and control of chemical processes and in oil exploitation.

Leuven, Belgium
January 2004

Jairo Espinosa
Joos Vandewalle
Vincent Wertz

Contents

Part I Fuzzy Modeling

1 Fuzzy Modeling 3
1.1 Function Approximation 4
1.1.1 System Description 4
1.1.2 Approximation Error 6
1.1.3 Constructing Units in the Fuzzy Models 8
1.2 Approximation Capabilities of Takagi–Sugeno Fuzzy Models 18
1.3 Conclusion and Summary 20

2 Constructing Fuzzy Models from Input-Output Data 21
2.1 Mosaic or Table Lookup Scheme 22
2.1.1 Illustrative Example 25
2.2 Using Gradient Descent 26
2.2.1 Gradient Updating for Trapezoidal Membership Functions 30
2.2.2 Gradient Updating for Triangular Membership Functions with Overlap $\frac{1}{2}$ 31
2.2.3 Gradient Updating for Polynomial Membership Functions 32
2.2.4 Gradient Updating for Polynomial Membership Functions with Overlap $\frac{1}{2}$ and $b_j^i = c_j^i = m_j^i$ 33
2.2.5 Gradient Updating for Gaussian Membership Functions 34
2.2.6 Illustrative Example 34
2.3 Using Clustering and Gradient Descent 38
2.3.1 Algorithm for Mamdani Models 38
2.3.2 Algorithm for Takagi–Sugeno Models 39
2.3.3 Illustrative Example 39
2.4 Using Evolutionary Strategies 46

2.5 Generalization and Consequences Estimation 50
2.5.1 Consequence Initialization 50
2.5.2 Consequence Estimation 51
2.6 Example of an Industrial Application 55
2.7 Conclusions 58

3 Fuzzy Modeling with Linguistic Integrity: A Tool for Data Mining 59
3.1 Introduction 59
3.2 Structure of the Fuzzy Model 61
3.3 The AFRELI Algorithm 63
3.4 The FuZion Algorithm 68
3.5 Examples 71
3.5.1 Modeling a Two-Input Nonlinear Function 71
3.5.2 Modeling of a Three-Input Nonlinear Function 77
3.5.3 Predicting Chaotic Time Series 80
3.5.4 Modeling of the Quality Properties on a High-Density Polyethylene (HDPE) Reactor 85
3.6 Complexity of the AFRELI Algorithm 89
3.7 Conclusions 89

4 Nonlinear Identification Using Fuzzy Models 91
4.1 System Identification 92
4.2 Basic Structure of the Fuzzy System 93
4.3 Experiment Design for System Identification 95
4.4 Choosing the Regressors 98
4.4.1 Search Methods 98
4.4.2 Regressors Evaluation 101
4.5 Choosing the Structure 105
4.6 Calculating the Parameters 106
4.7 Validation 108
4.8 Example: Identification of the Box and Jenkins Gas Furnace Data Set 109
4.9 Identification of Takagi–Sugeno Fuzzy Models Using Local Linear Identification 118
4.10 Conclusions 119

Part II Fuzzy Control

5 Fuzzy Control 123
5.1 Model-Free Fuzzy Control 124
5.1.1 Heuristic Trial-and-Error Design 124
5.1.2 Design of PID-like Fuzzy Controllers 124
5.2 Model Based Fuzzy Control 129

5.2.1 Using Adaptive Methods ... 129
5.2.2 Using Direct Synthesis ... 134
5.3 Conclusions and Future Perspectives ... 146

6 Predictive Control Based on Fuzzy Models ... 151
6.1 The Predictive Control Strategy ... 152
6.2 Unconstrained Nonlinear Predictive Control ... 154
6.2.1 Estimation of the Step Response to Construct $\mathbf{G}(t)$ ... 157
6.2.2 Example Predictive Control of a CSTR Using a Fuzzy Model ... 158
6.3 Constrained Nonlinear Predictive Control ... 167
6.3.1 The Constrained Nonlinear Predictive Control Problem ... 168
6.3.2 Approach Using Estimated Step Response ... 173
6.3.3 Approach Using Takagi–Sugeno Fuzzy Models ... 177
6.3.4 Approach Using Takagi–Sugeno Fuzzy Models and Multiple Models in the Predictor ... 178
6.3.5 Example Predictive Control of a Steam Generator Using a Fuzzy Model ... 181
6.3.6 Example: Nonlinear Predictive Control of a Gas-Phase High-Density Polyethylene (HDPE) Reactor ... 187
6.4 Conclusions ... 191

7 Robust Nonlinear Predictive Control Using Fuzzy Models ... 195
7.1 Introduction ... 195
7.2 Robust Quadratic Programming ... 196
7.3 Problem Description ... 198
7.4 Nominal Solution ... 199
7.5 Formulation of the MPC Problem as a Robust QP ... 201
7.6 The Control Algorithm ... 202
7.7 Uncertainty Description in Fuzzy Models ... 202
7.7.1 Local Uncertainty Described on Each Rule ... 203
7.7.2 Using the Active Rules ... 204
7.7.3 Using All the Rules ... 204
7.7.4 Using the Reachable Set ... 205
7.8 Conclusions and Perspectives ... 205

8 Conclusions and Future Perspectives ... 207
8.1 Conclusions and Summary ... 207
8.2 Perspectives and Future Work ... 210

Part III Appendices

A Fuzzy Set Theory 215
A.1 Introduction 215
A.2 Fuzzy Sets 215
A.2.1 Some Examples of Membership Functions 215
A.3 Basic Definitions of Fuzzy Sets 216
A.3.1 Support 216
A.3.2 Core 216
A.3.3 Height 217
A.3.4 Normal Fuzzy Set 217
A.3.5 α-Cut 217
A.3.6 Strict α-Cut 217
A.3.7 Convexity 217
A.4 Operations on Fuzzy Sets 217
A.4.1 A Is Contained in B 217
A.4.2 Complement, Negation 217
A.4.3 Intersection 218
A.4.4 Union 218
A.5 Fuzzy relations 219
A.5.1 Projection of Fuzzy Relations 220
A.5.2 Composition of Relations 220
A.6 Approximate Reasoning 220
A.6.1 Introduction 220
A.6.2 Linguistic Variables 221
A.7 General Structure of a Fuzzy Inference System 221
A.7.1 Control Rules as a Knowledge Representation 221
A.7.2 Defuzzification 223

B Clustering Methods 225
B.1 Fuzzy C-Means [2] 225
B.2 Using Fuzzy Covariance Matrix: Gustafson and Kessel Algorithm [3] 226
B.3 Mountain Clustering [4] 228

C Gradients Used in Identification with Fuzzy Models 231
C.1 Gradient for the Singleton Consequences 231
C.1.1 With Trapezoidal Membership Functions 232
C.1.2 With Polynomial Membership Functions 233
C.1.3 With Gaussian Membership Functions 233
C.2 Gradient for the Parameters of the Membership Functions 234
C.2.1 With Trapezoidal Membership Functions 234
C.2.2 With Polynomial Membership Functions 235
C.2.3 With Gaussian Membership Functions 237
C.2.4 With Triangular Membership Functions with 0.5 Overlap 237

C.2.5 With Polynomial Membership Functions with 0.5 Overlap . . . 239

D Discrete Linear Dynamical System Approximation Theorem . . . 243

E Fuzzy Control for a Continuously Variable Transmission . . . 247
E.1 Introduction and Process Description . . . 247
E.2 Performance Specifications . . . 248
E.3 A Physical Model for the CVT . . . 249
E.4 Design of the Controller . . . 251
E.5 Stability Analysis . . . 252
E.6 Conclusions . . . 253

References . . . 255

Index . . . 261

Part I

Fuzzy Modeling

1

Fuzzy Modeling

Fuzzy set theory can be used in the modeling of systems. The modeling task is carried by a so-called fuzzy inference system (FIS). Fuzzy Inference Systems are processing units that convert numerical information into linguistic variables by means of a *fuzzification* process, process the linguistic information using a *rule base* and generate a numerical result from the conclusions of the rules by means of the *defuzzification* process (see details in A.7).

Fuzzy inference systems are *universal approximators.* This property means that FISs are capable of approximating any continuous function into a compact domain with a certain level of accuracy (ϵ).

The universal approximation property of the fuzzy models is not the only remarkable property. Fuzzy models add a new dimension to the information that can be extracted from the model. The new dimension is the linguistic dimension, which provides intuitive (linguistic) descriptions over the behavior of the modeled system.

Fuzzy models can be dynamic or static. Different types of fuzzy models have been proposed in the literature. Perhaps the most used are the *rule-based fuzzy system* [5][6]. These models are characterized by having fuzzy propositions as antecedents and consequences of the rules (Mamdani models). Another important type of models are the *Takagi–Sugeno fuzzy models*[7], where the consequences of the rules are crisp functions of the antecedents.

After the model structures were proposed, many models were developed based on "pure" *empirical knowledge*, although in many applications this proved to be insufficient and not very efficient because most of the quantitative information was not used. Several data-driven techniques are mentioned in the literature. Some of them attempt to tune the parameters of the fuzzy systems once the structure was selected [8] [9] [10] [11]; others try to use the data to tune not only the parameters but also the structure [4] [12] [13] [14].

This chapter is divided into four sections, each of which presents important aspects of fuzzy modeling. The first section is devoted to presenting an extensive analysis of the capabilities for function approximation of different types of rule-based fuzzy systems. The analysis includes a proof of the universal approximation capabilities of a class of fuzzy systems and the geometrical properties of the approximating units that can be constructed with the fuzzy rule-based systems.

The second section presents the most important methods used to calculate the structure and the parameters of a fuzzy model, including gradient descent techniques, evolutionary strategies and cluster-based techniques.

The third section is focused on the issue of generalization. This chapter discusses this matter and shows a systematic method to improve the generalization especially under circumstances where rules are not completely excited by the data. The chapter closes with a discussion over linguistic modeling. The section presents an algorithm to construct fuzzy models. This algorithm can be used in data mining to extract linguistic knowledge from numerical data sets. The algorithm is based on the concepts of linguistic integrity [15] and optimal interface design [16]. The AFRELI (Autonomous Fuzzy Rule Extractor with Linguistic Integrity) algorithm selects the number of rules and the location of the membership functions to guarantee a trade-off between accuracy and comprehensibility of the rules.

Summary:
Fuzzy models can provide good numerical approximation of functions as well as linguistic information over the behavior of the functions.

1.1 Function Approximation

This section shows the approximation capabilities of some fuzzy systems. Initially the analysis is oriented to the approximation error and the universal approximation property of fuzzy systems. The second part of this section analyzes the geometric properties of the rules approximating a function. Fuzzy systems approach functions in a local way. This means that the information provided by each rule is restricted to a compact region and the union of these local descriptions achieves the approximation of the desired function. Each rule acts like a tile on a mosaic where the picture of the mosaic is the function.

1.1.1 System Description

For the analysis of approximation error, a class of fuzzy systems will be used. The class will be described as fuzzy systems with trapezoidal (or triangular) and normal ($\max \mu_i(x) = 1$) membership functions, with a maximum overlap of 0.5 with its neighboring membership functions. [1]

[1] For notation and a basic introduction to the fuzzy set theory, please see Appendix A.

$$\text{hgt}(\mu_i \cap \mu_{i\pm1}) \leq \frac{1}{2}$$

The center of the fuzzy set (modal value for triangular membership functions) will be given by m_i^j as shown in Figure 1.1. Using the product as the AND

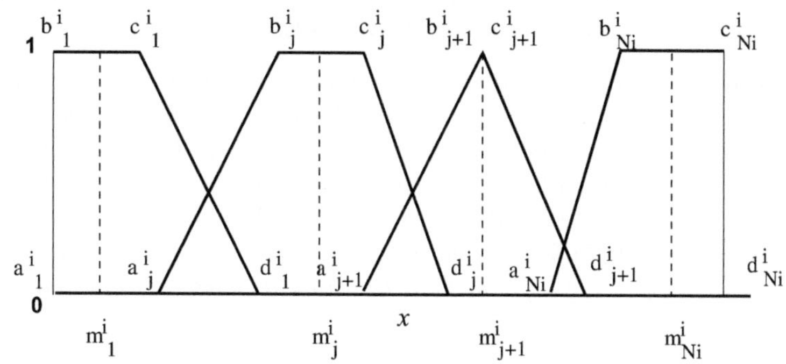

Figure 1.1. Membership function definition for the *i*th input

operation, singleton consequences and center average defuzzifier the inference process for a system with two inputs can be represented as

$$f(x) = \frac{\sum_{j_1=1}^{N_1} \sum_{j_2=1}^{N_2} \bar{y}^{j_1 j_2} (\mu_{j_1}^1(x_1)\mu_{j_2}^2(x_2))}{\sum_{j_1=1}^{N_1} \sum_{j_2=1}^{N_2} (\mu_{j_1}^1(x_1)\mu_{j_2}^2(x_2))} \tag{1.1}$$

where $x \in \Re^2$ and x_1 and x_2 are the components, μ_j^i is the membership function of the set A_j^i defined on the universe of the *i*th entry on the vector x and $\bar{y}^{j_1 j_2}$ represents a singleton consequence of the rule

IF x_1 is $A_{j_1}^1$ AND x_2 is $A_{j_2}^2$ THEN $f(x)$ is $\bar{y}^{j_1 j_2}$

A more general representation is given by

$$f(x) = \frac{\sum_{l=1}^{L} \bar{y}^l (\prod_{i=1}^{N} \mu_l^i(x_i))}{\sum_{l=1}^{L} (\prod_{i=1}^{N} \mu_l^i(x_i))} \tag{1.2}$$

where $x \in \Re^N$, L is the number of rules, and x_i represents each component of x. The membership functions are trapezoidal membership functions and for the case of the *i*th input are parameterized as follows (see Figure 1.1):

$$\mu_j^i(x) = \min\left[\max\left(\frac{x_i - a_j^i}{b_j^i - a_j^i}, 0\right), \max\left(1 - \frac{x_i - c_j^i}{d_j^i - c_j^i}, 0\right), 1\right] \tag{1.3}$$

Summary:
Fuzzy inference systems (FISs) can be represented in a compact mathematical expression [Equation (1.2)].

1.1.2 Approximation Error

This section shows the approximation capabilities of the fuzzy systems. The following theorem describes the capacity of fuzzy systems to approximate a continuous function in a compact domain. The theorem and the proof are limited to a prescribed type of fuzzy systems. However, the same reasoning can be extended to other types of fuzzy models.

Theorem 1.1. *Let $f(x)$ be a fuzzy inference system with an arbitrary number of normal membership functions (triangular or trapezoidal) with centers m_j^i, distributed over the intervals $[a_i, b_i]\,\forall i = 1, \ldots, N$ and covering the interval such that at least one and at most two membership functions are different from zero for a given value x_i and let $g(x) : \Re^N \to \Re$ be an unknown function. If $g(x)$ is continuously differentiable in the interval $U = [a_1, b_1] \times [a_2, b_2] \times \cdots \times [a_N, b_N]$, then the fuzzy system $f(x)$ can approximate the function $g(x)$ with an arbitrary bounded error ϵ*

$$||g(x) - f(x)||_\infty \leq \epsilon \tag{1.4}$$

where $||.||_\infty$ is defined as $||e(x)||_\infty = \sup_{x \in U} |e(x)|$.

Proof:
Let the consequences $\bar{y}$ of the rules of the fuzzy system $f(x)$ be calculated as

$$\bar{y}^{j_1 j_2 \ldots j_N} = g(m_{j_1}^1, m_{j_2}^2, \ldots, m_{j_N}^N)$$

Without lost of generality assume $g(x) : \Re^2 \to \Re$ defined and continuously differentiable in the interval $U = [a_1, b_1] \times [a_2, b_2]$. Define $U^{j_1 j_2} = [m_{j_1}^1, m_{j_1+1}^1] \times [m_{j_2}^2, m_{j_2+1}^2]$, where $j_1 = 1, 2, \ldots, N_1 - 1$ and $j_2 = 1, 2, \ldots, N_2 - 1$. The intervals $[a_i, b_i] = [m_1^i m_2^i] \cup [m_2^i m_3^i] \cup \ldots \cup [m_{N_i-1}^i m_{N_i}^i]$ with $i = 1, 2$. Then the input domain can be defined as

$$U = [a_1, b_1] \times [a_2, b_2] = \bigcup_{j_1=1}^{N_1-1} \bigcup_{j_2=1}^{N_2-1} U^{j_1 j_2} \tag{1.5}$$

This implies that for any $x \in U$ there is a "region" $U^{j_1 j_2}$ such that $x \in U^{j_1 j_2}$, so $x_1 \in [m_{j_1}^1 m_{j_1+1}^1]$ and $x_2 \in [m_{j_2}^2 m_{j_2+1}^2]$ (see Figure 1.2). Due to the selected description for the input membership functions, at least one and at most two membership functions will have a membership degree different from zero. The expression for the fuzzy system will be simplified as

$$f(x) = \frac{\sum_{i_1=j_1}^{j_1+1} \sum_{i_2=j_2}^{j_2+1} \bar{y}^{i_1 i_2} (\mu_{i_1}^1(x_1) \mu_{i_2}^2(x_2))}{\sum_{i_1=j_1}^{j_1+1} \sum_{i_2=j_2}^{j_2+1} (\mu_{i_1}^1(x_1) \mu_{i_2}^2(x_2))} \tag{1.6}$$

$$f(x) = \frac{\sum_{i_1=j_1}^{j_1+1} \sum_{i_2=j_2}^{j_2+1} g(m_{i_1}^1, m_{i_2}^2)(\mu_{i_1}^1(x_1) \mu_{i_2}^2(x_2))}{\sum_{i_1=j_1}^{j_1+1} \sum_{i_2=j_2}^{j_2+1} (\mu_{i_1}^1(x_1) \mu_{i_2}^2(x_2))} \tag{1.7}$$

The approximation error will be given by:

$$|g(x) - f(x)| \leq \left| g(x) - \frac{\sum_{i_1=j_1}^{j_1+1} \sum_{i_2=j_2}^{j_2+1} g(m_{i_1}^1, m_{i_2}^2)(\mu_{i_1}^1(x_1)\mu_{i_2}^2(x_2))}{\sum_{i_1=j_1}^{j_1+1} \sum_{i_2=j_2}^{j_2+1} (\mu_{i_1}^1(x_1)\mu_{i_2}^2(x_2))} \right|$$
$$\leq \max_{i_1=j_1,j_1+1;i_2=j_2,j_2+1} |g(x) - g(m_{i_1}^1, m_{i_2}^2)| \quad (1.8)$$

Applying the mean-value theorem,

$$|g(x) - f(x)| \leq \max_{i_1=j_1,j_1+1;i_2=j_2,j_2+1} (||\frac{\partial g}{\partial x_1}||_\infty |x_1 - m_{i_1}^1| + ||\frac{\partial g}{\partial x_2}||_\infty |x_2 - m_{i_2}^2|) \quad (1.9)$$

since $|x_1 - m_{i_1}^1| \leq |m_{j_1+1}^1 - m_{j_1}^1|$ and $|x_2 - m_{i_2}^2| \leq |m_{j_2+1}^2 - m_{j_2}^2|$, the local error becomes

$$|g(x) - f(x)| \leq ||\frac{\partial g}{\partial x_1}||_\infty |m_{j_1+1}^1 - m_{j_1}^1| + ||\frac{\partial g}{\partial x_2}||_\infty |m_{j_2+1}^2 - m_{j_2}^2| \quad (1.10)$$

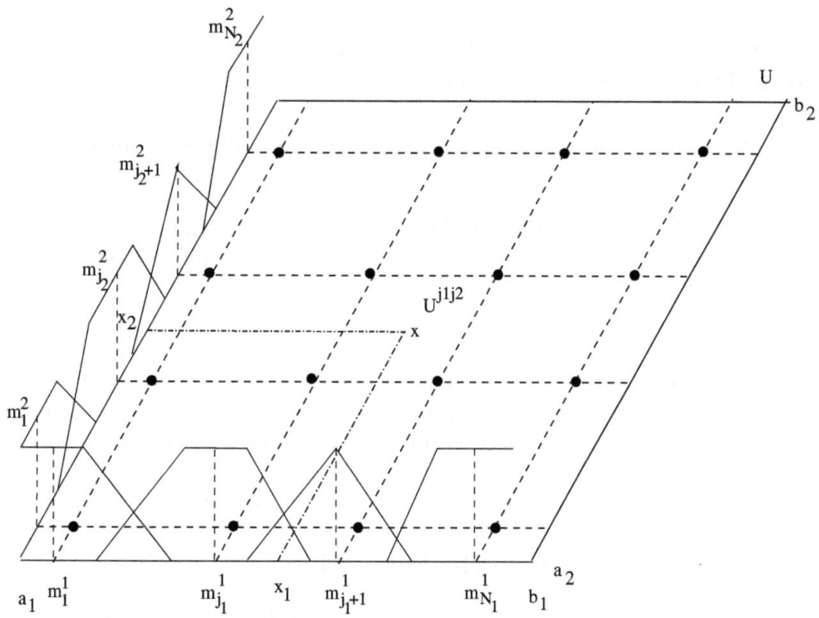

Figure 1.2. Input point in the rule space

The maximum global error will be given by

$$\begin{aligned} ||g(x) - f(x)||_\infty &= \sup_{x \in U} |g(x) - f(x)| \\ &\leq ||\frac{\partial g}{\partial x_1}||_\infty \max_{1 \leq j_1 \leq N_1 - 1} |m^1_{j_1+1} - m^1_{j_1}| \\ &\quad + ||\frac{\partial g}{\partial x_2}||_\infty \max_{1 \leq j_2 \leq N_2 - 1} |m^2_{j_2+1} - m^2_{j_2}| \\ &= ||\frac{\partial g}{\partial x_1}||_\infty e_1 + ||\frac{\partial g}{\partial x_2}||_\infty e_2 \end{aligned} \tag{1.11}$$

where $e_i = \max_{1 \leq j_i \leq N_i - 1} |m^i_{j_2+1} - m^i_{j_2}|$. Extending this result to N dimensions,

$$||g(x) - f(x)||_\infty \leq ||\frac{\partial g}{\partial x_1}||_\infty e_1 + ||\frac{\partial g}{\partial x_2}||_\infty e_2 + \cdots + ||\frac{\partial g}{\partial x_N}||_\infty e_N \tag{1.12}$$

This expression means that the maximum error in the approximation is bounded and the approximation error can be reduced to a certain $||g - f||_\infty \leq \epsilon$ value by playing with the distance e_i

In other words, the described fuzzy system $f(x)$ is a *universal approximator* to the function $g(x)$ within a finite domain $x \in U$ [17].

Observe that the accuracy of the approximation depends directly on two factors:

1. The value of the maximum gradient of the function $||\frac{\partial g}{\partial x_i}||_\infty$
2. The distance between the "centers" of the membership functions $|m^i_{j_2+1} - m^i_{j_2}|$.

Summary:
Fuzzy inference systems (FISs) can approximate any function in a compact domain. The accuracy of the approximation depends on the maximum *slope* of the function and the distance between the centers of the fuzzy sets.

1.1.3 Constructing Units in the Fuzzy Models

A clearer perspective on the approximation capabilities of fuzzy systems can be gained by making an analysis of the interpolation properties of neighboring rules. These elements constitute the constructing units of the fuzzy systems. These units work as "patches" that can be used to approximate a given function. The study of these units is important to analyze properties such as generalization and smoothness. It is important to remark that these constructing units are the "smallest" fuzzy model that can be defined on a given interval. The analysis of this section will be limited to three types of membership functions: triangular, polynomial and Gaussian. The interpolation properties of these types of membership functions, and especially the triangular, already cover the interpolation properties of the rules using trapezoidal membership functions. For this reason, these membership functions are not covered in this study.

Triangular Membership Functions with Overlap $\frac{1}{2}$

The triangular membership functions with overlap $\frac{1}{2}$ are plotted in Figure 1.3. These membership functions exhibit two important properties:

1. Overlap is equal to $\frac{1}{2}$.
2. $\mu^i_{j_i}(x_i) = 1 - \mu^i_{j_i+1}(x_i)$. Overlapping membership functions add up to 1.

Initially, a one-input–one-output system will be studied to simplify the analysis. The rules for the interval $x \in [m_{j_1}, m_{j_1+1}]$ are

$$\text{IF } x \text{ is } A_{j_1} \text{ THEN } \bar{y}^{j_1}$$
$$\text{IF } x \text{ is } A_{j_1+1} \text{ THEN } \bar{y}^{j_1+1}$$

The expression for the function defined by the fuzzy system when the input lies on the interval $x \in [m_{j_1}, m_{j_1+1}]$ will be

$$\begin{aligned} f(x) &= \frac{\bar{y}^{j_1}\mu_{j_1}(x) + \bar{y}^{j_1+1}\mu_{j_1+1}(x)}{\mu_{j_1}(x) + \mu_{j_1+1}(x)} \\ &= \bar{y}^{j_1}\mu_{j_1}(x) + \bar{y}^{j_1+1}\mu_{j_i+1}(x) \end{aligned} \tag{1.13}$$

where the membership functions are parameterized as

$$\mu_{j_1}(x) = \frac{m_{j_1+1} - x}{m_{j_1+1} - m_{j_1}} \tag{1.14}$$

$$\mu_{j_1+1}(x) = \frac{x - m_{j_1}}{m_{j_1+1} - m_{j_1}} \tag{1.15}$$

replacing this parameterization in (1.13):

$$\begin{aligned} f(x) &= \bar{y}^{j_1}\left(\frac{m^1_{j_1+1} - x}{m_{j_1+1} - m_{j_1}}\right) + \bar{y}^{j_1+1}\left(\frac{x - m_{j_1}}{m_{j_1+1} - m_{j_1}}\right) \\ &= \underbrace{\left(\frac{\bar{y}^{j_1+1} - \bar{y}^{j_1}}{m_{j_1+1} - m_{j_1}}\right)}_{a} x + \underbrace{\left(\frac{m_{j_1+1}\bar{y}^{j_1} - m_{j_1}\bar{y}^{j_1+1}}{m_{j_1+1} - m_{j_1}}\right)}_{b} \\ &= ax + b \end{aligned} \tag{1.16}$$

Equation (1.16) is clearly an affine function of x. For the one-dimensional case, the fuzzy system using the described fuzzy rules works as a linear interpolator between the consequences of the rules.

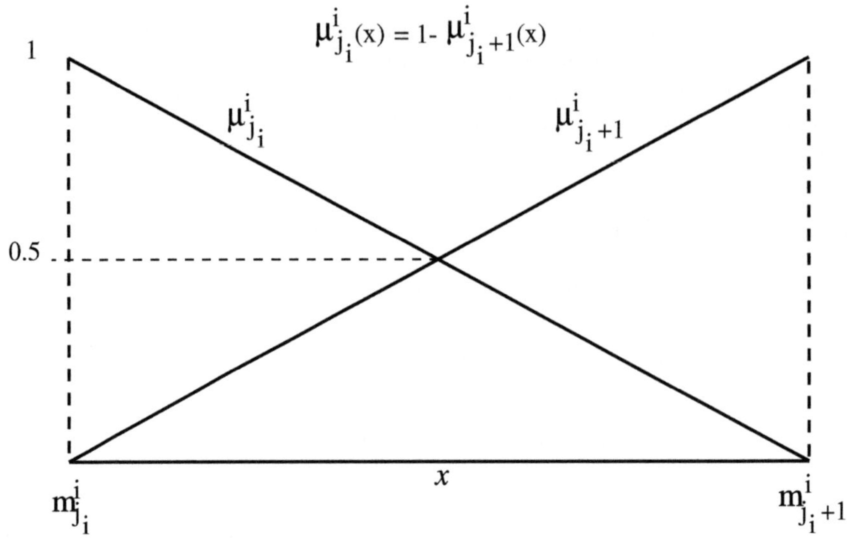

Figure 1.3. Triangular membership functions with overlap $\frac{1}{2}$

Extending the analysis to the case of functions with two inputs ($x \in \Re^2$): The function defined by the fuzzy system in the interval $x \in [m^1_{j_1}, m^1_{j_1+1}] \times [m^2_{j_2}, m^2_{j_1+1}]$ is given by

$$
\begin{aligned}
f(x) &= \\
&\frac{\bar{y}^{j_1 j_2}(\mu^1_{j_1}(x_1)\mu^2_{j_2}(x_2)) + \bar{y}^{(j_1+1)j_2}(\mu^1_{j_1+1}(x)\mu^2_{j_2}(x_2))}{\mu^1_{j_1}(x_1)\mu^2_{j_2}(x_2) + \mu^1_{j_1+1}(x)\mu^2_{j_2}(x_2) + \mu^1_{j_1}(x_1)\mu^2_{j_2+1}(x_2) + \mu^1_{j_1+1}(x)\mu^2_{j_2+1}(x_2)} \\
&+ \frac{\bar{y}^{j_1 j_2+1}(\mu^1_{j_1}(x_1)\mu^2_{j_2+1}(x_2)) + \bar{y}^{(j_1+1)(j_2+1)}(\mu^1_{j_1+1}(x)\mu^2_{j_2+1}(x_2))}{\mu^1_{j_1}(x_1)\mu^2_{j_2}(x_2) + \mu^1_{j_1+1}(x)\mu^2_{j_2}(x_2) + \mu^1_{j_1}(x_1)\mu^2_{j_2+1}(x_2) + \mu^1_{j_1+1}(x)\mu^2_{j_2+1}(x_2)} \\
&= \bar{y}^{j_1 j_2}(\mu^1_{j_1}(x_1)\mu^2_{j_2}(x_2)) + \bar{y}^{(j_1+1)j_2}(\mu^1_{j_1+1}(x)\mu^2_{j_2}(x_2)) + \\
&\quad + \bar{y}^{j_1 j_2+1}(\mu^1_{j_1}(x_1)\mu^2_{j_2+1}(x_2)) + \bar{y}^{(j_1+1)(j_2+1)}(\mu^1_{j_1+1}(x)\mu^2_{j_2+1}(x_2)) \qquad (1.17)
\end{aligned}
$$

where x_1 and x_2 are the components of x, and μ^i_j represents the jth membership function defined on the domain of x_i. Replacing the parameterization of the membership functions given by Equations (1.14) and (1.15), we can express the function of the system as

$$
\begin{aligned}
f(x) = &\frac{-m^2_{j_2+1}\bar{y}^{j_1 j_2} + m^2_{j_2+1}\bar{y}^{(j_1+1)j_2} + m^2_{j_2}\bar{y}^{j_1(j_2+1)} - m^2_{j_2}\bar{y}^{(j_1+1)(j_2+1)}}{(m^1_{(j_1+1)} - m^1_{j_1})(m^2_{(j_2+1)} - m^2_{j_2})} x_1 \\
&+ \frac{-m^1_{(j_1+1)}\bar{y}^{j_1 j_2} + m^1_{j_1}\bar{y}^{(j_1+1)j_2} + m^1_{j_1+1}\bar{y}^{j_1(j_2+1)} - m^1_{j_1}\bar{y}^{(j_1+1)(j_2+1)}}{(m^1_{(j_1+1)} - m^1_{j_1})(m^2_{(j_2+1)} - m^2_{j_2})} x_2 \\
&+ \frac{\bar{y}^{j_1 j_2} - \bar{y}^{(j_1+1)j_2} - \bar{y}^{j_1(j_2+1)} + \bar{y}^{(j_1+1)(j_2+1)}}{(m^1_{(j_1+1)} - m^1_{j_1})(m^2_{(j_2+1)} - m^2_{j_2})} x_1 x_2
\end{aligned}
$$

$$+\frac{m^1_{(j_1+1)}m^2_{(j_2+1)}\bar{y}^{j_1j_2}-m^2_{j_2+1}m^1_{j_1}\bar{y}^{(j_1+1)j_2}}{(m^1_{(j_1+1)}-m^1_{j_1})(m^2_{(j_2+1)}-m^2_{j_2})}$$
$$+\frac{-m^1_{j_1+1}m^2_{j_2}\bar{y}^{j_1(j_2+1)}+m^1_{j_1}m^2_{j_2}\bar{y}^{(j_1+1)(j_2+1)}}{(m^1_{(j_1+1)}-m^1_{j_1})(m^2_{(j_2+1)}-m^2_{j_2})} \quad (1.18)$$

Observe that even if Equation (1.18) looks very complex, it is nothing but a bilinear expression such as

$$f(x_1,x_2)=ax_1+bx_2+cx_1x_2+d \quad (1.19)$$

Figure 1.4 shows an example of a bilinear surface generated by a fuzzy system with two inputs and triangular membership functions with overlap $\frac{1}{2}$.

If the procedure is extended to higher dimensions, similar expressions

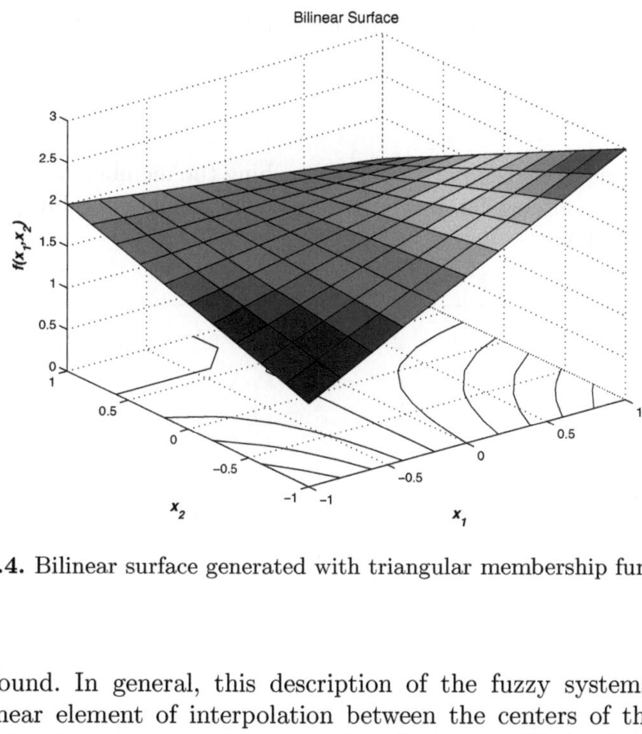

Figure 1.4. Bilinear surface generated with triangular membership functions with overlap $\frac{1}{2}$

will be found. In general, this description of the fuzzy systems produces a multilinear element of interpolation between the centers of the rules; in other words, the fundamental constructing element of this kind of fuzzy system is a finite multilinear hypersurface defined in the interval $U^{j_1j_2\cdots j_N}=[m^1_{j_1},m^1_{j_1+1}]\times[m^2_{j_2},m^2_{j_2+1}]\times\ldots\times[m^N_{j_N},m^N_{j_N+1}]$.

Summary:
Triangular membership functions with $\frac{1}{2}$ overlap generate linear interpolations among the consequences of their corresponding rules. If the system has multiple antecedents, the interpolation will be multilinear among the values of the consequences of the rules.

Polynomial Membership Function with Overlap $\frac{1}{2}$

Polynomial membership functions are characterized by two third-order polynomials describing the left and the right edge of the fuzzy set. The membership functions are shown in Figure - 1.5. Observe that these membership functions preserve the properties described in the previous section. Additionally, two new conditions are included. For the input domain x_i the conditions are

- Overlap is equal to $\frac{1}{2}$.
- $\mu^i_{j_i}(x_i) = 1 - \mu^i_{j_i+1}(x_i)$. Overlapping membership functions add up to 1.
- $\frac{\partial \mu^i_{j_i}(m_{j_i})}{\partial x_i} = 0.$
- $\frac{\partial \mu^i_{j_i}(m_{j_i+1})}{\partial x_i} = 0.$

With these conditions the polynomial describing the membership functions in the interval $[m_{j_i}, m_{j_i+1}]$ can be constructed. The membership function can be described as follows:

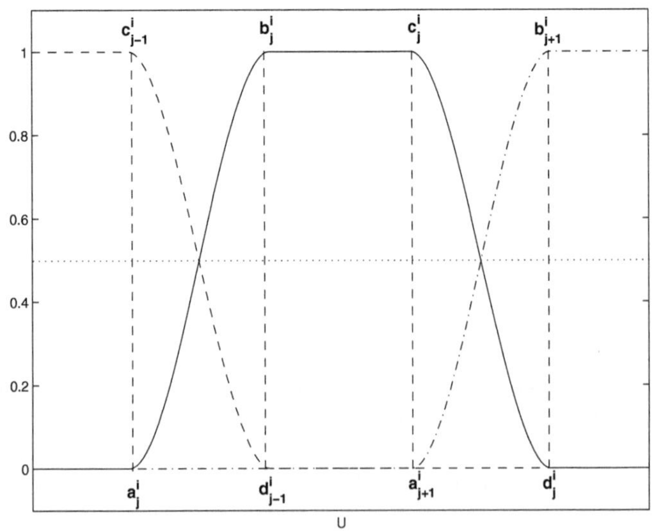

Figure 1.5. Polynomial membership functions

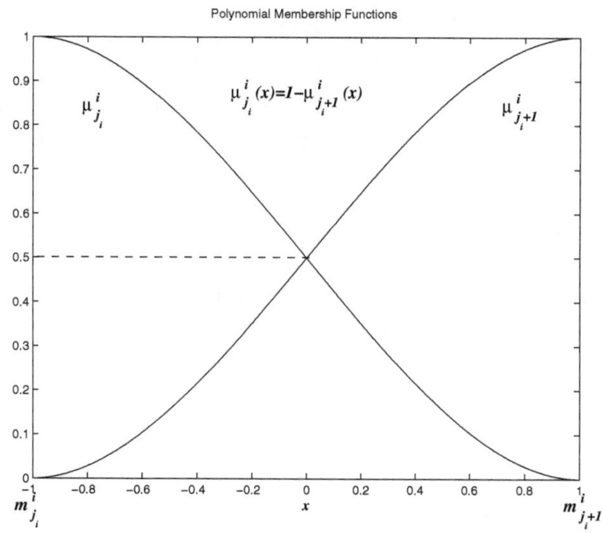

Figure 1.6. Polynomial membership functions with overlap $\frac{1}{2}$

$$\mu^i_{j_i}(x_i) = \begin{cases} 0 & \text{if } x_i < a^i_{j_i} \\ (x_i - a^i_{j_i})^2(2x_i + a^i_{j_i} - 3b^i_{j_i})/(a^i_{j_i} - b^i_{j_i})^3 & \text{if } a^i_{j_i} < x_i < b^i_{j_i} \\ 1 & \text{if } b^i_{j_i} \le x_i \le c^i_{j_i} \\ -(x_i - d^i_{j_i})^2(2x_i + d^i_{j_i} - 3c^i_{j_i})/(c^i_{j_i} - d^i_{j_i})^3 & \text{if } c^i_{j_i} < x_i < d^i_{j_i} \\ 0 & \text{if } x_i > d^i_{j_i} \end{cases} \tag{1.20}$$

Initially, a one-input–one-output system will be studied to simplify the analysis. The rules for the interval $x \in [m_{j_1}, m_{j_1+1}]$ are

$$\text{IF } x \text{ is } A_{j_1} \text{ THEN } \bar{y}^{j_1}$$
$$\text{IF } x \text{ is } A_{j_1+1} \text{ THEN } \bar{y}^{j_1+1}$$

The expression for the function generated by a system with one input defined on the interval $x \in [m_{j_1}, m_{j_1+1}]$ and membership functions are shown in Figure 1.6.

$$\begin{aligned} f(x) &= \frac{\bar{y}^{j_1}\mu_{j_1}(x) + \bar{y}^{j_1+1}\mu_{j_1+1}(x)}{\mu^1_{j_1}(x) + \mu_{j_1+1}(x)} \\ &= \bar{y}^{j_1}\mu_{j_1}(x) + \bar{y}^{j_1+1}\mu_{j_i+1}(x) \\ &= -\bar{y}^{j_1}(x - d^i_{j_i})^2(2x + m_{j_1+1} - 3m_{j_1})/(m_{j_1} - m_{j_1+1})^3 \\ &\quad + \bar{y}^{j_1+1}(x_i - m_{j_1})^2(2x_i + m_{j_1} - 3m_{j_1+1})/(m_{j_1} - m_{j_1+1})^3 \end{aligned} \tag{1.21}$$

Grouping the terms, we get

$$f(x) = -2(\bar{y}^{j_1} - \bar{y}^{j_1+1})x^3 + 3(\bar{y}^{j_1} - \bar{y}^{j_1+1})(m_{j_1} + m_{j_1+1})x^2$$

$$- 6(\bar{y}^{j_1} - \bar{y}^{j_1+1})(m_{j_1} m_{j_1+1})x$$
$$- \bar{y}^{j_1}(m_{j_1}^3 - 3m_{j_1}^2 m_{j_1+1}) + \bar{y}^{j_1+1}(m_{j_1+1}^3 - 3m_{j_1+1}^2 m_{j_1}) \quad (1.22)$$

It is clear from (1.22) that the interpolation given by this type of membership functions is a third-order polynomial. Observe that the interpolation is monotonic. Another very interesting property of this interpolation is that it has a continuous derivative at the extremes. Such a property is not present when triangular functions are used.

If the analysis is extended to more dimensions in the domain of the function, the interpolating unit will be a third-order polynomial on each of the inputs. A graphical representation of this constructing unit is shown in Figure-1.8.

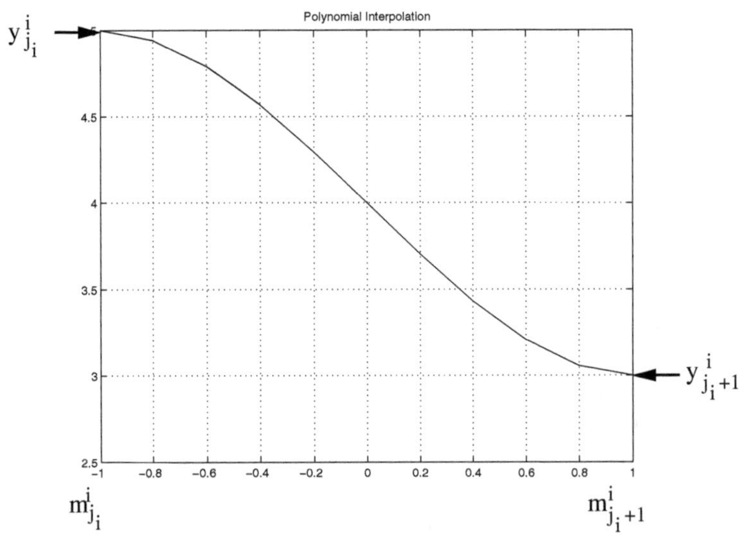

Figure 1.7. Polynomial interpolation generated with polynomial membership functions with overlap $\frac{1}{2}$

Summary: Polynomial membership functions with $\frac{1}{2}$ overlap generate smooth interpolations among the consequences of their corresponding rules. These interpolations are characterized by a third-order polynomial that is monotonic along the interpolation interval.

Gaussian Membership Functions with $\sigma = 0.6(m^i_{j_i+1} - m^i_{j_i})$

These membership functions are parameterized by a function shaped as a "Gauss' bell". For this type of membership functions the parameterization is

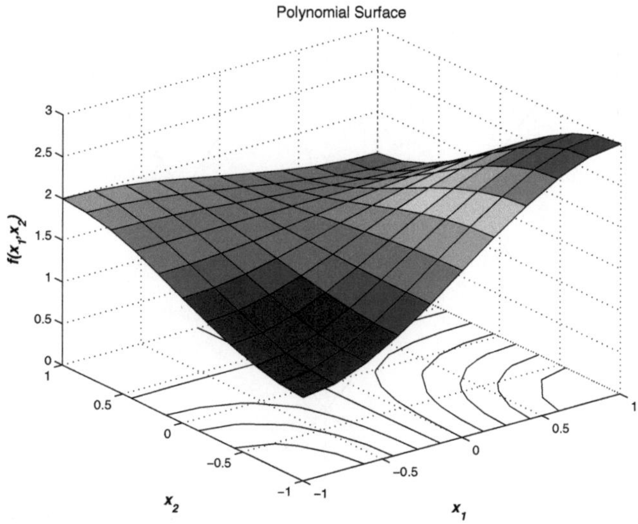

Figure 1.8. Polynomial surface generated with polynomial membership functions with overlap $\frac{1}{2}$ in a system with two inputs

given as

$$\mu^i_{j_i}(x_i) = e^{-\left(\frac{x(i)-m^i_{j_i}}{\sigma^i_{j_i}}\right)^2} \tag{1.23}$$

Assuming a simple fuzzy system with one input and one output and only two membership functions and two rules, the system will generate the following interpolation in the interval $x \in [m_{j_1}, m_{j_1+1}]$:

$$f(x) = \frac{\bar{y}^{j_1} e^{-\left(\frac{x-m_{j_1}}{\sigma}\right)^2} + \bar{y}^{j_1+1} e^{-\left(\frac{x-m_{j_1+1}}{\sigma}\right)^2}}{e^{-\left(\frac{x-m_{j_1}}{\sigma}\right)^2} + e^{-\left(\frac{x-m_{j_1+1}}{\sigma}\right)^2}} \tag{1.24}$$

The interpolation generated by this expression is shown in Figure- 1.10. Observe that this interpolation is smooth but the membership functions are not complementary. For this reason, the denominator in (1.24) is not equal to 1. This fact will make the interpolation less accurate because more than one rule will be activated when the inputs correspond to a modal value and the function will not pass exactly by the values given by the consequences.

The expression for the function generated by the fuzzy system when the domain of $x \in U$ is defined as $U = [m^1_{j_1}, m^1_{j_1+1}] \times [m^2_{j_2}, m^2_{j_2+1}]$ is

$$f(x) =$$

$$\frac{\bar{y}^{j_1 j_2}(\mu^1_{j_1}(x_1)\mu^2_{j_2}(x_2)) + \bar{y}^{(j_1+1)j_2}(\mu^1_{j_1+1}(x)\mu^2_{j_2}(x_2))}{\mu^1_{j_1}(x_1)\mu^2_{j_2}(x_2) + \mu^1_{j_1+1}(x)\mu^2_{j_2}(x_2) + \mu^1_{j_1}(x_1)\mu^2_{j_2+1}(x_2) + \mu^1_{j_1+1}(x)\mu^2_{j_2+1}(x_2)}$$

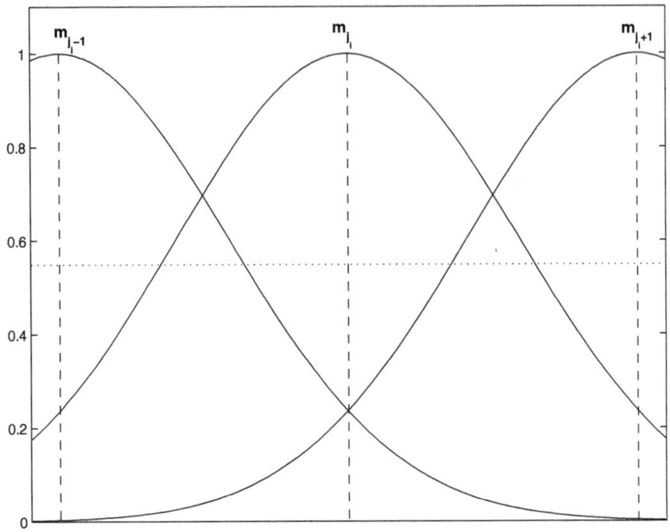

Figure 1.9. Gaussian membership functions with $\sigma = 0.6(m^i_{j_i+1} - m^i_{j_i})$

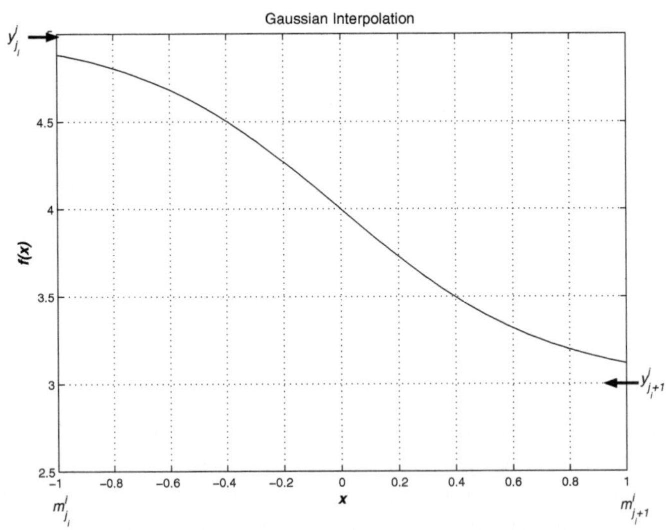

Figure 1.10. Gaussian interpolation generated with Gaussian membership functions with $\sigma = 0.6(m^i_{j_i+1} - m^i_{j_i})$

$$+ \frac{\bar{y}^{j_1 j_2+1}(\mu^1_{j_1}(x_1)\mu^2_{j_2+1}(x_2)) + \bar{y}^{(j_1+1)(j_2+1)}(\mu^1_{j_1+1}(x)\mu^2_{j_2+1}(x_2))}{\mu^1_{j_1}(x_1)\mu^2_{j_2}(x_2) + \mu^1_{j_1+1}(x)\mu^2_{j_2}(x_2) + \mu^1_{j_1}(x_1)\mu^2_{j_2+1}(x_2) + \mu^1_{j_1+1}(x)\mu^2_{j_2+1}(x_2)} \tag{1.25}$$

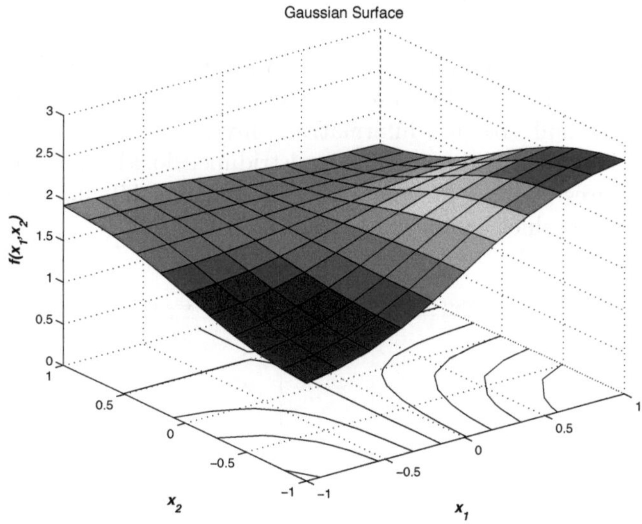

Figure 1.11. Gaussian surface generated with Gaussian membership functions with $\sigma = 0.6(m^i_{j_i+1} - m^i_{j_i})$

The product of two membership functions of this type can be parameterized as follows:

$$\mu^1_{j_1}(x_1)\mu^2_{j_2}(x_2) = e^{-\left(\frac{\sigma_2^2(x(1)-m^1_{j_1})^2+\sigma_1^2(x(2)-m^2_{j_2})^2}{\sigma_1^2\sigma_2^2}\right)}$$
$$\mu^1_{j_1}(x_1)\mu^2_{j_2}(x_2) = \mu_{j_1j_2}(x) = e^{-\frac{d^{j_1j_2\,T}\Sigma d^{j_1j_2}}{(\sigma_1\sigma_2)^2}} \tag{1.26}$$

with $\Sigma = \mathrm{diag}(\sigma_1, \sigma_2)$ and $d^{j_1j_2} = [x_1 - m^1_{j_1}, x_2 - m^2_{j_2}]^T$. Then the expression for the function will be

$$f(x) = \frac{\bar{y}^{j_1j_2}\mu_{j_1j_2}(x) + \bar{y}^{(j_1+1)j_2}\mu_{(j_1+1)j_2}(x)}{\mu_{j_1j_2}(x) + \mu_{(j_1+1)j_2}(x) + \mu_{j_1j_2+1}(x) + \mu_{(j_1+1)(j_2+1)}(x)} + \frac{\bar{y}^{j_1j_2+1}\mu_{j_1j_2+1}(x) + \bar{y}^{(j_1+1)(j_2+1)}\mu_{(j_1+1)(j_2+1)}(x)}{\mu_{j_1j_2}(x) + \mu_{(j_1+1)j_2}(x) + \mu_{j_1j_2+1}(x) + \mu_{(j_1+1)(j_2+1)}(x)} \tag{1.27}$$

The surface generated by this type of systems is shown in Figure- 1.11. The surface generated is a weighted sum of Gaussians. For a domain of n dimensions, the function is a weighted sum of Gaussians calculated using n dimensional distances.

Summary:
Gaussian membership functions generate smooth interpolations among points near the consequences of their corresponding rules. These interpolations are characterized as a sum of weighted Gaussian functions.

Other Approximating Units

Some other approximating units can be obtained with different distribution of the membership functions. Because the analytical expression for such "units" is very complex and not very informative, only the graphical representation is presented here. The bidimensional and tridimensional representations are shown in Figures- 1.13, 1.14, and 1.15. The membership functions used to generate these graphics are shown in Figure 1.12.

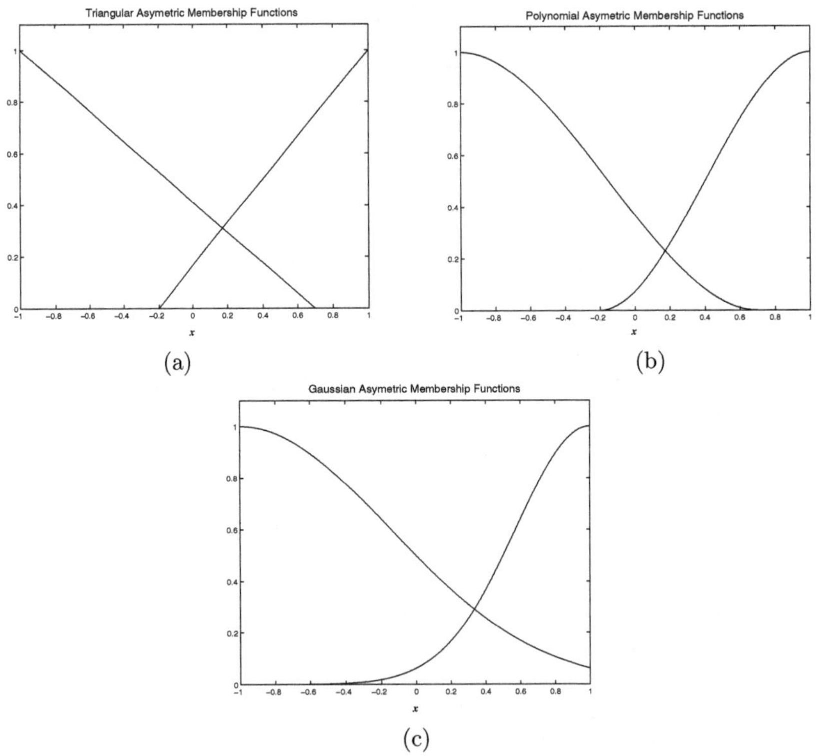

Figure 1.12. (a) Triangular membership functions with overlap different than $\frac{1}{2}$. (b) Polynomial membership functions with overlap different than $\frac{1}{2}$. (c) Gaussian membership functions with different σ

1.2 Approximation Capabilities of Takagi–Sugeno Fuzzy Models

The study of the approximation capabilities of Takagi–Sugeno (T-S) fuzzy models is more complex, because the consequences of the rules are no longer

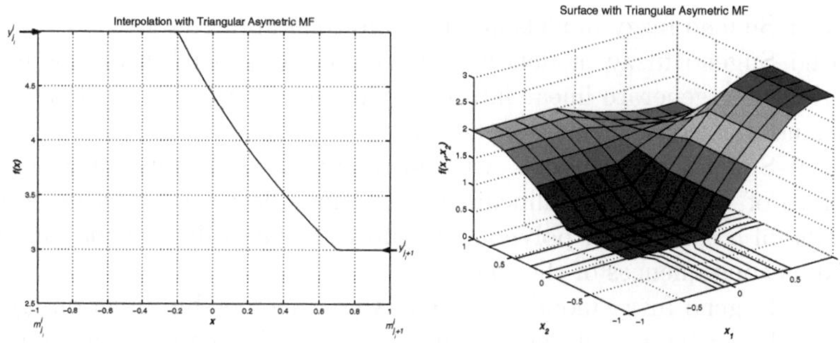

Figure 1.13. Interpolation and surfaces generated with triangular membership functions with overlap different than $\frac{1}{2}$

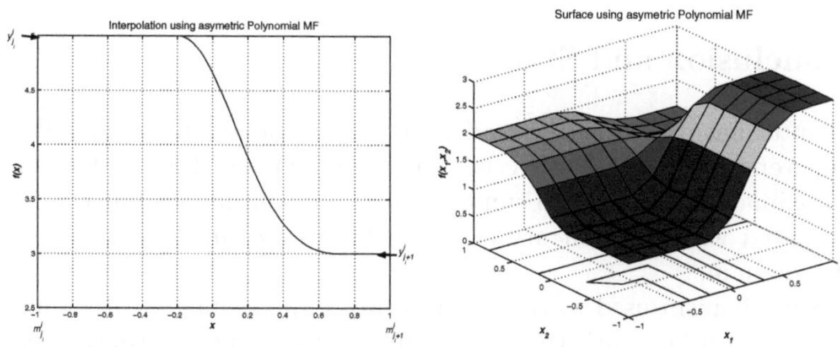

Figure 1.14. Interpolation and surfaces generated with polynomial membership functions with overlap different than $\frac{1}{2}$

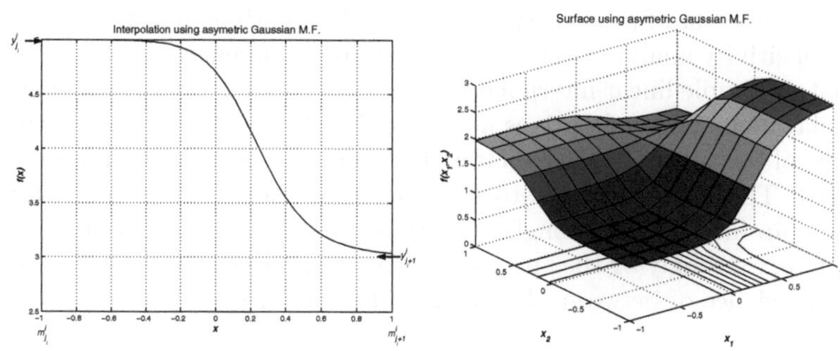

Figure 1.15. Interpolation and surfaces generated with Gaussian membership functions with different σ

fixed values but are functions of the antecedents or other variables. However, the reader can extend the reasonings of previous sections to T-S fuzzy models bearing in mind this important fact. The reasonings can be summarized in the following lines.

- Takagi–Sugeno fuzzy models are also universal approximators.
- Takagi–Sugeno fuzzy models with triangular membership functions and overlap of $\frac{1}{2}$ generate linear interpolations among the values of the functions used as consequences of the rules.
- Takagi–Sugeno fuzzy models with polynomial membership functions and overlap of $\frac{1}{2}$ generate nonlinear interpolations described by a monotonic section of a third-order polynomial connecting the values of the functions used as consequences of the rules.
- Takagi–Sugeno fuzzy models with Gaussian membership functions generate nonlinear interpolations described by a monotonic function that connects the "neighborhoods" of the functions used as consequences of the rules.

1.3 Conclusion and Summary

This chapter has presented the reader with the approximation capabilities of fuzzy inference systems. The section has presented the theorem for "universal approximation" showing in summary that any continuous function can be approximated with an arbitrary accuracy by a fuzzy inference system on a compact domain.

The mechanism used to construct such an approximation resembles the construction of a "mosaic" where each group of neighboring rules works like a "tile" helping to shape the function like a picture.

The use of one or other type of membership functions depends on many aspects. For instance, differentiability will favor the use of Gaussian and polynomial membership functions since they exhibit continuous derivatives facilitating sensitivity analysis over the obtained fuzzy inference system. If the goal is to obtain simple linear interpolations and simple numerical evaluations, the triangular membership functions are favored. If the goal is to guarantee local coverage of the rules, the triangular and polynomial membership functions are preferred. These examples try to illustrate that the selection of the type of membership functions is strongly conditioned by the application of the model and not very strongly by the approximation capabilities. The same can be said about the choice of between Mamdani models and Takagi–Sugeno models.

2

Constructing Fuzzy Models from Input-Output Data

In the previous chapter we presented a discussion of the approximation capabilities of fuzzy models. In summary, we have shown that fuzzy models can be used to reproduce the behavior of any continuous function. This chapter presents some of the methods used to construct fuzzy models that replicate the behavior of a given function. The information about the function is presented in the form of input–output data, which means that a set of points over the domain of the function (input) is selected and then evaluated in the function (output).

The construction of fuzzy models involves the selection of several parameters: position, shape and the distribution of the membership functions, rule base construction, selection of the logical operations, consequences of the rules, etc. This large number of "degrees of freedom" makes it very difficult to implement a unique method to select all these parameters at once. A typical approach is to set in advance the logical operations and the type of membership functions using certain criteria (differentiability, linguistic integrity, implementability, etc.). The remaining parameters can be estimated from the data using different strategies, but in general all are based on a single objective, which is to minimize the approximation error between the output values and the values given by the fuzzy model.

According to the tuned parameters and the strategies, different methods have been proposed in the literature. This chapter presents the following strategies:

- Mosaic or table lookup scheme [18]
- Using gradient descent [18] [19]
- Using clustering and gradient descent [12] [4]
- Using evolutionary strategies [20] [21]

The *mosaic or table lookup scheme* fixes in advance the type, number and position of the membership functions and calculates only the consequences of the rules. The methods based on *gradient descent* fix in advance the type and number of the membership functions and calculate their positions and the value of the consequences. The methods based on *clustering and gradient descent* fix only the type of membership functions and by means of clustering algorithms select the number and initial positions of the membership functions. Consequences and refined positions of the membership functions are found by means of a gradient descent algorithm.

Evolutionary strategies deserve a different comment since they can be used to optimize all possible aspects integrated in a fuzzy model including the set of inputs used to construct the model. Some interesting features of the *evolutionary strategies* are the fact that they can introduce complex constraints to enforce some desired features into the model and also the fact that they perform a gradient-free optimization.

Table 2.1 summarizes the methods and the parameters that are adjusted by the method. The following sections are dedicated to explain these methods. Finally, the chapter closes with an example of an industrial application of the fuzzy models constructed from input–output data.

Table 2.1. Parameters Adjusted by the Different Training Methods

Method	**Type of MFs**	**Number of MFs**	**Location of the MFs**	**Consequences**
Mosaic scheme	Fixed	Fixed	Fixed	Adjusted
Gradient descent	Fixed	Fixed	Adjusted	Adjusted
Clustering + gradient descent	Fixed	Adjusted	Adjusted	Adjusted
Evolutionary strategies (1)	Adjusted	Adjusted	Adjusted	Adjusted

Summary:
Fuzzy inference systems (FIS) can be systematically constructed from "pure" input–output data. All methods are based on the optimization of a cost function to minimize the "distance" between the predictions of the FIS and the output data. The main differences among the methods are the initialization and the adjusted parameters.

2.1 Mosaic or Table Lookup Scheme

The basic scheme of the method was proposed by Wang [18]. Here some simple modifications are introduced, and these modifications are related to the consequence calculation. In this method the position, the shape and the distribution of the membership functions are choices for the designer. The rule base is composed and the method finds only the consequences of the rules.

Assume a sequence of input–output $\{x^i, y^i\}\ i = 1, \ldots, N$ data is collected, the inputs $x^i \in U \subset \Re^p$ and the output $y^i \in V \subset \Re$. The subset U is a portion of the space $\Re^p$ and is defined as $U = [a_1, b_1] \times \ldots \times [a_p, b_p]$. The procedure to construct the model is laid out in the following.

- For each of the p inputs of the system distribute over the interval $[a_i, b_i]$ N_i membership functions. The shape, the position and the distribution is a user's choice. The only condition is that the full interval is covered and at least two membership functions are placed on each point of the input domain. As shown in the previous sections, the shape and the distribution affect the smoothness and the accuracy of the approximation.
- Generate the rule base using all possible combinations among the antecedents and the AND operator (choosing in advance "min" or "product" operator). The rule l of the rule base for Mamdani fuzzy systems is

$$\text{IF } x_1^i \text{ is } A_1^l \text{ AND } \ldots \text{ AND } x_p^i \text{ is } A_p^l \text{ THEN } y \text{ IS } \bar{y}^l$$

and for Takagi–Sugeno fuzzy systems

$$\text{IF } x_1^i \text{ is } A_1^l \text{ AND } \ldots \text{ AND } x_p^i \text{ is } A_p^l \text{ THEN } y = a_1^l x_1^i + \ldots + a_p^l x_p^i + b^l$$

- Calculate the inference of each rule. For rule l of the form

$$\mu_l(x^i) = \min\{\mu_l^1(x_1^i), \mu_l^2(x_2^i), \ldots, \mu_l^p(x_p^i)\} \tag{2.1}$$

or

$$\mu_l(x^i) = \mu_l^1(x_1^i).\mu_l^2(x_2^i).\ldots.\mu_l^p(x_p^i) \tag{2.2}$$

the general expressions for these fuzzy system with L rules will be given by

$$f(x^i) = \frac{\sum_{l=1}^{L} \bar{y}^l \mu_l(x^i)}{\sum_{l=1}^{L} \mu_l(x^i)} \tag{2.3}$$

for the Mamdani models and

$$f(x^i) = \frac{\sum_{l=1}^{L} (a_1^l x_1^i + \ldots + a_p^l x_p^i + b^l)\mu_l(x^i)}{\sum_{l=1}^{L} \mu_l(x^i)} \tag{2.4}$$

for Takagi–Sugeno models.
- Calculate the consequence parameters
  - In the Mamdani model the parameter to be calculated is $\bar{y}^l$ $l = 1, \ldots, L$ such that $f(x^i) \approx y^i$. Observe that Equation (2.3) can be written as

$$f(x^i) = \sum_{l=1}^{L} \bar{y}^l w_l(x^i) \tag{2.5}$$

$$w_l(x^i) = \frac{\mu_l(x^i)}{\sum_{l=1}^{L} \mu_l(x^i)} = w_l^i \tag{2.6}$$

The N output values can be represented as the vector Y in terms of the inference process:

$$\underbrace{\begin{bmatrix} y^1 \\ y^2 \\ \vdots \\ y^N \end{bmatrix}}_{Y} = \underbrace{\begin{bmatrix} w_1^1 & w_2^1 & \dots & w_L^1 \\ w_1^2 & w_2^2 & \dots & w_L^2 \\ \vdots & \vdots & \ddots & \vdots \\ w_1^N & w_2^N & \dots & w_L^N \end{bmatrix}}_{W} \underbrace{\begin{bmatrix} \bar{y}^1 \\ \bar{y}^2 \\ \vdots \\ \bar{y}^L \end{bmatrix}}_{\theta} + \underbrace{\begin{bmatrix} e_1 \\ e_2 \\ \vdots \\ e_N \end{bmatrix}}_{E} \tag{2.7}$$

– In the Takagi–Sugeno model the parameters to be calculated are $a_1^l \dots a_p^l$ and b^l $l = 1, \dots, L$ such that $f(x^i) \approx y^i$. Using the reasoning applied for the Mamdani models, Equation (2.4) can be written as

$$f(x^i) = \sum_{l=1}^{L} (a_1^l x_1^i + \dots + a_p^l x_p^i + b^l) w_l(x^i) \tag{2.8}$$

where $w_l(x^i)$ has the same form shown in Equation (2.6).
The N output values can be represented as the vector Y in terms of the inference process

$$\underbrace{\begin{bmatrix} y^1 \\ y^2 \\ \vdots \\ y^N \end{bmatrix}}_{Y} = \underbrace{\begin{bmatrix} w_1^1 x_1^1 & \dots & w_1^1 x_p^1 & w_1^1 & w_2^1 x_1^1 & \dots & w_L^1 x_p^1 & w_L^1 \\ w_1^2 x_1^1 & \dots & w_1^2 x_p^1 & w_1^2 & w_2^2 x_1^1 & \dots & w_L^2 x_p^1 & w_L^2 \\ \vdots & \ddots & \vdots & \vdots & \vdots & \ddots & \vdots & \vdots \\ w_1^N x_1^1 & \dots & w_1^N x_p^1 & w_1^N & w_2^N x_1^1 & \dots & w_L^N x_p^1 & w_L^N \end{bmatrix}}_{W} \underbrace{\begin{bmatrix} a_1^1 \\ a_2^1 \\ \vdots \\ a_p^1 \\ b^1 \\ a_1^2 \\ \vdots \\ a_p^L \\ b^L \end{bmatrix}}_{\theta}$$

$$+ \underbrace{\begin{bmatrix} e_1 \\ e_2 \\ \vdots \\ e_N \end{bmatrix}}_{E} \tag{2.9}$$

In both cases the vector E is the approximation error and the aim is to reduce the norm of this vector as much as possible. Using the quadratic norm to measure the approximation error, we obtain

$$\min_{\theta} ||E||_2 = \min_{\theta} ||Y - W\theta||_2 \tag{2.10}$$

It is a least-squares problem and the consequences can be calculated using least squares. The solution to this least-squares problem is

$$\theta = \arg\min_{\theta} ||E||_2 = (W^T W)^{-1} Y^T W \tag{2.11}$$

This solution is applicable as far as the rank$(W^T W) = \dim(\theta)$; otherwise other methods must be applied to guarantee a reliable set of consequences for the rules. In Section 2.5, a method based on recursive least squares is detailed.

Summary:
A mosaic or table lookup scheme is probably the simplest method to construct fuzzy models from data. The method demands from the user the definition of the antecedent of the rules and finds the consequences by using least squares.

2.1.1 Illustrative Example

In this example we show a simple application of the method *mosaic or table lookup scheme* to approximate the function $f(x) = sin(x)$ over the interval $[0, 2\pi]$ using 629 points equidistant along the domain of x. In this case we illustrate the results using six membership functions over the input domain. Four models are presented: three of the Mamdani type and one Takagi–Sugeno. The three Mamdani models are created with three different types of membership functions: *triangular*, *polynomial* and *Gaussian*. For the model using Takagi–Sugeno rules only the results using *triangular* membership are illustrated.

Observe that the interpolations generated by the Mamdani models are mentioned in previous Chapter: linear for the *triangular* membership functions (see Figure 2.1), *polynomial* for the polynomial membership functions (see Figure 2.2) and between the *neighborhoods* of the consequences for the Gaussian membership functions.

The best model is by far the Takagi–Sugeno model (see Figure 2.4). In fact, in the figure it is difficult to distinguish the approximation from the original function. Figure 2.4 shows some straight segments corresponding to the consequences of the rules. The successful result of the Takagi–Sugeno can be explained in part because the model exhibits 12 degrees of freedom (two adjustable parameters per rule a_1^l and b^l) in contrast with the Mamdani models with only 6 degrees of freedom (only one adjustable parameter per rule $\bar{y}^l$). Having more degrees of freedom can be beneficial as long as the number of points is large enough and as long as they are well spread over the input domain (*persistent excitation*). Otherwise the generalization capabilities of the model can be compromised.

Among the Mamdani models the best model is the model generated using *Gaussian* membership functions. The result is explained in part for the resemblance between the shapes of the sine function and the Gaussian membership functions. Also, it is interesting to observe how the "flat" sections of the polynomial membership functions affect the approximation by generating local plateaus in the function approximation.

Observe that the results of this example are only an illustration of the method and are by no means a benchmark to judge the capabilities of some membership functions or model types.

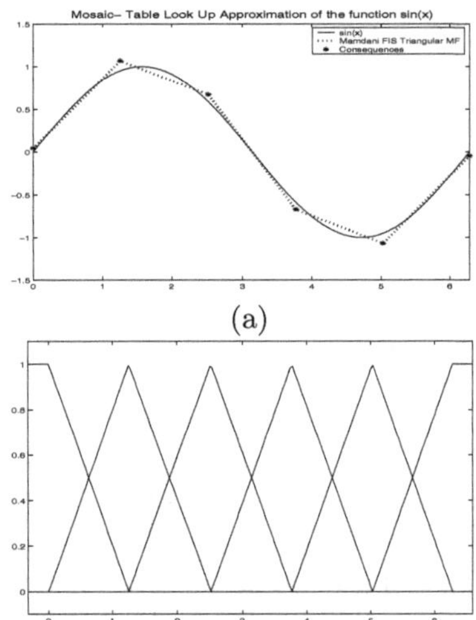

(a) (b)

Figure 2.1. (a) Approximation generated by a Mamdani fuzzy model trained using the *mosaic or table lookup scheme* using triangular membership functions with 6 membership functions. (-) Original function (- -) Approximation generated by the fuzzy model (*) Consequences of the rules (b) Membership functions

2.2 Using Gradient Descent

This method requires the definition of the number of membership functions and their shape. Normally the AND function is fixed to be the "product" because an analytical expression for the gradient of the cost function is needed. The initial position of the membership functions is another element that must be chosen. The method proceeds as follows:

- For each of the p inputs of the system, distribute over the interval $[a_i, b_i]$, N_i membership functions. The shape, the initial positions and the distribution are user's choices. The membership functions must cover the input

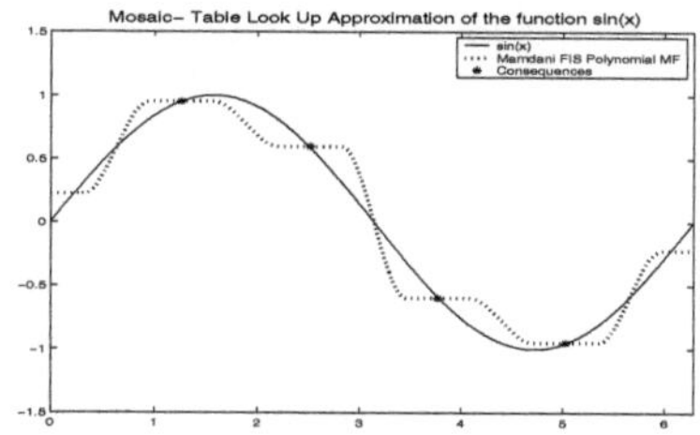

(a)

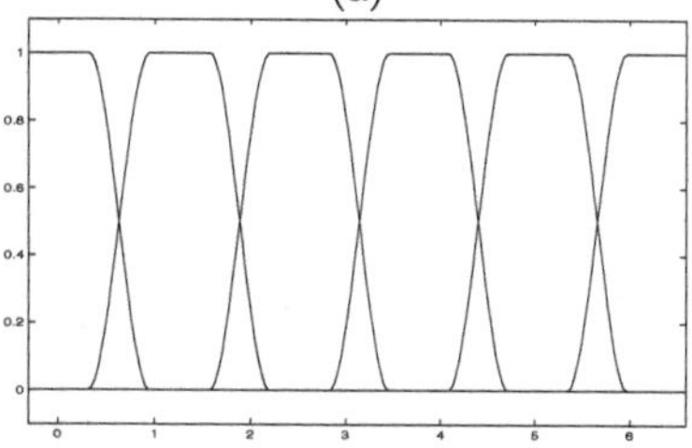

(b)

Figure 2.2. (a) Approximation generated by a Mamdani fuzzy model trained using the *mosaic or table lookup scheme* using polynomial membership functions with 6 membership functions. (-) Original function (- -) Approximation generated by the fuzzy model (*) Consequences of the rules (b) Membership functions

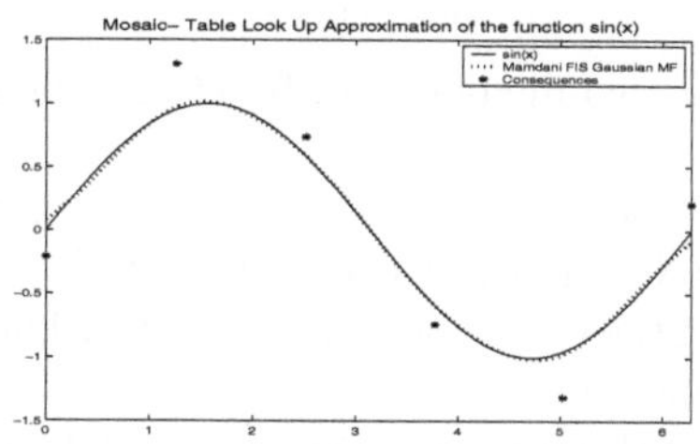

(a)

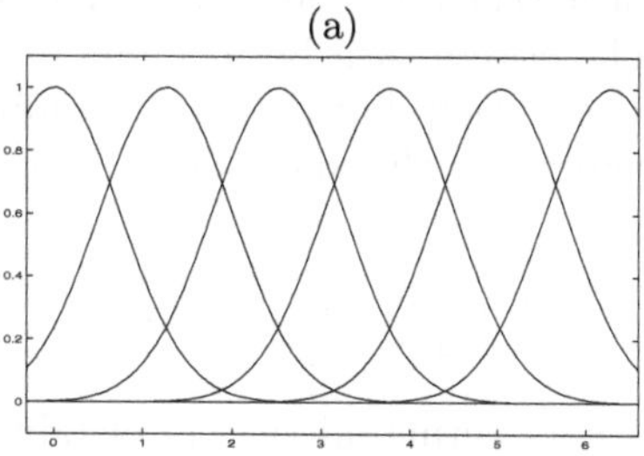

(b)

Figure 2.3. (a) Approximation generated by a Mamdani fuzzy model trained using the *mosaic or table lookup scheme* using polynomial membership functions with 6 membership functions. (-) Original function (- -) Approximation generated by the fuzzy model (*) Consequences of the rules (b) Membership functions

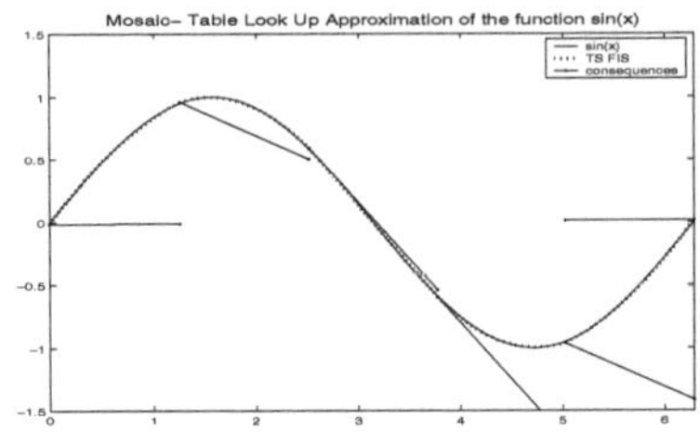

(a)

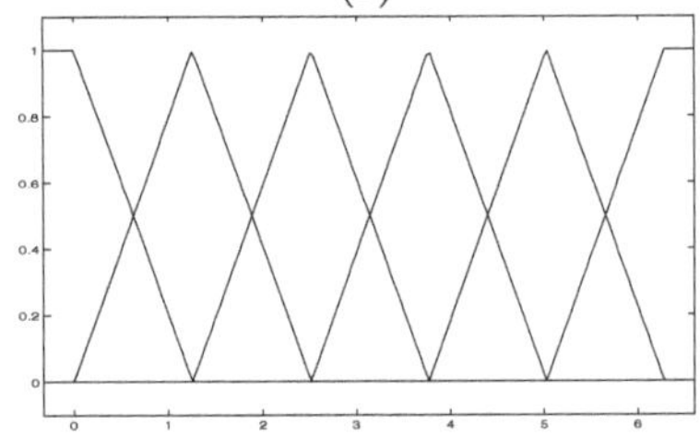

(b)

Figure 2.4. (a) Approximation generated by a Takagi–Sugeno fuzzy model trained using the *mosaic or table lookup scheme* using triangular membership functions with 6 membership functions. (-) Original function (- -) Approximation generated by the fuzzy model (.-) Consequences of the rules (b) Membership functions

interval, and at least two membership functions should be placed on each input domain.

- Generate the rule base using all possible combinations among the antecedents and the AND operator using "product".
- Initialize the value of the consequences using prior knowledge, least squares or recursive least squares.
- Optimize the value of the consequences $\bar{y}^l$ and the parameters of the membership functions. The criteria will be to minimize the cost function described in the previous section, but now the optimization will also adjust the membership functions of the antecedents. The cost function can be described as

$$J = \frac{1}{2}\sum_{i=1}^{N}(y^i - f(x^i, \theta))^2 \tag{2.12}$$

where θ is a vector representing all the "adjustable" parameters (consequences, parameters of the membership functions) of the fuzzy system $f(.,.)$. The problem will be the minimization of the cost function J. This minimization is a nonlinear, nonconvex optimization problem. The objective is to obtain an "acceptable" solution and not necessarily "the global minima" of this cost function. Different schemes for optimization can be applied to find this solution. Probably the simplest one will be the gradient descent method. This method consists of an iterative calculation of the parameters oriented to the negative direction of the gradient. The explanation behind this method is that by taking the negative direction of the gradient, the steepest route toward the minimum will be taken. This descent direction does not guarantee convergence of the scheme; for this reason, the α parameter is introduced and it can be modified to improve the convergence rate and properties. Some choices of α are given by Newton and quasi-Newton methods [22].

$$\theta(k+1) = \theta(k) + \alpha \frac{\partial J}{\partial \theta} \tag{2.13}$$

α is sometimes called the "learning rate." The gradient descent method can be modified; for example, by calculating the consequences by means of least squares (see ANFIS scheme [19]).

The gradient of the cost function will be in general

$$\frac{\partial J}{\partial \theta} = \sum_{i=1}^{N} (y^i - f(x^i, \theta)) \frac{\partial f(x^i, \theta)}{\partial \theta} = \sum_{i=1}^{N} e^i \frac{\partial f(x^i, \theta)}{\partial \theta} \tag{2.14}$$

Using the general expression of the fuzzy system can be parameterized as

$$f(x^i) = \frac{\sum_{l=1}^{L} \bar{y}^l \mu_l(x^i)}{\sum_{l=1}^{L} \mu_l(x^i)} = \frac{A}{B} \tag{2.15}$$

The updating of the consequence parameters will be independent of the parameterization of the membership functions and will be given by

$$\bar{y}^l(k+1) = \bar{y}^l(k) + \alpha \sum_{i=1}^{N} (y - f(x^i)) \frac{\mu_l(x^i)}{B} \tag{2.16}$$

The expressions to update the parameters of the membership functions are different for each parameterization. Special attention is devoted to the gradient calculation to guarantee a 0.5 overlap between contiguous membership functions. The updating formulas for some of the membership functions are shown in the next sections.

Summary:
The *gradient descent* method calculates parameters on the antecedents and the consequences of the fuzzy inference system. The method demands from the user the definition of the initial location of the membership functions of the antecedents. The method can be combined with a calculation of the consequences by using least squares. In this case the method is known as the ANFIS scheme. Such a method exhibits faster convergence, especially for Takagi–Sugeno models.

2.2.1 Gradient Updating for Trapezoidal Membership Functions

Assuming the parameterization given in the expression

$$\mu_j^i(x_i, a_j^i, b_j^i, c_j^i, d_j^i) = \min\left[\max\left(\frac{x_i - a_j^i}{b_j^i - a_j^i}, 0\right), \max\left(1 - \frac{x_i - c_j^i}{d_j^i - c_j^i}, 0\right), 1\right] \tag{2.17}$$

the updating formulas will be

$$a_j^i(k+1) = a_j^i(k) + \alpha \sum_{t=1}^{N} \frac{(y^t - f(x^t))}{B} \sum_{l \in \mathcal{U}} (\bar{y}^l - f(x^t)) \frac{\mu_l(x^t)}{\mu_j^i(x_i^t)} \frac{\partial \mu_j^i(x_i^t)}{\partial a_j^i} \tag{2.18}$$

$$b_j^i(k+1) = b_j^i(k) + \alpha \sum_{t=1}^{N} \frac{(y^t - f(x^t))}{B} \sum_{l \in \mathcal{U}} (\bar{y}^l - f(x^t)) \frac{\mu_l(x^t)}{\mu_j^i(x_i^t)} \frac{\partial \mu_j^i(x_i^t)}{\partial b_j^i} \tag{2.19}$$

$$c_j^i(k+1) = c_j^i(k) + \alpha \sum_{t=1}^{N} \frac{(y^t - f(x^t))}{B} \sum_{l \in \mathcal{U}} (\bar{y}^l - f(x^t)) \frac{\mu_l(x^t)}{\mu_j^i(x_i^t)} \frac{\partial \mu_j^i(x_i^t)}{\partial c_j^i} \tag{2.20}$$

$$d_j^i(k+1) = d_j^i(k) + \alpha \sum_{t=1}^{N} \frac{(y^t - f(x^t))}{B} \sum_{l \in \mathcal{U}} (\bar{y}^l - f(x^t)) \frac{\mu_l(x^t)}{\mu_j^i(x_i^t)} \frac{\partial \mu_j^i(x_i^t)}{\partial d_j^i} \tag{2.21}$$

where the set $\mathcal{U}$ is the set of rules that includes the function $\mu_j^i(x)$ in the antecedents and

$$\frac{\partial \mu_j^i(x_i^t)}{\partial a_j^i} = \begin{cases} 0 & \text{if } x_i^t < a_j^i \\ \frac{x_i^t - b_j^i}{(b_j^i - a_j^i)^2} & \text{if } a_j^i < x_i^t < b_j^i \\ 0 & \text{if } x_i^t > b_j^i \end{cases} \tag{2.22}$$

$$\frac{\partial \mu_j^i(x_i^t)}{\partial b_j^i} = \begin{cases} 0 & \text{if } x_i^t < a_j^i \\ \frac{a_j^i - x_i^t}{(b_j^i - a_j^i)^2} & \text{if } a_j^i < x_i^t < b_j^i \\ 0 & \text{if } x_i^t > b_j^i \end{cases} \tag{2.23}$$

$$\frac{\partial \mu_j^i(x_i^t)}{\partial c_j^i} = \begin{cases} 0 & \text{if } x_i^t < c_j^i \\ \frac{d_j^i - x_i^t}{(d_j^i - c_j^i)^2} & \text{if } c_j^i < x_i^t < d_j^i \\ 0 & \text{if } x_i^t > d_j^i \end{cases} \tag{2.24}$$

$$\frac{\partial \mu_j^i(x_i^t)}{\partial d_j^i} = \begin{cases} 0 & \text{if } x_i^t < c_j^i \\ \frac{x_i^t - c_i^j}{(d_i^j - c_j^i)^2} & \text{if } c_j^i < x_i^t < d_j^i \\ 0 & \text{if } x_i^t > d_j^i \end{cases} \tag{2.25}$$

It is important to remark that the updating should preserve the condition $a_j^i \leq b_j^i \leq c_j^i \leq d_j^i$. This adaptation rule can be applied to triangular membership functions by just making $b_j^i = c_j^i$.

2.2.2 Gradient Updating for Triangular Membership Functions with Overlap $\frac{1}{2}$

The membership functions are parameterized by using only their modal values. This parameterization not only preserves the overlap but also reduces the number of parameters to be tuned. Triangular membership functions are parameterized by the position of their three vertices; but the condition of overlap $\frac{1}{2}$ makes the lower right vertex of one membership function to be at the same position as the modal value of the next membership function. So, instead of tuning three parameters (the vertices), only one parameter is tuned for each membership function.

The parameterization for a triangular membership function using the modal values as parameters is

$$\mu_j^i(x_i, m_{j-1}^i, m_j^i, m_{j+1}^i) = \max\left[0, \min\left(\frac{x_i - m_{j-1}^i}{m_j^i - m_{j-1}^i}, 1 - \frac{x_i - m_j^i}{m_{j+1}^i - m_j^i}\right)\right] \tag{2.26}$$

The updating formula will be

$$\begin{aligned} m_j^i(k+1) &= m_j^i(k) + \alpha \sum_{t=1}^{N} \frac{(y^t - f(x^t))}{B} \\ &\left[\sum_{l \in \mathcal{U}} (\bar{y}^l - f(x^t)) \frac{\mu_l(x^t)}{\mu_{j-1}^i(x_i^t)} \frac{\partial \mu_{j-1}^i(x_i^t)}{\partial m_j^i} + \sum_{l \in \mathcal{V}} (\bar{y}^l - f(x^t)) \frac{\mu_l(x^t)}{\mu_j^i(x_i^t)} \frac{\partial \mu_j^i(x_i^t)}{\partial m_j^i} \right. \\ &\left. + \sum_{l \in \mathcal{W}} (\bar{y}^l - f(x^t)) \frac{\mu_l(x^t)}{\mu_{j+1}^i(x_i^t)} \frac{\partial \mu_{j+1}^i(x_i^t)}{\partial m_j^i} \right] \end{aligned} \tag{2.27}$$

where the sets $\mathcal{U}$, $\mathcal{V}$ and $\mathcal{W}$ are the set of rules that includes the functions $\mu_{j-1}^i(.)$, $\mu_j^i(.)$ and $\mu_{j+1}^i(.)$, respectively, and with

$$\frac{\partial \mu_{j-1}^i(x_i^t)}{\partial m_j^i} = \begin{cases} 0 & \text{if } x_i^t < m_{j-1}^i \\ \frac{x_i^t - m_j^i}{(m_j^i - m_{j-1}^i)^2} & \text{if } m_{j-1}^i < x_i^t < m_j^i \\ 0 & \text{if } x_i^t > m_j^i \end{cases} \tag{2.28}$$

$$\frac{\partial \mu_j^i(x_i^t)}{\partial m_j^i} = \begin{cases} 0 & \text{if } x_i^t < m_{j-1}^i \\ \frac{m_{j-1}^i - x_i^t}{(m_j^i - m_{j-1}^i)^2} & \text{if } m_{j-1}^i < x_i^t < m_j^i \\ \frac{m_{j+1} - x_i^t}{(m_{j+1}^i - m_j^i)^2} & \text{if } m_j^i < x_i^t < m_{j+1}^i \\ 0 & \text{if } x_i^t > m_{j+1}^i \end{cases} \tag{2.29}$$

$$\frac{\partial \mu_{j+1}^i(x_i^t)}{\partial m_j^i} = \begin{cases} 0 & \text{if } x_i^t < m_j^i \\ \frac{x_i^t - m_{j+1}^i}{(m_{j+1}^i - m_j^i)^2} & \text{if } m_j^i < x_i^t < m_{j+1}^i \\ 0 & \text{if } x_i^t > m_{j+1}^i \end{cases} \tag{2.30}$$

Here the adaptation must be constrained such that the condition $m_j^i \leq m_{j+1}^i$ is preserved.

2.2.3 Gradient Updating for Polynomial Membership Functions

Assuming the parameterization given in the expression

$$\mu_{j_i}^i(x_i) = \begin{cases} 0 & \text{if } x_i < a_{j_i}^i \\ (x_i - a_{j_i}^i)^2(2x_i + a_{j_i}^i - 3b_{j_i}^i)/(a_{j_i}^i - b_{j_i}^i)^3 & \text{if } a_{j_i}^i < x_i < b_{j_i}^i \\ 1 & \text{if } b_{j_i}^i \leq x_i \leq c_{j_i}^i \\ -(x_i - d_{j_i}^i)^2(2x_i + d_{j_i}^i - 3c_{j_i}^i)/(c_{j_i}^i - d_{j_i}^i)^3 & \text{if } c_{j_i}^i < x_i < d_{j_i}^i \\ 0 & \text{if } x_i > d_{j_i}^i \end{cases} \tag{2.31}$$

the expressions to update the parameters a_j^i, b_j^i, c_j^i and d_j^i are similar to the ones used for the trapezoidal membership functions. Only the expression for the gradient of the membership functions changes.

$$a_j^i(k+1) = a_j^i(k) + \alpha \sum_{t=1}^{N} \frac{(y^t - f(x^t))}{B} \sum_{l \in \mathcal{U}} (\bar{y}^l - f(x^t)) \frac{\mu_l(x^t)}{\mu_j^i(x_i^t)} \frac{\partial \mu_j^i(x_i^t)}{\partial a_j^i} \tag{2.32}$$

$$b_j^i(k+1) = b_j^i(k) + \alpha \sum_{t=1}^{N} \frac{(y^t - f(x^t))}{B} \sum_{l \in \mathcal{U}} (\bar{y}^l - f(x^t)) \frac{\mu_l(x^t)}{\mu_j^i(x_i^t)} \frac{\partial \mu_j^i(x_i^t)}{\partial b_j^i} \tag{2.33}$$

$$c_j^i(k+1) = c_j^i(k) + \alpha \sum_{t=1}^{N} \frac{(y^t - f(x^t))}{B} \sum_{l \in \mathcal{U}} (\bar{y}^l - f(x^t)) \frac{\mu_l(x^t)}{\mu_j^i(x_i^t)} \frac{\partial \mu_j^i(x_i^t)}{\partial c_j^i} \tag{2.34}$$

$$d_j^i(k+1) = d_j^i(k) + \alpha \sum_{t=1}^{N} \frac{(y^t - f(x^t))}{B} \sum_{l \in \mathcal{U}} (\bar{y}^l - f(x^t)) \frac{\mu_l(x^t)}{\mu_j^i(x_i^t)} \frac{\partial \mu_j^i(x_i^t)}{\partial d_j^i} \tag{2.35}$$

with

$$\frac{\partial \mu_j^i(x_i^t)}{\partial a_j^i} = \begin{cases} 0 & \text{if } x_i^t < a_j^i \\ 6\frac{(b_j^i - x_i^t)^2(a_j^i - x_i^t)}{(a_j^i - b_j^i)^4} & \text{if } a_j^i < x_i^t < b_j^i \\ 0 & \text{if } x_i^t > b_j^i \end{cases} \tag{2.36}$$

$$
\frac{\partial \mu_j^i(x_i^t)}{\partial b_j^i} = \begin{cases} 0 & \text{if } x_i^t < a_j^i \\ -6\frac{(a_j^i - x_i^t)^2 (b_j^i - x_i^t)}{(a_j^i - b_j^i)^4} & \text{if } a_j^i < x_i^t < b_j^i \\ 0 & \text{if } x_i^t > b_j^i \end{cases} \tag{2.37}
$$

$$
\frac{\partial \mu_j^i(x_i^t)}{\partial c_j^i} = \begin{cases} 0 & \text{if } x_i^t < c_j^i \\ -6\frac{(d_j^i - x_i^t)^2 (c_j^i - x_i^t)}{(c_j^i - d_j^i)^4} & \text{if } c_j^i < x_i^t < d_j^i \\ 0 & \text{if } x_i^t > d_j^i \end{cases} \tag{2.38}
$$

$$
\frac{\partial \mu_j^i(x_i^t)}{\partial d_j^i} = \begin{cases} 0 & \text{if } x_i^t < c_j^i \\ 6\frac{(c_j^i - x_i^t)^2 (d_j^i - x_i^t)}{(c_i^j - d_j^i)^4} & \text{if } c_j^i < x_i^t < d_j^i \\ 0 & \text{if } x_i^t > d_j^i \end{cases} \tag{2.39}
$$

The adaptation algorithm should preserve the condition $a_j^i \leq b_j^i \leq c_j^i \leq d_j^i$.

2.2.4 Gradient Updating for Polynomial Membership Functions with Overlap $\frac{1}{2}$ and $b_j^i = c_j^i = m_j^i$

The parameterization of the membership functions is made using only their modal values. This parameterization guarantees the overlap of $\frac{1}{2}$ with the neighboring membership functions. The number of adjusted parameters is reduced: instead of adjusting four parameters $(a_j^i, b_j^i, c_j^i, d_j^i)$, for each membership function, only one parameter m_j^i is adjusted. Observe that the overlap $\frac{1}{2}$ is preserved only if the parameters $a_{j+1}^i, b_j^i, c_j^i, d_{j-1}^i$ describing the polynomial membership function are equal among each other and equal to the modal value m_j^i. The parameterization using the modal values is as follows:

$$
\mu_{j_i}^i(x_i) = \begin{cases} 0 & \text{if } x_i < m_{j-1}^i \\ \frac{(x_i - m_{j-1}^i)^2 (2x_i + m_{j-1}^i - 3m_j^i)}{(m_{j-1}^i - m_j^i)^3} & \text{if } m_{j-1}^i < x_i < m_j^i \\ \frac{-(x_i - m_{j+1}^i)^2 (2x_i + m_{j+1}^i - 3m_j^i)}{(m_j^i - m_{j+1}^i)^3} & \text{if } m_j^i < x_i < m_{j+1}^i \\ 0 & \text{if } x_i > m_{j+1}^i \end{cases} \tag{2.40}
$$

The parameters m_j^i are updated with a similar formula as the one used for the triangular membership functions with overlap $\frac{1}{2}$:

$$
\begin{aligned}
m_j^i(k+1) = m_j^i(k) + \alpha \sum_{t=1}^{N} \frac{(y^t - f(x^t))}{B} \\
\Bigg[\sum_{l \in \mathcal{U}} (\bar{y}^l - f(x^t)) \frac{\mu_l(x^t)}{\mu_{j-1}^i(x_i^t)} \frac{\partial \mu_{j-1}^i(x_i^t)}{\partial m_j^i} + \sum_{l \in \mathcal{V}} (\bar{y}^l - f(x^t)) \frac{\mu_l(x^t)}{\mu_j^i(x_i^t)} \frac{\partial \mu_j^i(x_i^t)}{\partial m_j^i} \\
+ \sum_{l \in \mathcal{W}} (\bar{y}^l - f(x^t)) \frac{\mu_l(x^t)}{\mu_{j+1}^i(x_i^t)} \frac{\partial \mu_{j+1}^i(x_i^t)}{\partial m_j^i} \Bigg]
\end{aligned} \tag{2.41}
$$

where the sets $\mathcal{U}$, $\mathcal{V}$ and $\mathcal{W}$ are the set of rules that includes the functions $\mu^i_{j-1}(.)$, $\mu^i_j(.)$ and $\mu^i_{j+1}(.)$, respectively, and with

$$\frac{\partial \mu^i_{j-1}(x^t_i)}{\partial m^i_j} = \begin{cases} 0 & \text{if } x^t_i < m^i_{j-1} \\ 6\frac{(m^i_{j-1}-x^t_i)^2(m^i_j-x^t_i)}{(m^i_{j-1}-m^i_j)^4} & \text{if } m^i_{j-1} < x^t_i < m^i_j \\ 0 & \text{if } x^t_i > m^i_j \end{cases} \tag{2.42}$$

$$\frac{\partial \mu^i_j(x^t_i)}{\partial m^i_j} = \begin{cases} 0 & \text{if } x^t_i < m^i_{j-1} \\ -6\frac{(m^i_{j-1}-x^t_i)^2(m^i_j-x^t_i)}{(m^i_{j-1}-m^i_j)^4} & \text{if } m^i_{j-1} < x^t_i < m^i_j \\ -6\frac{(m^i_{j+1}-x^t_i)^2(m^i_j-x^t_i)}{(m^i_j-m^i_{j+1})^4} & \text{if } m^i_j < x^t_i < m^i_{j+1} \\ 0 & \text{if } x^t_i > m^i_{j+1} \end{cases} \tag{2.43}$$

$$\frac{\partial \mu^i_{j+1}(x^t_i)}{\partial m^i_j} = \begin{cases} 0 & \text{if } x^t_i < m^i_j \\ 6\frac{(m^i_{j+1}-x^t_i)^2(m^i_j-x^t_i)}{(m^i_j-m^i_{j+1})^4} & \text{if } m^i_j < x^t_i < m^i_{j+1} \\ 0 & \text{if } x^t_i > m^i_{j+1} \end{cases} \tag{2.44}$$

Observe that the adaptation must preserve the condition $m^i_j \leq m^i_{j+1}$.

2.2.5 Gradient Updating for Gaussian Membership Functions

The parameterization of the membership functions is given by

$$\mu^i_j(x^t_i) = \exp(-(\frac{x^t_i - \bar{x}^i_j}{\sigma^i_j})^2) \tag{2.45}$$

The updating formula for the parameters $\bar{x}^i_j$ and σ^i_j will be given by

$$\bar{x}^i_j(k+1) = \bar{x}^i_j(k) + \alpha \sum_{t=1}^{N} \frac{(y^t - f(x^t))}{B} \sum_{l \in \mathcal{U}} 2(\bar{y}^l - f(x^t))\mu_l(x^t)\frac{x^t_i - \bar{x}^i_j}{{\sigma^i_j}^2} \tag{2.46}$$

$$\sigma^i_j(k+1) = \sigma^i_j(k) + \alpha \sum_{t=1}^{N} \frac{(y^t - f(x^t))}{B} \sum_{l \in \mathcal{U}} 2(\bar{y}^l - f(x^t))\mu_l(x^t)\frac{(x^t_i - \bar{x}^i_j)^2}{{\sigma^i_j}^3} \tag{2.47}$$

where $\mathcal{U}$ is the set of rules with the antecedent term $\mu^i_j(.)$.

2.2.6 Illustrative Example

This example uses the same simple sine function presented in Section 2.1.1. The same 629 equidistant points were used to approximate the function $f(x) = sin(x)$ over the interval $[0, 2\pi]$. In this case the number of membership functions is 6 and they were initially equally distributed along the input domain in the same way as in the example of Section 2.1.1. Again, we prepared four models: three of the Mamdani type and one Takagi–Sugeno. The three Mamdani models were created with three different types of membership functions *triangular*, *polynomial* and *Gaussian*, and they were trained during 400 iterations (epochs) using pure gradient descent. For the model using Takagi–Sugeno rules, only the results using *triangular* membership are illustrated. The Takagi–Sugeno model was trained during 400 iterations using a combination of *gradient descent* and *least squares* (ANFIS Scheme [19]). The ANFIS scheme was more efective in the Takagi–Sugeno scheme showing a faster convergence. For the Mamdani models, the use of ANFIS or "pure" *gradient descent* did not show major differences.

Observe that all the approximations are better than the approximations given by the models obtained with the method of *mosaic or table lookup*. The Mamdani model with *triangular* membership functions improves the approximation by extending the overlap of the most external membership functions (see Figure 2.5). Observe that the function no longer crosses the points of the consequences and the interpolation is no longer linear, all because the overlap of the membership functions is no longer $\frac{1}{2}$.

On the other hand, the Mamdani models using *polynomial* and *Gaussian* membership functions improve the approximation by narrowing the central membership functions and putting their centers closer (see Figures 2.6 and 2.7). The improvement shown by the approximation using polynomial membership functions (see Figure 2.6) is very remarkable compared with the approximation obtained with the simple *mosaic or table lookup* method.
The Takagi–Sugeno model shows again a good approximation with some improvement as shown in Table 2.2, but compared with the other models the improvement brought by the gradient descent method was not as significant as it was for the other models. However, observe that even that the membership functions did not have significant changes; the functions describing the consequences show completely different slopes.

In general, the improvement in the approximation provided by the tuning of the membership functions using the *gradient descent* method is clear. The observed improvement, which in one case (Mamdani polynomial model) was of almost two orders of magnitude, is partially explained by the increased number of degrees of freedom (consequences + parameters of the membership functions) introduced in the gradient descent method (see Table 2.2).

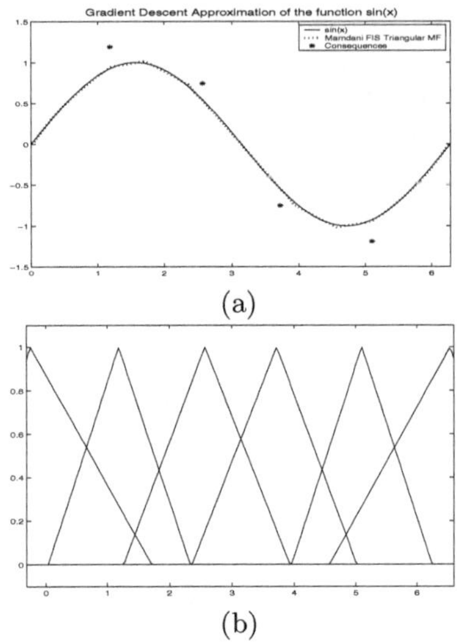

Figure 2.5. (a) Approximation generated by a Mamdani fuzzy model trained using the *gradient descent method* using triangular membership functions with 6 membership functions initially equally spaced. (-) Original function (- -) Approximation generated by the fuzzy model (*) Consequences of the rules (b) Membership functions after training

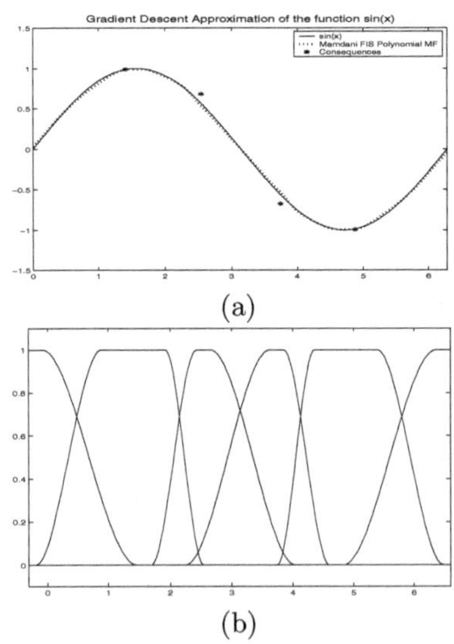

Figure 2.6. (a) Approximation generated by a Mamdani fuzzy model trained using the *gradient descent method* using polynomial membership functions with 6 membership functions initially equally spaced. (-) Original function (- -) Approximation generated by the fuzzy model (*) Consequences of the rules (b) Membership functions after training

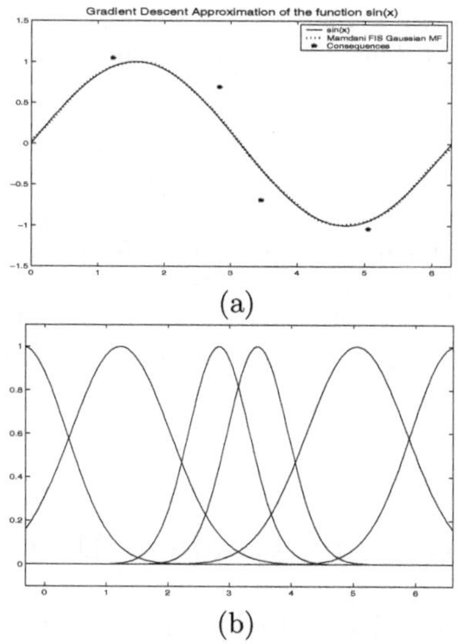

Figure 2.7. (a) Approximation generated by a Mamdani fuzzy model trained using the *gradient descent method* using polynomial membership functions with 6 membership functions initially equally spaced. (-) Original function (- -) Approximation generated by the fuzzy model (*) Consequences of the rules (b) Membership functions after training

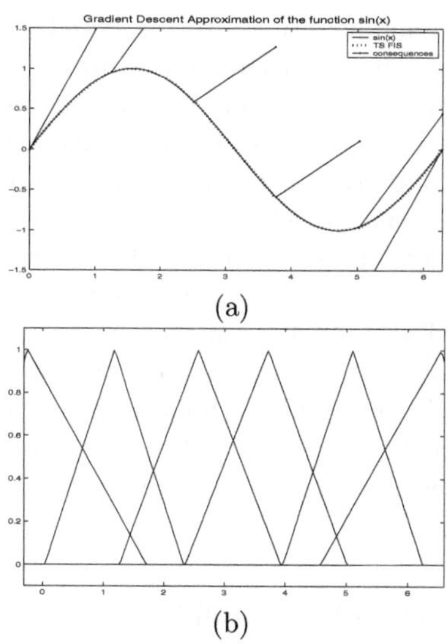

Figure 2.8. (a) Approximation generated by a Takagi–Sugeno fuzzy model trained using the *mosaic or table lookup scheme* using triangular membership functions with 6 membership functions equally spaced. (-) Original function (- -) Approximation generated by the fuzzy model (.-) Consequences of the rules (b) Membership functions after training

Table 2.2. Approximation Error ($\sum_{i=1}^{629}[f(x) - \hat{f}(x)]^2$)of the Different Models Trained with the *Mosaic or Table Lookup* Scheme and the *Gradient Descent* Method for 400 Steps.

	Table look up		Gradient Descent	
Model	Error	DOF	Error	DOF
Mamdani triangular M.F.	1.3392	6	0.0615	24
Mamdani Gaussian M.F.	0.2378	6	0.0449	18
Mamdani polynomial M.F.	12.9666	6	0.1755	30
Tak.Sug. triangular M.F.	0.0206	12	0.0117	30

DOF=degrees of freedom number of adjustable parameters

2.3 Using Clustering and Gradient Descent

The methods studied so far had placed the fuzzy sets of the input domains on their initial positions according to the choice made by the designer (typically equally distributed). Two choices has been made by the designer – the number of membership functions and their initial distribution. The methods based on clustering aim to obtain both parameters at the same time, the number of fuzzy sets needed to make the function approximation and their distribution along the input domains.

The methods based on clustering are considered as data-driven methods. The main idea of these methods is to find structures (clusters) among the data according to their distribution in the space of the function and assimilate each cluster as a multidimensional fuzzy set representing a rule. The cluster prototypes can be either a point (to construct *Mamdani* models) or a hyperplane (to construct *Takagi–Sugeno* models).

The fuzzy inference system is constructed by means of projecting the clusters into the input space and approximating the projected cluster with a one-dimensional fuzzy set. The advantage of these methods is that they generate automatically the membership functions, leaving as the user's choices only the parameters of the clustering algorithms (number of clusters and distance function). According to the type of model to be constructed the method will be slightly different. Here is a summary of the methods:

2.3.1 Algorithm for Mamdani Models

- Collect the data and construct a set of vectors $Z^t = \{x^{t^T}, y^t\}$ where x^t and y^t are, respectively, the inputs and the output of the function. Observe that here we assume $x^t \in \Re^n$ and $y^t \in \Re$.
- Search for clusters using the Fuzzy C-means algorithm [2] or the mountain-clustering algorithm [4] for problems where the dimension of the input

space is small. Appendix B includes a description of the mentioned algorithms.

- Project the membership functions from the partition matrix U into the input space.
- Approximate the projected membership function using convex membership functions (triangular, Gaussian, polynomial, trapezoidal, etc.)
- Construct the rules with the projected membership functions.
- Calculate the consequences using recursive least squares.
- Adjust the parameters of the antecedents (if needed) using gradient descent.

2.3.2 Algorithm for Takagi–Sugeno Models

- Collect the data and construct a set of vectors $Z^t = \{x^{t^T}, y^t\}$ where x^t and y^t are, respectively, the inputs and the output of the function. Observe that here we assume $x^t \in \Re^n$ and $y^t \in \Re$.
- Search for clusters using the Gustafson and Kessel (G-K) algorithm [3]. Appendix B describes the G-K algorithm.
- Check for similarities among the clusters. Do two clusters describe a similar hyperplane?
- Project the membership functions from the partition matrix U into the input space.
- Approximate the projected membership function using convex membership functions (triangular, Gaussian, polynomial, trapezoidal, etc.)
- Construct the rules with the projected membership functions.
- Generate the consequences using the covariance matrices of each cluster.
- Calculate the consequences that are not covered by the clusters using recursive least squares.
- Adjust the parameters of the antecedents (if needed) using gradient descent.

Summary:
The *clustering + gradient descent* method calculates the initial location of the membership function by projecting the partition matrices obtained from a clustering applied to the input–output data. The consequences are generated from the centers of the clusters and for the Takagi–Sugeno models from the centers and their covariance matrices. The parameters can be refined to improve the approximation by applying gradient descent.

2.3.3 Illustrative Example

This example uses the same function ($f(x) = sin(x)$) presented in Sections 2.1.1 and 2.2.6. The data are composed of 629 equidistant points that were used to approximate the function $f(x) = sin(x)$ over the interval $[0, 2\pi]$. In this case the models were constructed based in two clustering procedures.

Fuzzy C-Means to construct Mamdani models and Gustafson and Kessel to construct a Takagi–Sugeno model. For both procedures the number of clusters selected *a priori* was 6 and the stopping criteria $\epsilon = 5 \times 10^{-5}$. This selection was made such that the results are comparable with the ones shown in previous examples. Both clustering algorithms were executed and they generated the clusters shown in Figures 2.9(a) and 2.10(a). Observe that the cluster of the G-K algorithm are characterized by their center and "main direction" of its covariance matrix. The partition matrix was projected over the input domain obtaining the membership functions shown in Figures 2.9(b) and 2.10(b).

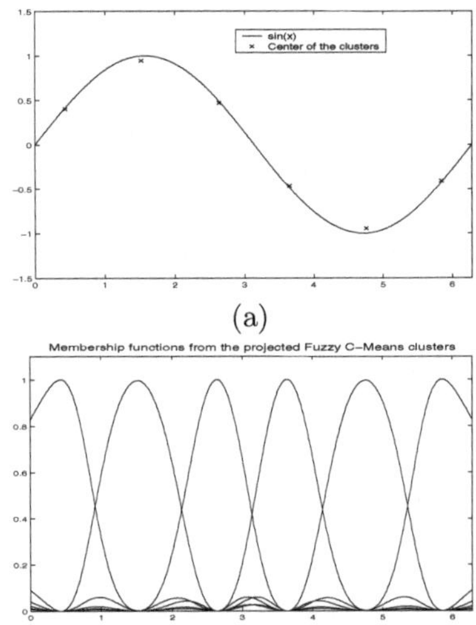

(b)

Figure 2.9. (a) Center of the clusters found by the Fuzzy C-means algorithm. Original function (-) Center of the clusters (*) (b) Membership functions projected from the partition matrix U

The projected membership functions obtained from the partition matrix U are approximated by convex membership functions as they are shown in Figures 2.11 and 2.12.

The rule base was constructed and the models were further optimized using gradient descent for 400 steps. Figures 2.13 and 2.14 show the approximation of the function. It is important to comment that for the Mamdani models there are little differences with the models shown in previous examples, but it is not the case of the Takagi–Sugeno models. Observe the orientation of the consequences of the rules, which are almost tangent to the function.

Table 2.3 summarizes the results obtained with the three methods shown. Perhaps the most remarkable results are the improvement of the models using Gaussian functions. The reason for such a benefit from the clustering can be

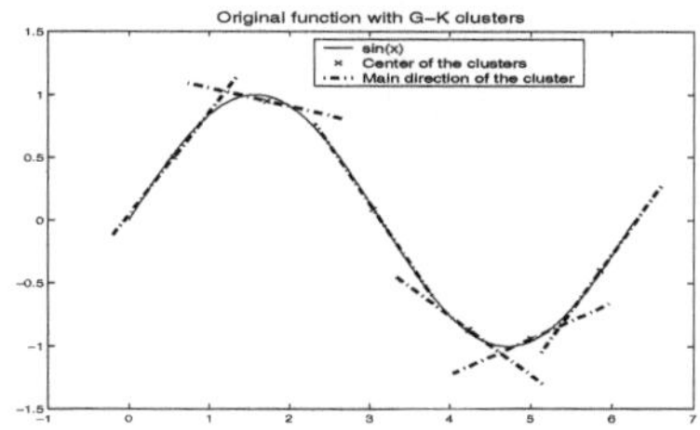

(a)

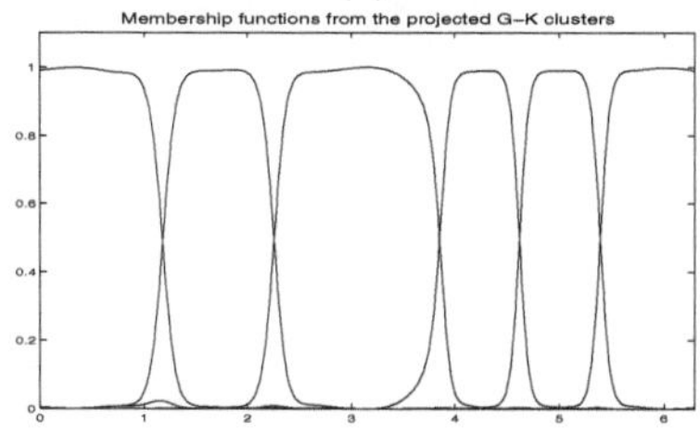

(b)

Figure 2.10. (a) Clusters found by the G-K algorithm. Original function (-) Center of the clusters (*) Main direction of the covariance matrix (.-) (b) Membership functions projected from the partition matrix U

explained by the strong similarity between the projected partition function from the clusters and the Gaussian membership functions. Observe that these results are simple illustrations of the methods and do not represent an absolute benchmark. For other functions the performance exhibit by the models will be different.

Table 2.3. Approximation error ($\sum_{i=1}^{629}[f(x) - \hat{f}(x)]^2$)of the Different Models Trained with the *Mosaic or Table Lookup* Scheme, the *Gradient Descent* Method for 400 Steps and *Clustering Gradient Descent* Method for 400 steps)

Model	Table lookup app. error	Gradient descent app. error	Clustering + GD app. error
Mamdani triangular M.F.	1.3392	0.0615	0.0858
Mamdani Gaussian M.F.	0.2378	0.0449	0.0037
Mamdani polynomial M.F.	12.9666	0.1755	0.1959
Takagi–Sugeno model.	0.0206	0.0117	0.5058

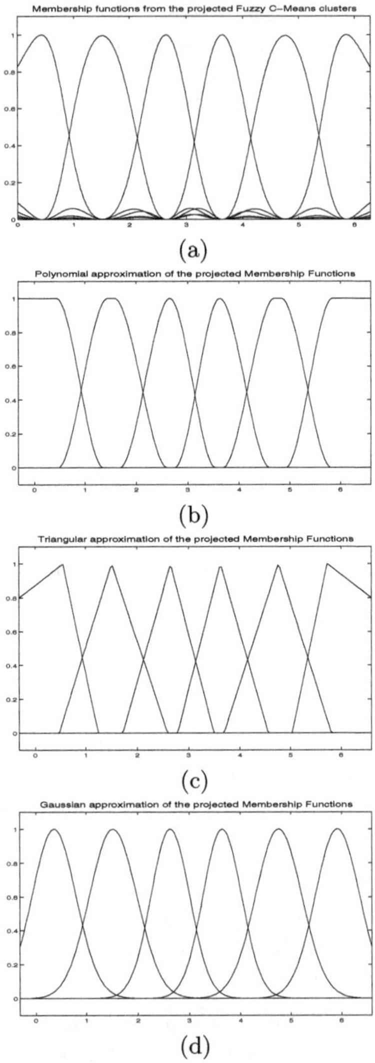

Figure 2.11. Membership functions for Mamdani models.(a) Membership functions projected from the partition matrix U (b) Approximation with polynomial M.Fs. (c)Approximation with triangular M.Fs. (d) Approximation with Gaussian M.Fs.

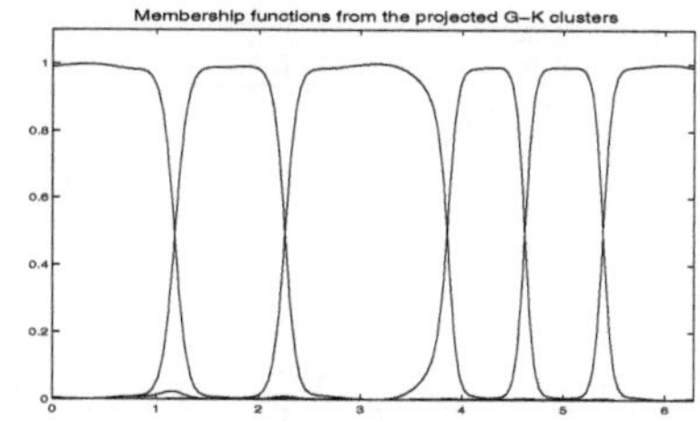

(a)

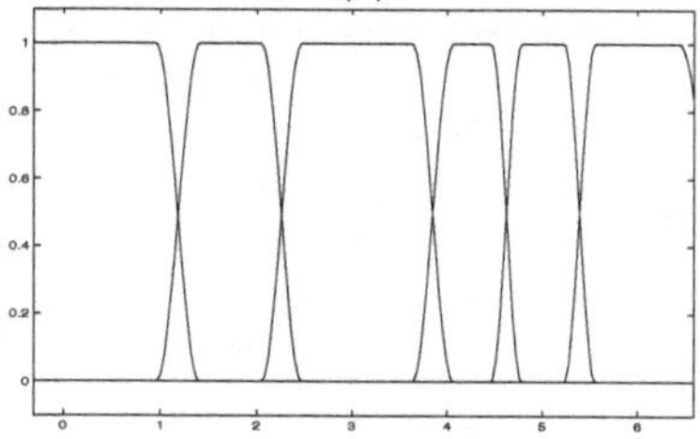

(b)

Figure 2.12. Membership functions for Takagi–Sugeno models.(a) Membership functions projected from the partition matrix U from the G–K clustering (b) Approximation with polynomial M.Fs.

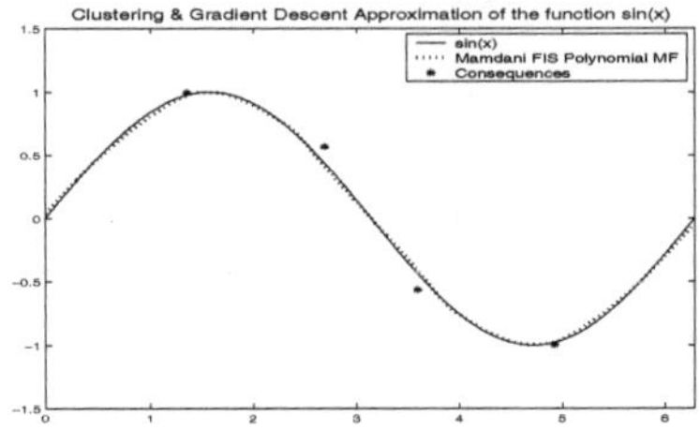

(a)

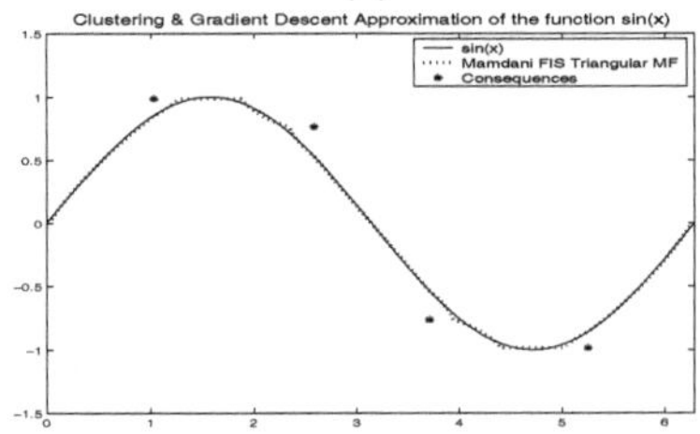

(b)

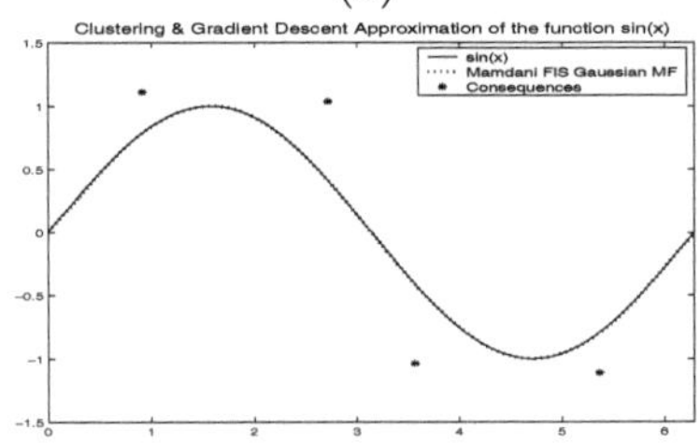

(c)

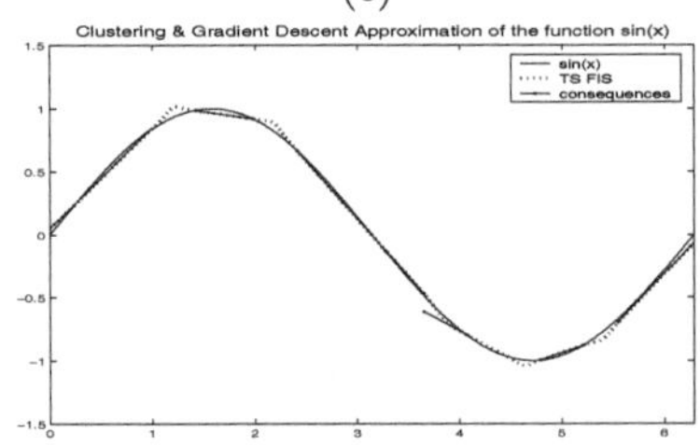

(d)

Figure 2.13. Function approximation of the models obtained using clustering and gradient. Original function (-) Approximated function (–) Consequences (*) Consequence of the TS model (-.). (a) Membership functions projected from the partition matrix U (b) Approximation with polynomial M.Fs. (c)Approximation with triangular M.Fs. (d) Approximation with Gaussian M.Fs.

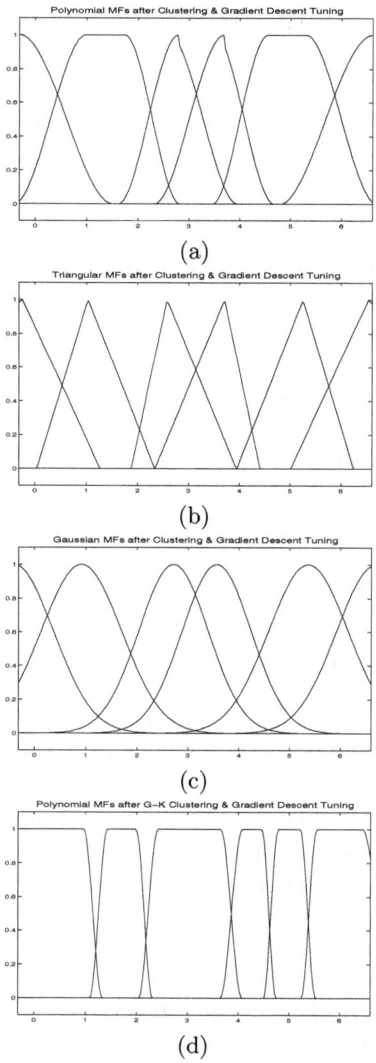

Figure 2.14. Membership functions for the models obtained by clustering and gradient descent optimization.(a) Mamdani model with polynomial M.Fs. (b) Mamdani model triangular M.Fs. (c) Mamdani model with Gaussian M.Fs.(d) Takagi–Sugeno model with polynomial M.Fs.

2.4 Using Evolutionary Strategies

The evolutionary strategies are computational algorithms that use methods derived from the concept of "natural evolution." Some of the methods include reproduction, mutation and selection. The use of these algorithms has been oriented to the search of parameters such that a certain computational entity can achieve some goals.

In this case the computational entity will be a fuzzy system, the goal will be to approximate a function with certain accuracy and a limited complexity and the parameters could be the number of membership functions, their distribution, etc.

Basic methods in these strategies are the so-called genetic algorithms [23]. In genetic algorithms, the data are represented as binary strings. The parameters are encoded on these binary strings. It is important to remark that the efficiency of these techniques is strongly affected by the "code book" used to construct the strings [24]. Initiallty, a group of these strings is generated as the initial "population." The fulfillment of the goal is tested for each element of the population (cost evaluation) and a "fitness" value is generated such that, if the value is larger, the objective is better achieved. The procedure can be outlined as follows:

- Take the initial population N and evaluate the "fitness" of the individuals (binary strings).
- Reproduce the population according to the "fitness," such that those individuals with higher values of fitness will have a higher probability of being reproduced.
- Make random couples among the individuals of the reproduced population and apply the "crossover" operation. The crossover operation takes two individuals and generates a random number $l \leq L$ where L is the length of the string. This operation generates two new individuals by taking the first l elements of one string and the remaining $L - l$ element from the other string. For example, take the first string $A_1A_2A_3A_4A_5A_6$ and the second string $B_1B_2B_3B_4B_5B_6$. In this case $L = 6$. Suppose $l = 2$. The crossover will be represented as

$$
\begin{array}{c}
A_1A_2A_3A_4A_5A_6 \\
B_1B_2B_3B_4B_5B_6 \\
------- \\
A_1A_2B_3B_4B_5B_6 \\
B_1B_2A_3A_4A_5A_6
\end{array}
$$

- Finally, some members of the population are selected for "mutation." A random number l is generated such that $0 < l \leq L$ for each selected member and the bit l is flipped. Suppose the string $A_1A_2A_3A_4A_5A_6 = 101100$ is selected for mutation and $l = 4$. The string after mutation is $A_1A_2A_3A_4A_5A_6 = 101000$.

- The "fitness" of the generated population is evaluated and the procedures of reproduction, crossover and mutation are repeated for a given number of times (generations).

These algorithms are very powerful for the search of "global" solutions in the search space, and there is a probability equal to 1 that the algorithm will find the "global solution" after a number of generations given by [20]

$$\frac{1}{1-(1-p_M\frac{N_{opt}}{2^L})^N} \tag{2.48}$$

where p_M is the probability of mutation, N_{opt} is the number of global solutions in the final population, L is the length of the strings and N is the number of strings in the population.

The application of these algorithms to the design of fuzzy systems is mainly oriented to the generation of the number and distribution of the membership functions. One example of codification is: Assume a number of triangular, trapezoidal or polynomial membership functions with overlap $\frac{1}{2}$ have been fixed for each input. Then the string will represent the distance from the previous point, as shown in Figure- 2.15. The length of the string is $L = 28$, four groups of seven bits. An example of mutation is shown in Figure- 2.16, where

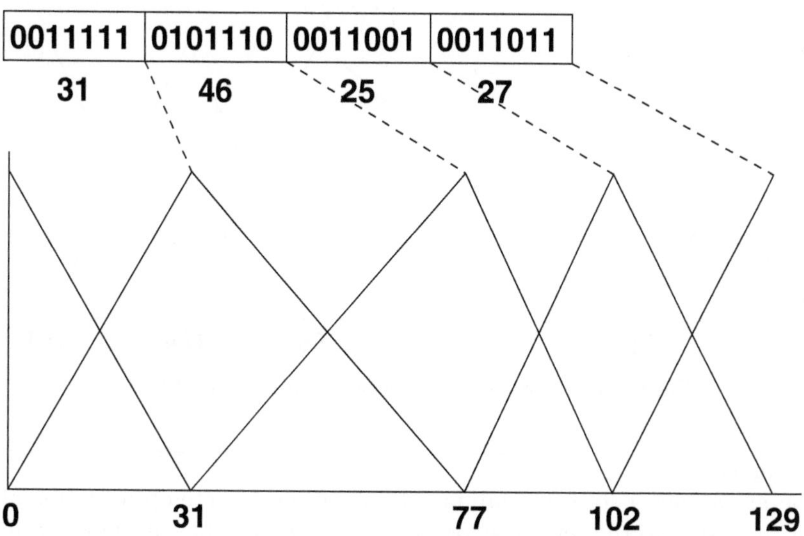

Figure 2.15. Codification of the membership functions

$l = 10$. Finally, an example of the effect of the crossover operation is shown in Figure- 2.17, where $l = 10$. Other codification methods and details can be seen in [20].

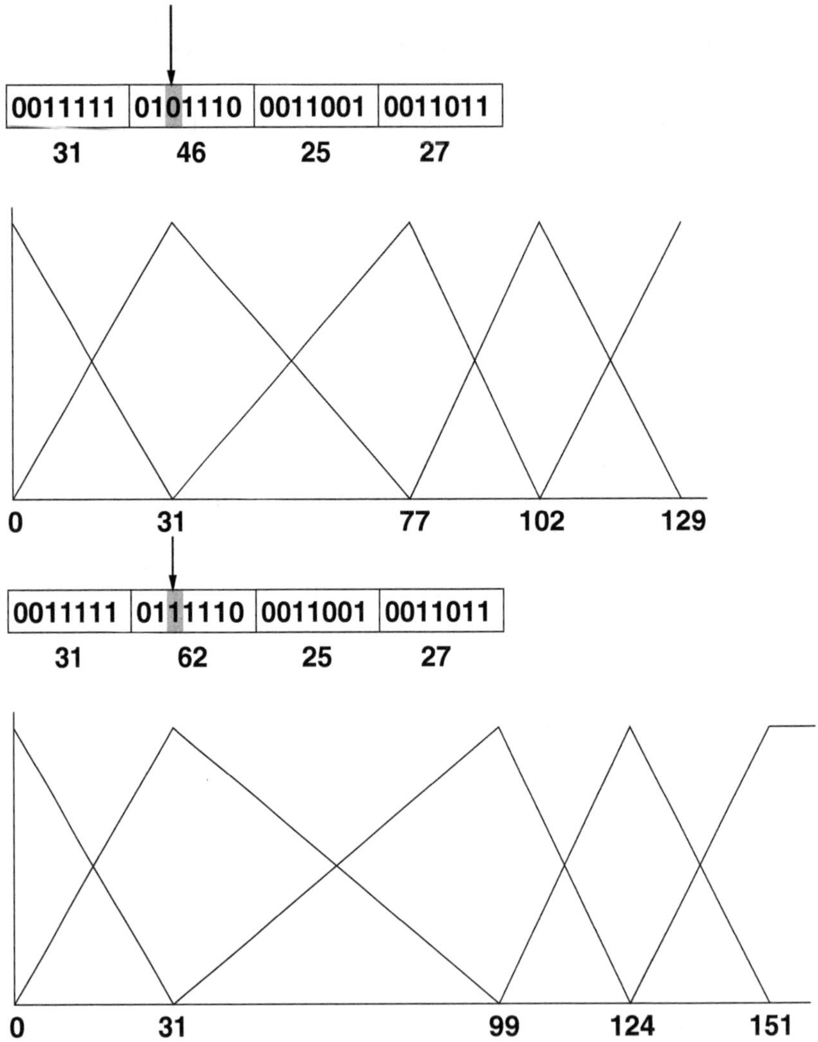

Figure 2.16. Mutation operation in a fuzzy partitions

Summary:
The *evolutionary strategies* are mainly based on discrete optimization algorithms such as the genetic algorithms. The method can calculate parameters such as number and location of the membership functions.

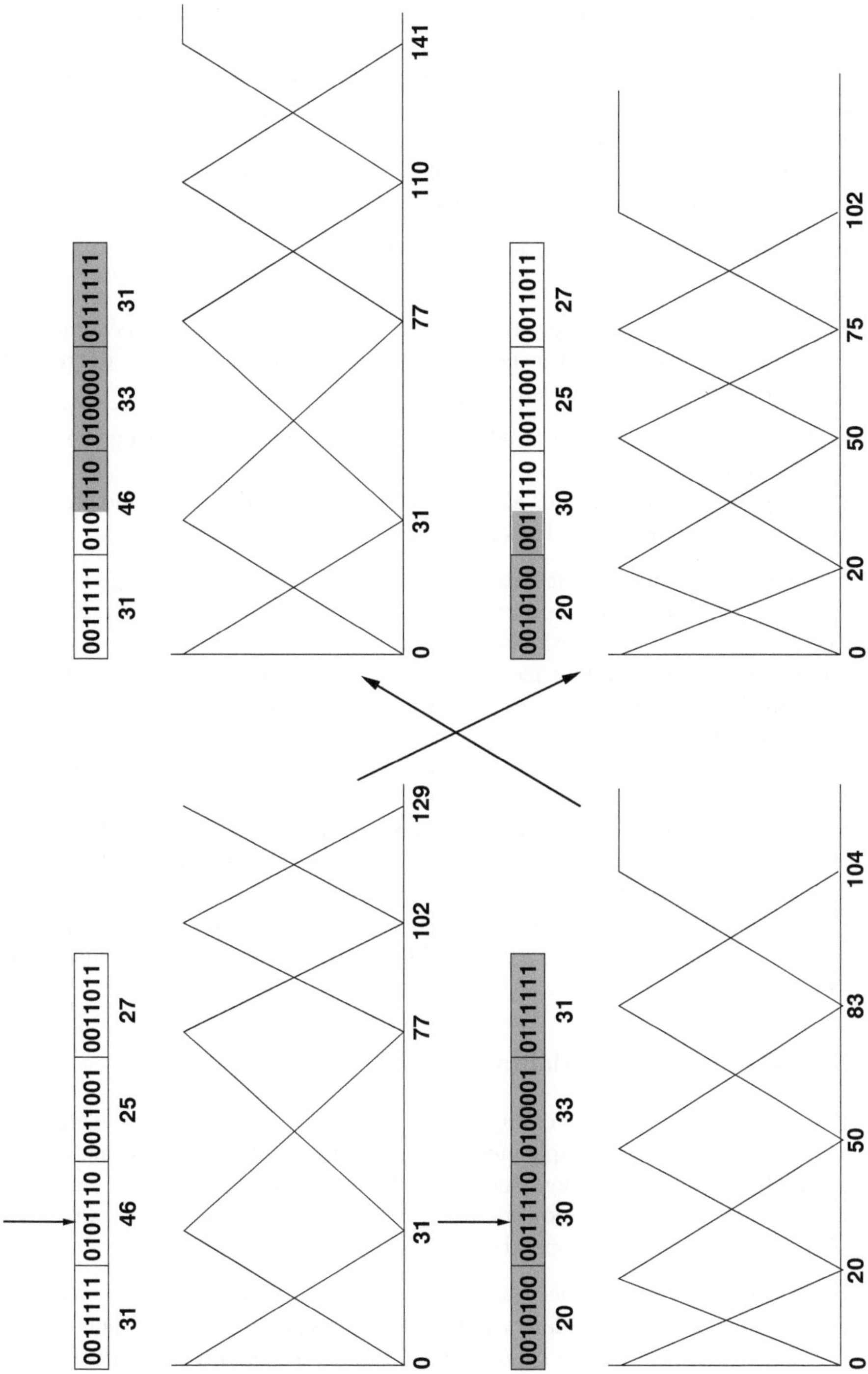

Figure 2.17. Crossover operation between two fuzzy partitions

2.5 Generalization and Consequences Estimation

The issue of generalization is quite related with the issue of consequence estimation in fuzzy systems, as will be explained in the following lines. Generalization is the capacity of the designed system (the fuzzy inference system) to generate "good" output values when new inputs are presented to the system. Two issues limit the capacity of generalization:

- Lack of excitation during model construction
- Too many degrees of freedom in the model

These two issues are strongly related because as the degrees of freedom grow the data must excite all the new modes introduced by the new degrees of freedom. There are two ways to improve the generalization:

- Reducing the degrees of freedom with the drawback of reducing the accuracy of the approximation.
- Generating many data for all possible operating modes. For some practical cases, this is almost an impossible task.

Assuming the input–output data are given in advance, the challenge is to design a system with good approximation properties and good generalization. The application of the methods reviewed in previous sections postulates the generation of the consequences of the rules by means of least squares. As mentioned in Section 2.1, the calculation of the least squares using Equation (2.11) will be possible only if $\mathrm{rank}(W^T W) = \dim\theta$. In cases where $\mathrm{rank}(W^T W) \leq \dim\theta$, the estimation will be very poor and the consequences of those unexcited rules will be very far from their real value. A reasonable solution is to initialize the rules using information given by a simpler model (with very few degrees of freedom) and to improve the estimation of the consequences of those rules that have been excited using recursive least squares. The advantage of the recursive least-squares algorithm is that it only updates those terms that have been excited. The procedure can be detailed in two steps.

2.5.1 Consequence Initialization

The initialization of the consequences can be done in two ways using the information given by a simple model with sufficient excitation or using expert knowledge. The use of expert knowledge demands, the designer that initialize those rules with empirical knowledge. The initialization using a simple model with sufficient excitation operates as follows:

- The smallest fuzzy model $\hat{f}(x^t)$ is constructed by placing only two membership functions (triangular or polynomial) on each input with their modal values placed, respectively, in the maximum and the minimum values of the universe of discourse and fixing the overlap value in $\frac{1}{2}$. This distribution of the membership functions will generate a fuzzy system with 2^N rules where N is the number of inputs. This fuzzy system has the property that any input presented will excite the whole set of rules. This property guarantees enough excitation such that the 2^N consequences can be estimated using the least-squares solution given in the Equation (2.11).

- If the constructed model uses triangular or polynomial membership functions with overlap $\frac{1}{2}$, the consequences of the rules can be initialized using the model $\hat{f}(x^t)$ as follows:

$$\bar{y}^{j_1 j_2 \ldots j_N} = \hat{f}(M^{j_1 j_2 \ldots j_N}) \tag{2.49}$$

with

$$M^{j_1 j_2 \ldots j_N} = \{m^1_{j_1}, m^2_{j_2}, \ldots, m^N_{j_N}\}^T$$

where $m^i_{j_i}$ are the modal values of the membership functions of the fuzzy model $f(x^t)$. If the model is not constructed as described above, the initial consequences can be estimated by using a data set generated from the model $\hat{f}(x^t)$, so that the condition of sufficient excitation is guaranteed. This can be done just by generating input data regularly distributed and with "enough" density over the input space U.

This initialization method guarantees that the constructed fuzzy model will be at least as good as the best multilinear model, if the smaller model is constructed with triangular membership functions, or at least as good as the best third-order multipolynomial model. These bounds guarantee the quality of the generalization even if the training data have no information about some of the regions described in the rule base.

2.5.2 Consequence Estimation

Once the consequences have been initialized, the recursive least-squares algorithm can be applied to improve the estimation. The algorithm is described as follows using the notation presented in Section 2.1:

$$\theta(k+1) = \theta(k) + \gamma(k)[y^t - W^t\theta(k)] \tag{2.50}$$

with $W^t = \{w^t_1, w^t_2, \ldots, w^L_t\}$, $\theta(k) = \{\bar{y}^1(k), \bar{y}^2(k), \ldots, \bar{y}^L(k)\}$ and:

$$\gamma(k) = P(k+1)W_{k+1} \tag{2.51}$$

$$= \frac{1}{W^t P(k) W^{t^T} + 1} P(k) W^t \tag{2.52}$$

$$P(k+1) = [I - \gamma(k)W^t]P(k) \tag{2.53}$$

with the initial value $P(0) = \alpha I$, where α is a large scalar value. The procedure is repeated and each time the index k is incremented. Also, the index t is incremented until it reaches the value N, and then the value of t is reset to

$t = 1$. The initial values of $\theta(0)$ are the initialization values given by the procedure described before. The following example is presented in order to illustrate how the present method improves the generalization.

Example 2.1. The objective is to approximate the function of two variables $f(x, y) = 6x + 4y + \cos(\pi x) + \cos(\pi y) + 50$ on the interval $(x, y) \in U$ $U = [-2, 2] \times [-2, 2]$. The function is plotted in Figure- 2.18.

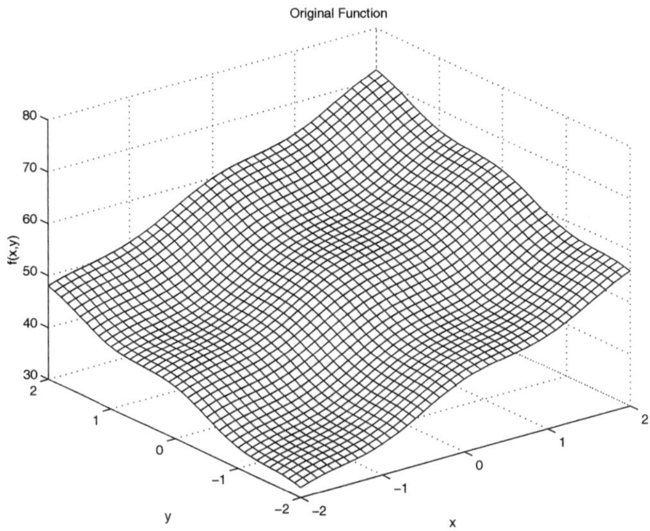

Figure 2.18. Function to be approximated $f(x, y) = 6x+4y+\cos(\pi x)+\cos(\pi y)+50$

The function is sampled at 153 points. The sampling was done such that only one point falls in the interval $V = [-2, 0] \times [-2, 0]$. The data points are depicted in Figure 2.19.

The function will be approximated by a fuzzy system using five triangular membership functions equally distributed on each domain with overlap 0.5. A total of 25 rules is generated and the consequences will be estimated using three methods: least squares (LS), recursive least squares (RLS) and RLS with the consequence initialization method explained in Section 2.5.1. Observe in Figure 2.20 that the LS method and the "pure" RLS fail to approximate the function in the domain V and even the LS solution fails to make a good approximation in the region where the "training" points were selected.

The third method as was explained in Section 2.5.1 first calculates the smallest fuzzy model $\hat{f}(x, y)$ with only two membership functions with overlap 0.5 covering the whole domain on each input. The model has four rules that are excited by all the points such that the consequence estimation does not represent any numerical problem. The approximation generated by this model

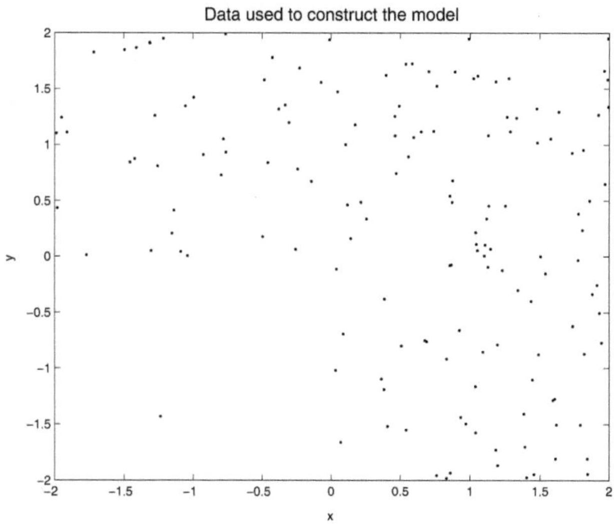

Figure 2.19. Sampled points to approximate the function $f(x, y) = 6x + 4y + \cos(\pi x) + \cos(\pi y) + 50$

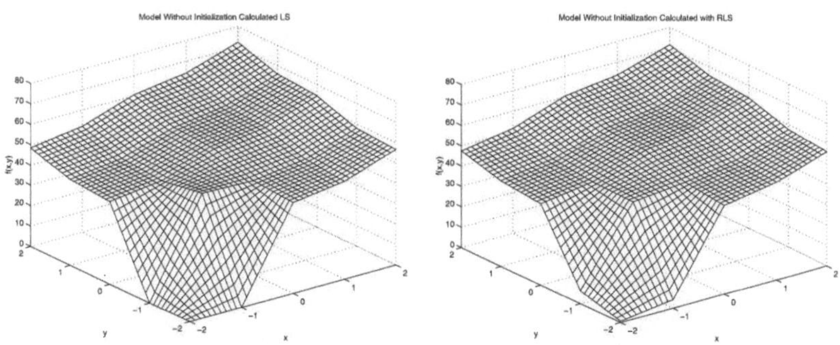

(a) (b)

Figure 2.20. (a) Approximation obtained with the consequences calculated using LS (b)Approximation obtained with the consequences calculated using RLS

is shown in Figure 2.21. Equation (2.49) is used to initialize the consequences of the rules in the model with 25 rules such that it generates an approximation perfectly equivalent to the approximation given by the model $\hat{f}(x, y)$.

Then the consequences are estimated using RLS. The results are shown in Figure 2.22. Observe that the approximation is good in the whole domain U and there are no big changes in the region V where almost no data exist during the training. A final comparison was performed by generating 141 points in the domain U but excluding the region V (the same conditions used for the training) to observe the approximation error in the "well-excited" region. The results are presented in Table 2.4, and the error index is calculated as $E =$

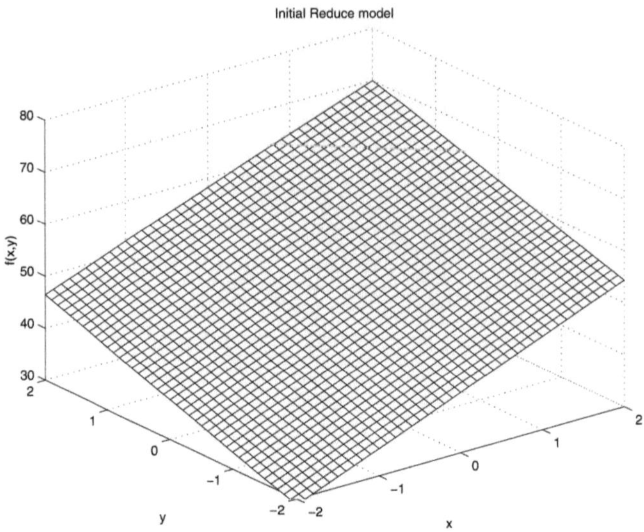

Figure 2.21. Approximation generated by the "smallest" fuzzy model $\hat{f}(x, y)$ with only 4 rules

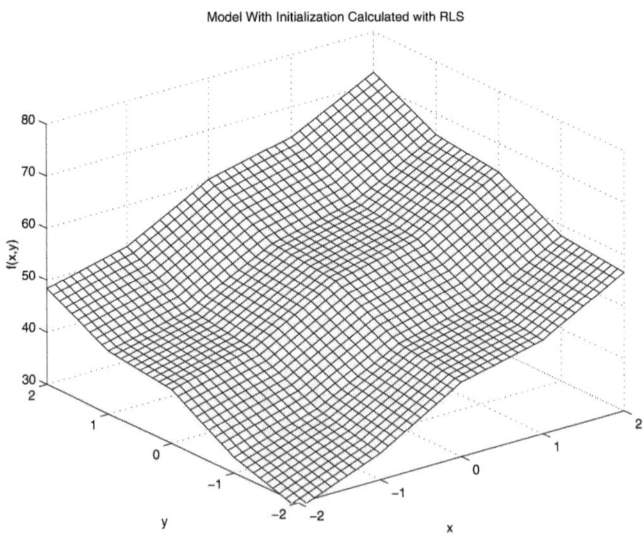

Figure 2.22. Approximation generated by the fuzzy model initialized with $\hat{f}(x, y)$ and with the consequences calculated using RLS

$\frac{1}{141}\sum_{i=1}^{141} e^2$, where E is the difference between the real and the estimated values. From these results, it is clear that the LS method is the worst. The reason is that the inputs selected for the estimation did not excite some of the rules and in this specific case the $\text{rank}(W^T W) = 22$; therefore, the LS solution is badly conditioned. The RLS solution is better because it updates only the excited rules but the badly excited rules are not updated, making a bad generalization on "poorly" excited regions. Finally, the best performance is by far the one of the proposed method. The reason is that this method assumes the generalization given by a "well"-excited model ($\hat{f}(x, y)$) and the tuning will only improve the approximation on these regions where there is enough excitation.

Table 2.4. Example: Comparison Between Methods for Consequences Calculation

Method	Approx. error
Least Squares	0.3455
Recursive Least Squares	0.1659
RLS with initialization using $\hat{f}(x, y)$	0.0146

Summary:
Fuzzy models should make good predictions even when they are asked to predict on regions that were not excited during the construction of the model. The generalization capabilities can be controlled by an appropriate initialization of the consequences (prior knowledge) and the use of the recursive least squares to improve the prior choices. The prior knowledge can be obtained from the data.

2.6 Example of an Industrial Application

This section presents an industrial application of a static model. In this case the system helps to supply hot water for domestic use. The water is heated using steam coming from the cooling circuit of an electric power plant. The heat is transferred to the cold water by a heat exchanger (see Figure 2.23). Since the demand of hot water (F_{hw}) and the supply of steam change (F_{steam} and T_{steam}), the system must be commanded by a control system to guarantee a supply of hot water at a constant temperature (T_{hw}) (Set-point = 60°C). This objective is achieved by combining a feedback controller constructed with a PID (proportional integral derivative) and a feed-forward controller constructed using a fuzzy model (see Figure 2.24). The fuzzy model is constructed using experimental data supplied by the manufacturer of the heat exchanger. The fuzzy model is constructed to map the flow of water (F_{hw}), the temperature of the steam (T_{steam}) and the temperature of the cold water (T_{cw}) into a steam flow (F_{steam}) to guarantee that the hot water is supplied at the correct temperature (60°C).

$$F_{steam} = f(F_{hw}, T_{steam}, T_{cw})$$

Since the function is constructed using nominal data and the controller is not supposed to be "fine-tuned" on each installation, the feed-forward action will be insufficient to guarantee the supply of the water at the correct temperature. For this reason an additional feedback controller is put in place.

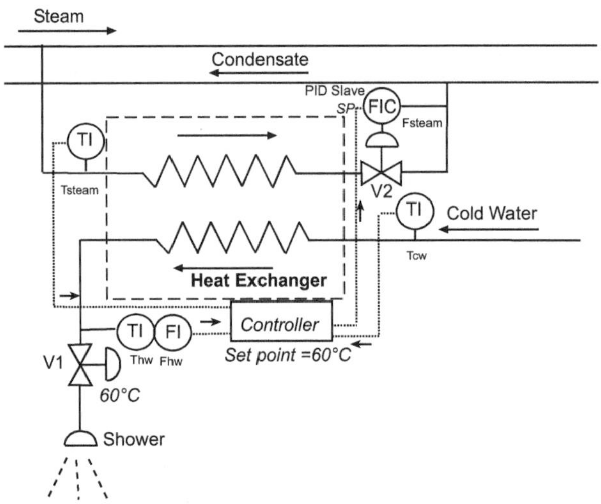

Figure 2.23. Diagram of the installation of the heat exchanger including the instrumentation and the control system

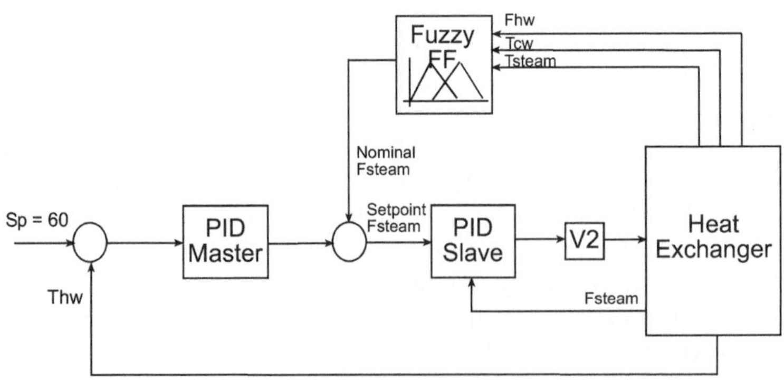

Figure 2.24. Diagram of the control system for the heat exchanger

The data supplied by the manufacturer of the heat exchanger are shown in Figure 2.25 together with the result of the approximation [see Figure 2.25(d)].

The system was implemented using triangular membership functions since the memory and the computational time available in the microcontroller were limited. Figure 2.26 shows the membership functions of the feed-forward controller.

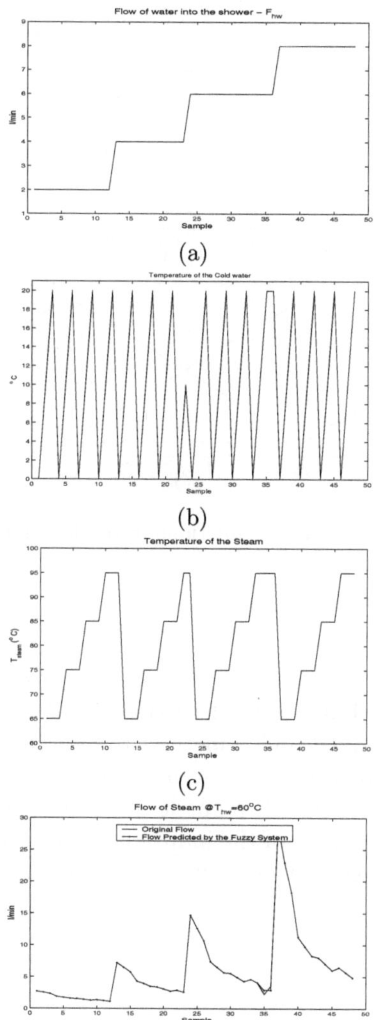

Figure 2.25. Signal collected from the heat exchanger to guarantee a nominal temperature of 60°C (a) Flow of hot water F_{hw} (b) Temperature of the cold water T_{cw} (c) Temperature of the steam T_{steam} (d) Flow of steam F_{steam} (-) Original value (.-) Value generated by the fuzzy system

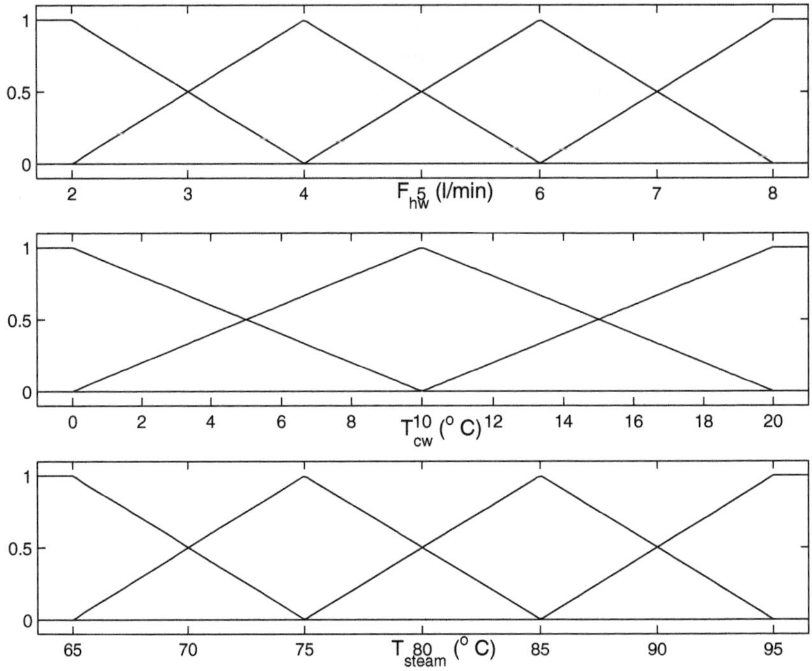

Figure 2.26. Membership functions of the feed-forward controller

2.7 Conclusions

This chapter has presented different methods to construct fuzzy models that approximate nonlinear functions. The issue of lack of excitation and generalization has been analyzed and a method to guarantee good generalization has been proposed. This method guarantees a lower bound in the quality of the model (the fuzzy model will be at least as good as the best multilinear approximation). The example of the industrial application shows a method to construct feed-forward controllers using fuzzy inference systems for function approximation.

3

Fuzzy Modeling with Linguistic Integrity: A Tool for Data Mining

3.1 Introduction

The use of models is an essential element of human behavior. When a human being predicts the impact of his actions, he is using a model. This causality analysis conditions our capacity to act and our scheme for decision making. Causality is a paramount assumption that makes models useful. Causality is reflected in language as IF–THEN rules (IF **cause-happens** THEN **a consequence is foreseen**). A set of IF–THEN rules is a linguistic representation of a mental model created inside the brain.

New instrumentation and data acquisition systems have expanded the capacity of human beings beyond the five senses. This expanded sensorial capacity has been accompanied with an increase in the storage capacity, but the capacity of the human brain to interact with this information remains limited. This situation motivates the development of computer techniques that can extract the "knowledge" and represent it in a linguistic way using IF–THEN rules. This is one of the goals of data mining, to discover causal relations among features in large databases.

There is a trade-off between numerical accuracy and linguistic interpretability. This trade-off is a consequence of a limitation of the human brain to represent a limited number of categories on a given domain. A consequence of this limitation is reflected in language. The number of linguistic labels that a human being can generate to represent categories on a given domain is limited to as much as nine and it will typically be seven.

On the other hand, the numerical accuracy is important in the implementation of policies and control actions based on the information given by model. This issue of accuracy is especially relevant when the models are used in a dynamic way where the predicted value is fed back.

In the previous chapter, some methods for model construction are explained. Initially the algorithms based on gradient descent techniques, also known as neurofuzzy models [18] [19], have been oriented to minimize the numerical error. The use of this technique generates in some cases fuzzy sets with "too much" or "no" overlap, making the interpretation of the model a difficult task.

To overcome the drawback of the initial selection of the fuzzy sets, several methods have been proposed, some of them based on local error approximation [13] [14] and others based on clustering techniques [11] [4] [12]. These methods generate multidimensional fuzzy sets and project them into the input spaces. The projections also exhibit unsatisfactory overlap, making the interpretation and the labeling of the fuzzy sets a difficult task.

This section presents the AFRELI algorithm (**A**utonomous **F**uzzy **R**ule **E**xtractor with **L**inguistic **I**ntegrity); the algorithm is able to fit input–output data while maintaining the semantic integrity of the rule base. The AFRELI algorithm uses clustering and projection techniques to find a good initial position for the fuzzy set in the input domains. A FuZion algorithm is introduced to reduce the complexity of the projected fuzzy sets. A rule base is constructed using the reduced representation of the fuzzy sets and the consequences are initialized and calculated with a method that improves generalization and avoids the lack of excitation in some rules. Finally, the consequences of the rules are represented by two fuzzy sets with different strength. The number of terms in the consequences of the fuzzy rules is reduced again using the FuZion algorithm.

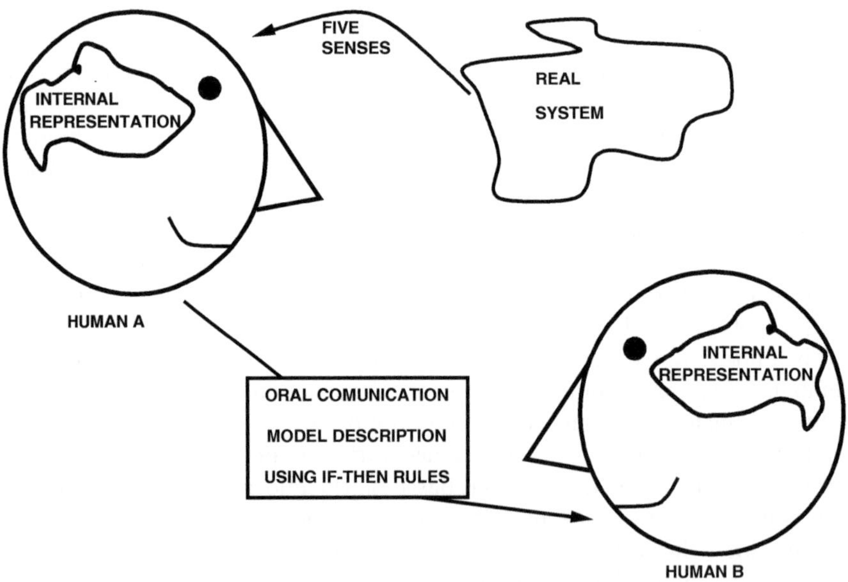

Figure 3.1. Traditional knowledge acquisition (Courtesy of Springer-Verlag [25])

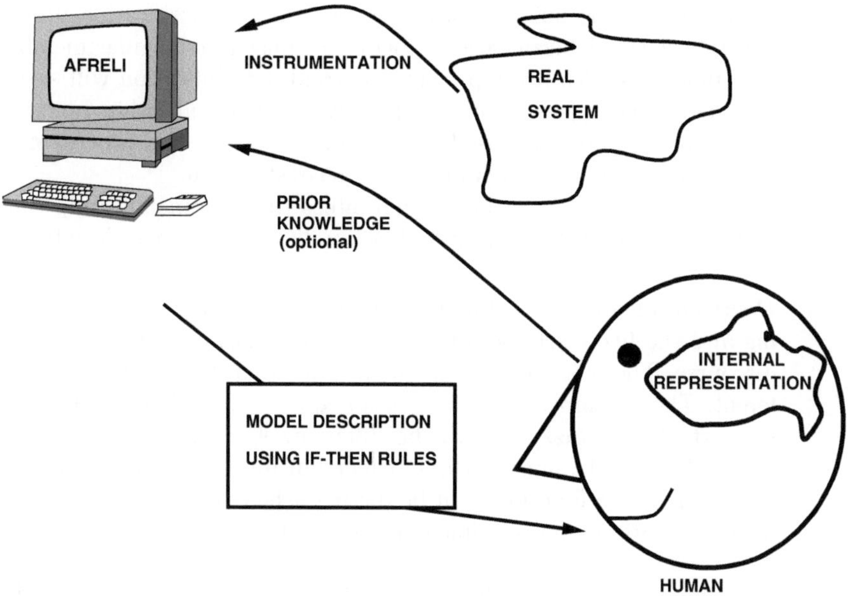

Figure 3.2. Knowledge Acquisition using AFRELI (Courtesy of Springer-Verlag [25])

Summary:
Fuzzy models with embedded linguistic interpretability are useful to extract knowledge from data. This knowledge is represented as a set of IF–THEN rules where the antecedents and the consequences are semantically meaningful.

3.2 Structure of the Fuzzy Model

The high number of degrees of freedom in a fuzzy inference system (shape and number of membership functions, T-norms, aggregation methods, etc.) gives high flexibility to the fuzzy system but also demands systematic criteria to select these parameters. Some parameters can be fixed taking into account the following issues: optimal interface design [16] and semantic integrity [15].

- Optimal interface design
  - *Errorfree reconstruction*: In a fuzzy system a numerical value is converted into a linguistic value by means of fuzzification. A defuzzification method should guaranteed that this linguistic value can be reconstructed in the same numerical value

$$\forall x \in [a, b]: \qquad \mathfrak{L}^{-1}[\mathfrak{L}(x)] = x \tag{3.1}$$

where $[a, b]$ is the universe of discourse, $\mathfrak{L}$ is the fuzzification process and $\mathfrak{L}^{-1}$ is the defuzzification process. The use of triangular membership functions with overlap $\frac{1}{2}$ and centroid defuzzification will satisfy this requirement (see proof: [16]). Polynomial membership functions with overlap $\frac{1}{2}$ also are used, but a new defuzzification process must be designed to guarantee an errorfree reconstruction.

- Semantic integrity: This property guarantees that the membership functions represent linguistic concepts. The conditions for semantic integrity are
  - *Distinguishability:* Each linguistic label should have semantic meaning and the fuzzy set should clearly define a range in the universe of discourse. Therefore, the membership functions should be clearly different. The assumption of the overlap equal to $\frac{1}{2}$ makes sure that the support of each fuzzy set will be different. A minimum distance between the modal values of the membership functions makes sure that the membership functions can be distinguished. The modal value of a membership function is defined as the α-cut with $\alpha = 1$

$$m_i = \mu_{i(\alpha=1)}(x), \quad i = 1, \ldots, N \tag{3.2}$$

  - *Justifiable number of elements:* The number of sets on each domain should be compatible with the number of "quantifiers" that a human being can handle. This number should not exceed the limit of 7 ± 2 distinct terms [26]. The choice of the shape of the membership functions does not guarantee this property. To assure that this requirement is fulfilled, the FuZion algorithm is presented further in this chapter. This algorithm reduces the number of sets on each input or output domain.
  - *Coverage:* Any element from the universe of discourse should belong to at least one of the fuzzy sets. This concept is also mentioned in the literature as ϵ completeness [19].
  - *Normalization:* Due to the fact that each linguistic label has semantic meaning, at least one of the values in the universe of discourse should have a membership degree equal to 1. In other words, all the fuzzy sets should be normal.

Based on these criteria the selected *membership functions* will be triangular and normal $(\mu_1(x), \mu_2(x), \ldots, \mu_n(x))$ with a specific overlap of $\frac{1}{2}$. It means that the height of the intersection of two successive fuzzy sets is

$$\text{hgt}(\mu_i \cap \mu_{i\pm1}) = \frac{1}{2} \tag{3.3}$$

The choice of the AND and the OR operation will be conditioned by the need of constructing a continuous and differentiable nonlinear map. This property is important if optimization of the antecedent terms is needed. In this case AND

and OR operations using *product* and *probabilistic sum* will be preferred because their derivatives are continuous.

Summary:
The selection of the appropriate structure of the fuzzy models guarantees their linguistic interpretability.

3.3 The AFRELI Algorithm

The AFRELI (Automatic Fuzzy Rule Extractor with Linguistic Integrity)[27] [28] is an algorithm designed to obtain a good compromise between numerical approximation and linguistic meaning. This particular trade-off has been referenced for long time in science (for a compilation of remarks, see [29]). The main steps of this algorithm are (see Figure 3.3)

- Clustering
- Projection
- Reduction of terms in the antecedents(FuZion, see Section- 3.4)
- Consequence calculation
- (Optional step) Further antecedent optimization
- Reduction of terms in the consequences and rule modification (FuZion, see Section- 3.4)

The AFRELI algorithm proceeds as follows:

1. Collect N points from the inputs ($X = \{x^1, \ldots, x^N\}$) and the output ($Y = \{y^1, \ldots, y^N\}$)

$$x^k = \begin{bmatrix} x_1^k \\ \vdots \\ x_p^k \end{bmatrix} \tag{3.4}$$

where $x^k \in \Re^p$ and $y^k \in \Re$ represent the inputs and the output on instant k and construct the *feature vectors*

$$U^k = \begin{bmatrix} x_1^k \\ \vdots \\ x_p^k \\ y^k \end{bmatrix} \tag{3.5}$$

$u^k \in \Re^{p+1}$.

2. Using the N feature vectors find C clusters. Apply the mountain clustering method [4] [12] to initialize the centers and to obtain the number of clusters (C). Refine the clusters using the fuzzy C-means algorithm [2].

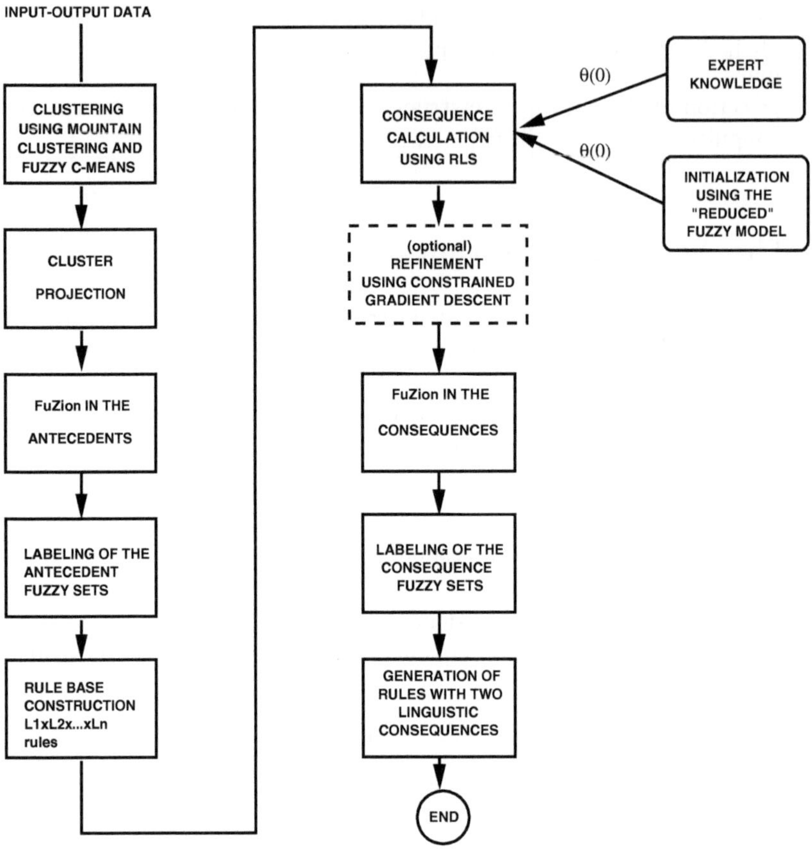

Figure 3.3. Flow diagram of AFRELI algorithm (Courtesy of IEEE [27])

$$Uc = \begin{pmatrix} \tilde{x}_1^1 & \tilde{x}_1^2 & \dots & \tilde{x}_1^C \\ \vdots & \vdots & \ddots & \vdots \\ \tilde{x}_p^1 & \tilde{x}_p^2 & \dots & \tilde{x}_p^C \\ \tilde{y}^1 & \tilde{y}^2 & \dots & \tilde{y}^C \end{pmatrix} \tag{3.6}$$

with $Uc \in \Re^{p+1 \times C}$ and $\tilde{x}_i^j$ represents the ith coordinate of the jth cluster. It is very important to remark that the use of *mountain clustering* will be limited to low-dimensional problems (up to four or five dimensions). Its inherent advantage is that it can guide good initial points and number of clusters. For high-dimensional problems, the alternative is to use only fuzzy c-means with an overestimated number of clusters; the subsequent steps (FuZion) will reduce the number of terms.

3. Project the C prototypes of the clusters into the input spaces, by converting the projected value of each prototype into the modal value of a triangular membership function.

$$m_j^i = \tilde{x}_i^j \tag{3.7}$$

where $i = 1, \ldots, p$, $j = 1, \ldots, C$

4. Sort the modal values on each domain such that

$$m_j^i \leq m_{j+1}^i \qquad \forall i = 1, \ldots, p \tag{3.8}$$

5. Add two more modal values to each input to guarantee full coverage of the input space.

$$m_0^i = \min_{k=1,\ldots,N} x_i^k \tag{3.9}$$

$$m_{C+1}^i = \max_{k=1,\ldots,N} x_i^k \tag{3.10}$$

6. Construct the triangular membership functions with overlap of $\frac{1}{2}$ as

$$\mu_j^i(x_i^k) = \max\left[0, \min\left(\frac{x_i^k - m_{j-1}^i}{m_j^i - m_{j-1}^i}, \frac{x_i^k - m_{j+1}^i}{m_j^i - m_{j+1}^i}\right)\right] \tag{3.11}$$

where: $j = 1, \ldots, C$, and the trapezoidal membership functions at the extremes of each universe of discourse

$$\mu_0^i(x_i^k) = \max\left[0, \min\left(\frac{x_i^k - m_1^j}{m_0^j - m_1^j}, 1\right)\right] \tag{3.12}$$

$$\mu_{C+1}^j(x_i^k) = \max\left[0, \min\left(\frac{x_i^k - m_C^j}{m_{C+1}^j - m_C^j}, 1\right)\right] \tag{3.13}$$

7. Apply FuZion algorithm (see Section 3.4) to reduce the number of membership functions on each input domain. This algorithm does a somewhat one-dimensional clustering among the modal values of the fuzzy sets.
8. Associate linguistic labels (*e.g.*BIG, MEDIUM, SMALL, etc.) to the resulting membership functions. This association will depend on the type of variable and the criteria of the designer. In fact, the association of a fuzzy set with a label will be the result of the agreement between the fuzzy set proposed by the algorithm and the "sense" that this set creates in the mind of the user.
9. Construct the rule base with all possible antecedents (all possible permutations) using rules of the form

...

IF x_1^k is μ_l^1 AND x_2^k is μ_l^2 AND ... AND x_p^k is μ_l^p THEN $\hat{y}_k = \bar{y}_l$

...

Equivalently, the evaluation of the antecedents of each rule can be expressed in terms of the operators *min* and *product*. Using *min* operator:

$$\mu_l(x^k) = \min\{\mu_l^1(x_1^k), \mu_l^2(x_2^k), \ldots, \mu_l^n(x_p^k)\} \tag{3.14}$$

Using *product* operator:

$$\mu_l(x^k) = \mu_l^1(x_1^k) \cdot \mu_l^2(x_2^k) \cdot \ldots \cdot \mu_l^n(x_p^k) \tag{3.15}$$

Observe that if the number of fuzzy sets on the input i is L_i and there are n inputs, the number of rules will be

$$L_1 \times L_2 \times \ldots \times L_n$$

This structure guarantees a complete description of the system in the space interval U, because every possible condition will be represented in the rule base. Observe that the number of rules will grow very fast as the number of input increases. This fact is a limitation in the sense that the comprehension of a set of rules with a large number of antecedents is difficult. In addition, the storage problem, generated by a large number of terms to be kept in the computer's memory. On the other hand, it does not represent a limitation in terms of execution time because the use of the described type of triangular membership functions will guarantee that at most 2^p rules will be evaluated during the inference process.

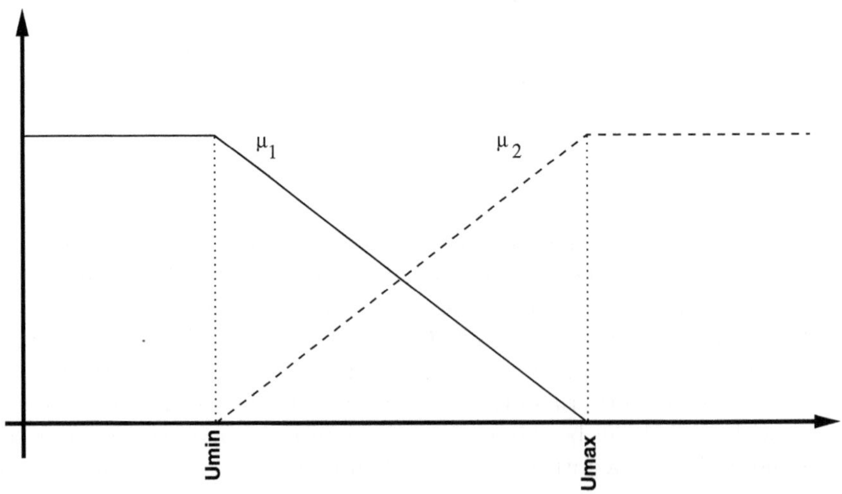

Figure 3.4. Input membership functions for the smallest fuzzy model

10. Propagate the N values of the inputs and calculate the consequences of the rules as singletons ($\bar{y}_l$). These singletons should be calculated with the method explained in Section 2.5. It is important to remember that this method guarantees that the "full" model will be at least as good as the "reduced" model. Because the "reduced" model is the best multilinear model that can be built with the given data, there is a guarantee that the "full" fuzzy model will be at least as good as the best multilinear model.

11. (**Optional step**) The approximation can be improved by fine-tuning the parameters using constrained optimization methods based on gradient descent. These methods will use the expressions for the gradient calculated in Section 2.2.2. This tuning can improve the location of the modal values of the antecedent membership functions. The main constraint applied in the optimization phase is the "distinguishability" constraint. This constraint can be represented as the minimum acceptable distance between consecutive modal values. The use of gradient descent methods will move the system parameters toward a "local minimum" close to the initial values. Because the improvement obtained by this step will not be very significant, this step is considered *optional* and will only be recommended when the numeric performance of the model does not satisfy the requirements of the user.
12. Because the singletons in the consequences are crisp sets, the linguistic meaning of the rules will be lost. The next step is to convert the singletons of the consequences to triangular membership functions with overlap $\frac{1}{2}$ and modal values equal to the position of the singleton $\bar{y}_l$. Consider the vector $\tilde{Y}$ whose entries are the L consequences of the rules but sorted in such a way that

$$\tilde{y}_1 \leq \tilde{y}_2 \leq \ldots \leq \tilde{y}_L \tag{3.16}$$

The triangular membership function of the ith consequence is

$$\mu_i^{\tilde{y}}(y) = \max\left[0, \min\left(\frac{y - \tilde{y}_{i-1}}{\tilde{y}_i - \tilde{y}_{i-1}}, \frac{y - \tilde{y}_{i+1}}{\tilde{y}_i - \tilde{y}_{i+1}}\right)\right] \tag{3.17}$$

and the two membership functions of the extremes:

$$\mu_1^{\tilde{y}}(y) = \max\left[0, \min\left(\frac{y - 2\tilde{y}_1 + \tilde{y}_2}{-\tilde{y}_1 + \tilde{y}_2}, \frac{y - \tilde{y}_2}{\tilde{y}_1 - \tilde{y}_2}\right)\right] \tag{3.18}$$

$$\mu_L^{\tilde{y}}(y) = \max\left[0, \min\left(\frac{y - \tilde{y}_{L-1}}{\tilde{y}_L - \tilde{y}_{L-1}}, \frac{y - 2\tilde{y}_L + \tilde{y}_{L-1}}{-\tilde{y}_L + \tilde{y}_{L-1}}\right)\right] \tag{3.19}$$

This description of the outer membership functions guarantees that their centers of gravity will be exactly on their modal values. This guarantees that the condition of errorfree reconstruction for optimal interface will be achieved.
13. Apply FuZion algorithm (see Section 3.4) to reduce the number of membership functions in the output universe. The FuZion process reduces groups of neighboring singletons to triangular membership functions whose modal values are representative for a group of singletons. It is optimal in a sense that the modal value of the "FuZioned" membership function is placed at the mean value of the neighboring singletons.
14. Associate linguistic labels to the resulting membership functions.
15. With the partition of the output universe, fuzzify the values of the singletons. Observe that each singleton will have a membership degree in at least one set and in as much as two.

16. Relate the fuzzified values with the corresponding rule. It means that each rule will have one consequence or two weighted consequences, where the weights are the nonzero membership values of the fuzzified singleton. This description of the consequences of the rules using two linguistic fuzzy sets and two strength values improves the interpretability of the consequences compared when only one singleton describes the consequence. The advantage of this description is that interpretability is gained without a cost in numerical precision. This strategy was independently proposed previously in [30] and [31].

Summary:
The AFRELI algorithm is a clustering-based algorithm to construct fuzzy models with linguistically meaningful parameters.

3.4 The FuZion Algorithm

The FuZion algorithm is a routine that merges consecutive triangular membership functions when their values are "too close" to each other. This merging process is needed to preserve the *distinguishability* and a *justifiable number of elements* on each domain guaranteeing the semantic integrity. A fundamental parameter of this algorithm is the minimum acceptable distance between modal values and it is given by M (see Figure 3.5). The FuZion algorithm goes as follows:

1. Take the triangular membership functions $\mu_1(x), \mu_2(x), \ldots, \mu_N(x)$ with $\frac{1}{2}$ overlap, and the modal values

$$m_i = \mu_{i(\alpha=1)}(x), \quad i = 1, \ldots, N \tag{3.20}$$

with

$$m_1 \leq m_2 \leq \ldots \leq m_N \tag{3.21}$$

2. Define the minimum distance acceptable M between the modal values.
3. Calculate the difference between successive modal values as

$$d_j = m_{j+1} - m_j, \quad j = 1, \ldots, N-1 \tag{3.22}$$

4. While $\exists d_j < M$ do steps 5–8
5. Find all the differences smaller than M.
6. Merge all the modal values corresponding to consecutive differences smaller than M using (3.23).

$$m_{new} = \frac{\sum_{i=a}^{b} m_i}{D} \tag{3.23}$$

$$D = b - a + 1 \tag{3.24}$$

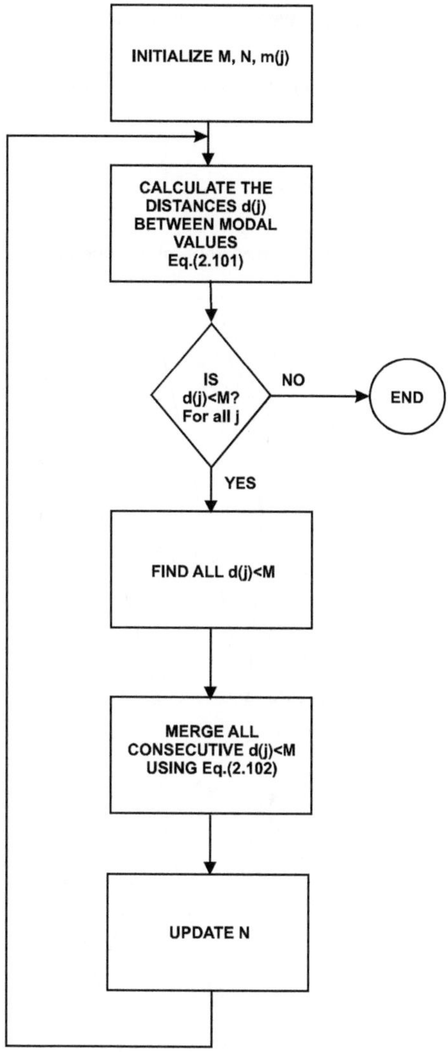

Figure 3.5. Flow diagram of FuZion algorithm (Courtesy of IEEE [27])

where a and b are, respectively, the index of the first and the last modal value of the fusioned sequence and D is the number of merged membership functions.

7. Update N.

8. Calculate the difference between the new successive modal values as:

$$d_j = m_{j+1} - m_j, \quad j = 1, \ldots, N-1 \tag{3.25}$$

9. end while
10. end

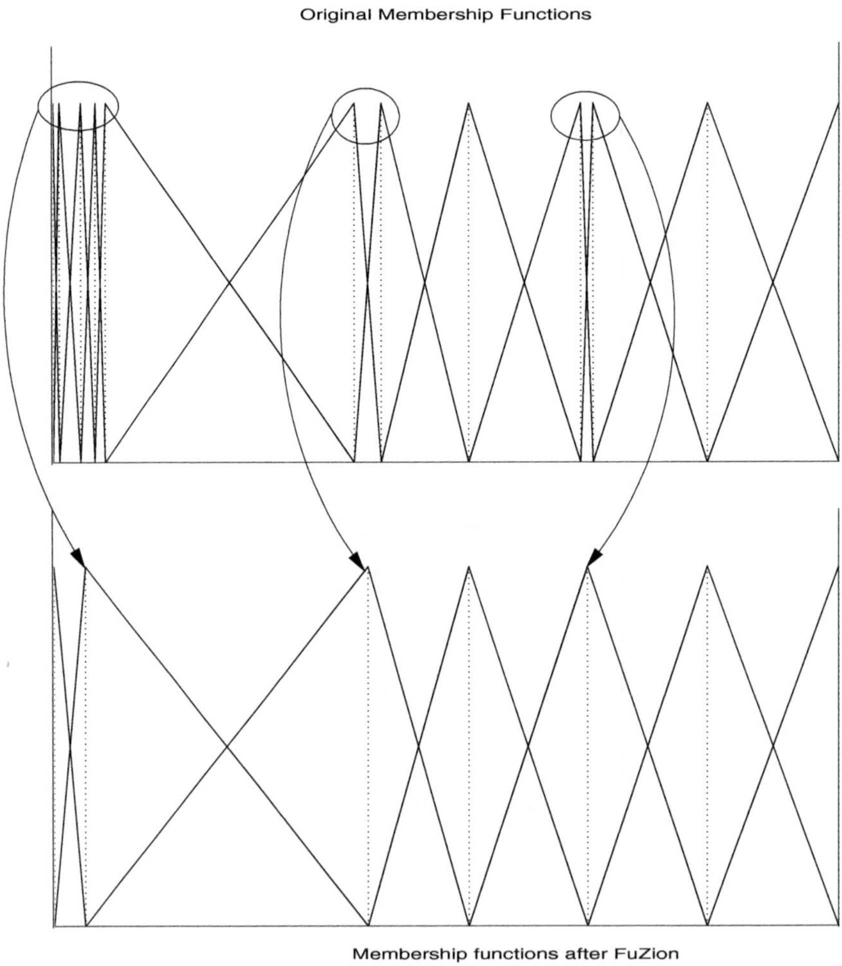

Figure 3.6. Effect of the FuZion algorithm

Summary:
The FuZion algorithm is an algorithm that reduces the number of membership functions obtained from the projection of the clusters into the input space. The method combines membership functions close to each other, reducing the number of membership functions on each input domain.

3.5 Examples

The present section shows four examples of applications of the AFRELI and FuZion algorithms. The first two examples are approximations of nonlinear static maps, the third one is the prediction of a chaotic time series and the last one is a practical example where the density of the polyethylene produced in a gas-phase HDPE reactor is predicted.

3.5.1 Modeling a Two-Input Nonlinear Function

In this example, we consider the function

$$f(x,y) = \sin\left(\frac{\pi x}{10}\right)\sin\left(\frac{\pi y}{10}\right) \tag{3.26}$$

The steps applied to the current example will be numbered using the same numbering as the one used in the FuZion algorithm in Section 3.4.

- **Step 1** 441 points regularly distributed were selected from the interval $[-10,10] \times [-10,10]$. The graph of the function is shown in Figure 3.7.
- **Step 2** Using mountain clustering and fuzzy C-means algorithm 26 clusters were found and are shown in Figure 3.8 represented with ‘x’.
- **Step 3** After the clusters were found, their center values were projected into the input domains as shown in Figure 3.9.
- **Steps 4, 5, 6** The modal values were sorted and two more modal values were added to each input domain on minus 10 and 10. The triangular membership functions were constructed. Figure 3.10 shows the projected membership functions.
- **Steps 7, 8** The FuZion algorithm was applied with M equal to 10% of the universe of discourse on each domain; observe that with this value of M five membership functions were generated, as shown in Figure 3.11. Figure 3.12 shows the membership functions obtained (seven) when the M parameter in the FuZion algorithm is chosen equal to 7% of the universe of discourse. It is clear that the smaller the value of M, the larger the number of membership functions. Linguistic labels were associated to the membership functions, as shown in Figure 3.11.
- **Step 9** A rule base was generated by combining all the membership functions present on each domain, $5 \times 5 = 25$ rules.

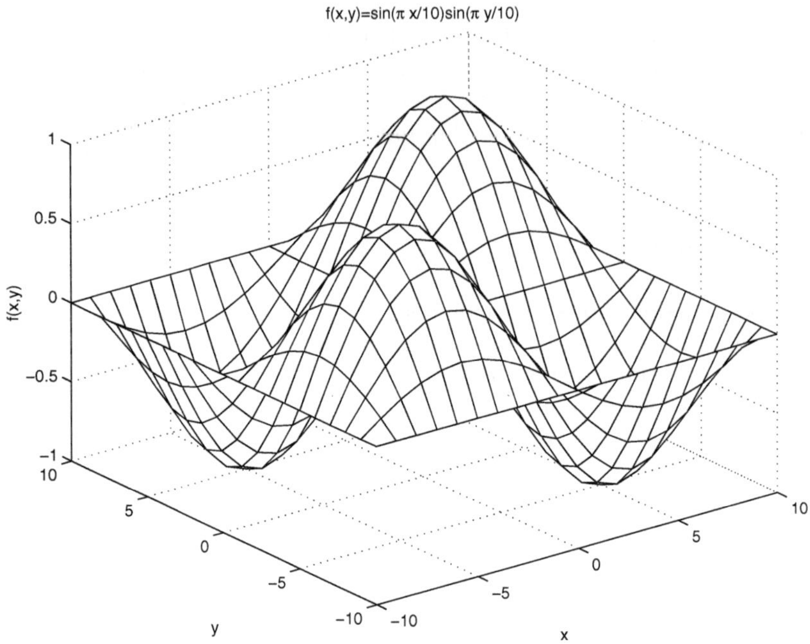

Figure 3.7. Function $f(x, y) = \sin\left(\frac{\pi x}{10}\right)\sin\left(\frac{\pi y}{10}\right)$ (Courtesy of IEEE [27])

- **Step 10** The 25 singletons of the consequences were calculated using RLS. The output membership functions are shown in Figure 3.13(a).
- **Step 11** The optional step was not applied because the approximation was considered acceptable.
- **Step 12** The singletons were converted into 25 triangular membership functions.
- **Steps 13, 14** The FuZion algorithm was applied to the 25 output triangular membership functions with M equal to 10% of the universe of discourse. Three membership functions were obtained and they received their linguistic values, as shown in Figure 3.13(b).
- **Steps 15, 16** The 25 singletons were fuzzified using the three membership functions obtained in the previous step. All the nonzero membership values were associated to the consequences of the rules, as shown in the following list.

1. **IF** x is Negative Large AND y is Negative Large **THEN** z is Negative with strength 0.01 AND Zero with strength 0.99
2. **IF** x is Negative Medium AND y is Negative Large **THEN** z is Zero with strength 0.92 AND Positive with strength 0.08
3. **IF** x is Zero AND y is Negative Large **THEN** z is Negative with strength 0.01 AND Zero with strength 0.99
4. **IF** x is Positive Medium AND y is Negative Large **THEN** z is Negative with strength 0.1 AND Zero with strength 0.9

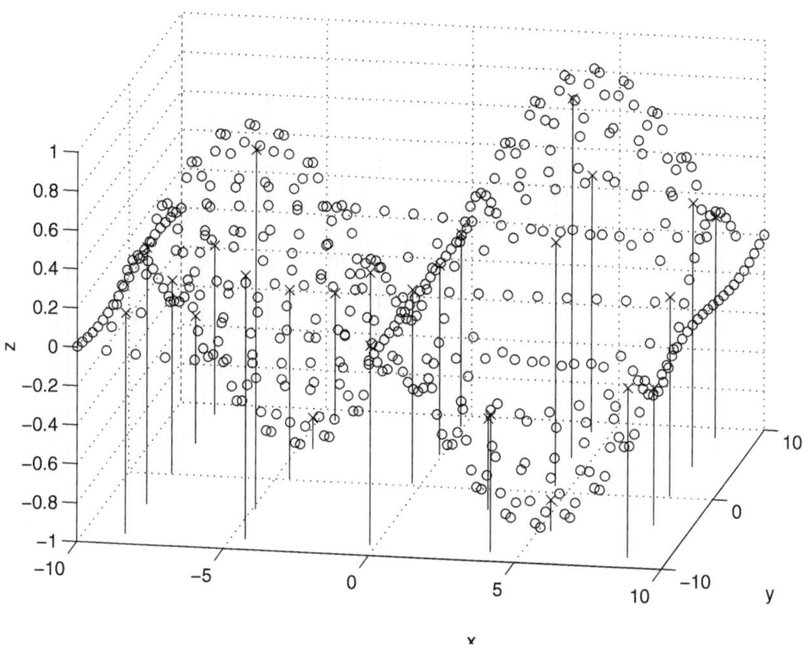

Figure 3.8. Clustering operation over the data collected from the function $f(x, y) = \sin\left(\frac{\pi x}{10}\right)\sin\left(\frac{\pi y}{10}\right)$. Extracted clusters ('x') from the data (o)

5. **IF** x is Positive Large AND y is Negative Large **THEN** z is Negative with strength 0.03 AND Zero with strength 0.97
6. **IF** x is Negative Large AND y is Negative Medium **THEN** z is Zero with strength 0.96 AND Positive with strength 0.04
7. **IF** x is Negative Medium AND y is Negative Medium **THEN** z is Zero with strength 0.01 AND Positive with strength 0.99
8. **IF** x is Zero AND y is Negative Medium **THEN** z is Zero with strength 0.92 AND Positive with strength 0.08
9. **IF** x is Positive Medium AND y is Negative Medium **THEN** z is Negative with strength 0.99 AND Zero with strength 0.01
10. **IF** x is Positive Large AND y is Negative Medium **THEN** z is Negative with strength 0.1 AND Zero with strength 0.90
11. **IF** x is Negative Large AND y is Zero **THEN** z is Negative with strength 0.02 AND Zero with strength 0.98
12. **IF** x is Negative Medium AND y is Zero **THEN** z is Negative with strength 0.11 AND Zero with strength 0.89
13. **IF** x is Zero AND y is Zero **THEN** z is Negative with strength 0.03 AND Zero with strength 0.97
14. **IF** x is Positive Medium AND y is Zero **THEN** z is Zero with strength 0.92 AND Positive with strength 0.08
15. **IF** x is Positive Large AND y is Zero **THEN** z is Negative with strength 0.01 AND Zero with strength 0.99

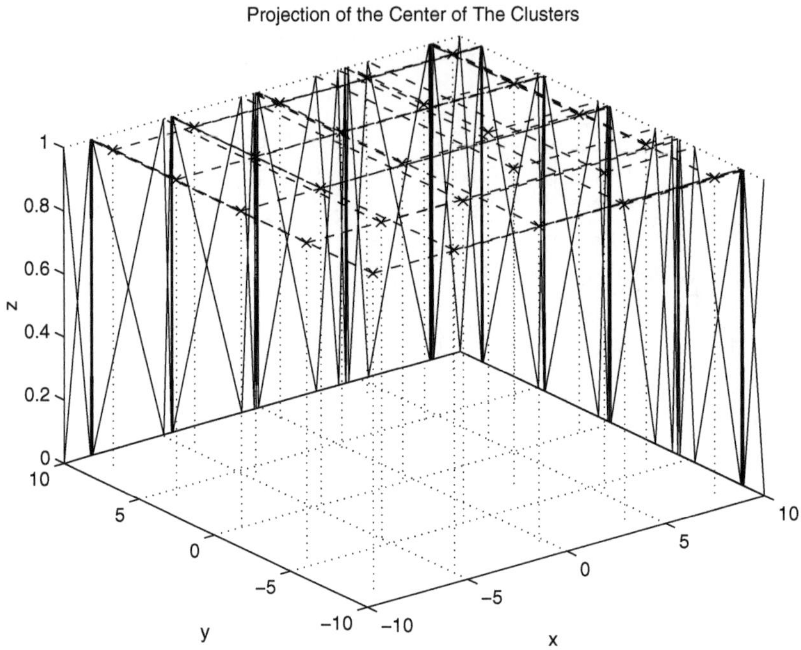

Figure 3.9. Projection of the centers of the clusters to approximate the function $f(x,y) = \sin\left(\frac{\pi x}{10}\right)\sin\left(\frac{\pi y}{10}\right)$

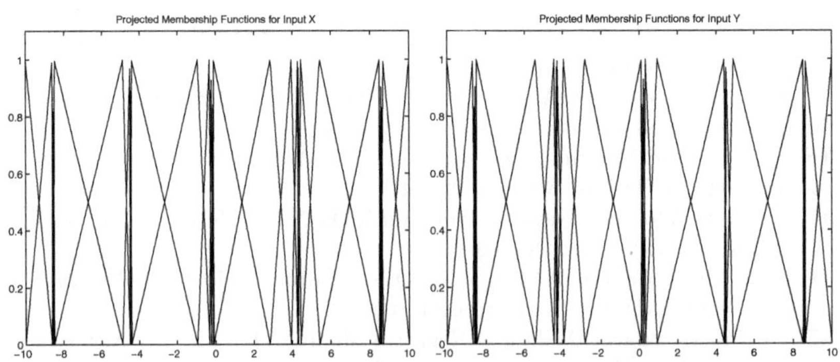

Figure 3.10. Projected membership functions over the input domains (Courtesy of IEEE [27])

16. **IF** x is Negative Large AND y is Positive Medium **THEN** z is Negative with strength 0.07 AND Zero with strength 0.93
17. **IF** x is Negative Medium AND y is Positive Medium **THEN** z is Negative with strength 1
18. **IF** x is Zero AND y is Positive Medium **THEN** z is Negative with strength 0.11 AND Zero with strength 0.89

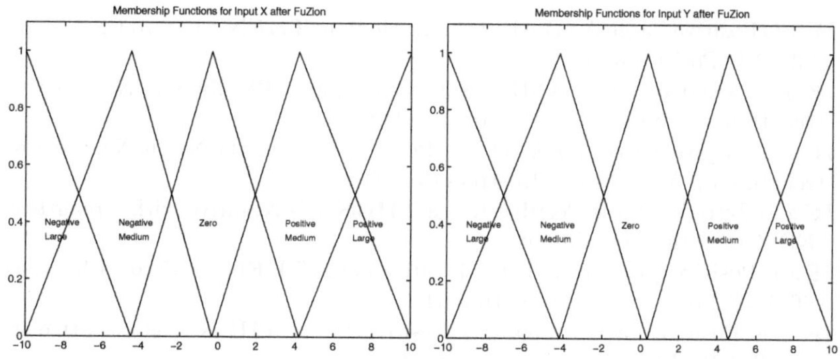

Figure 3.11. Membership functions after FuZion with $M = 10\%$ (Courtesy of IEEE [27])

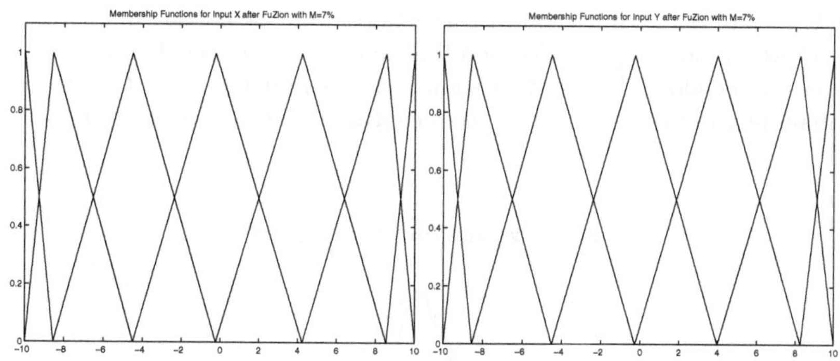

Figure 3.12. Membership functions after FuZion with $M = 7\%$ (Courtesy of IEEE [27])

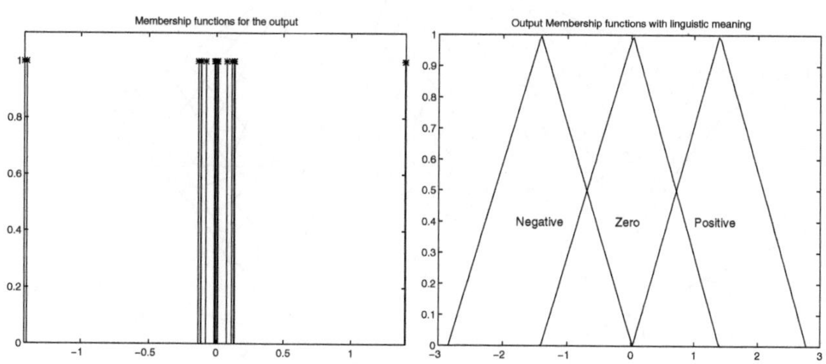

Figure 3.13. Consequences of the rules. (a) Singletons (b) Membership functions with linguistic meaning. (Courtesy of IEEE [27])

19. **IF** x is Positive Medium AND y is Positive Medium **THEN** z is Positive with strength 1

20. **IF** x is Positive Large AND y is Positive Medium **THEN** z is Zero with strength 0.93 AND Positive 0.07
21. **IF** x is Negative Large AND y is Positive Large **THEN** z is Negative with strength 0.02 AND Zero with strength 0.98
22. **IF** x is Negative Medium AND y is Positive Large **THEN** z is Negative with strength 0.07 AND Zero with strength 0.93
23. **IF** x is Zero AND y is Positive Large **THEN** z is Negative with strength 0.02 AND Zero with strength 0.98
24. **IF** x is Positive Medium AND y is Positive Large **THEN** z is Zero with strength 0.96 AND Positive with strength 0.04
25. **IF** x is Positive Large AND y is Positive Large **THEN** z is Negative with strength 0.01 AND Zero with strength 0.99

Observe that the obtained rules exhibit a clear dominance of one of the consequences. When this happens it will be possible to eliminate the consequence with the small strength without a major impact on the numerical approximation. However, this step is a decision that must be left to the designer because it is case-dependent. Figure 3.14 shows the identified surface. Observe that the main features of the function were captured by the fuzzy system.

Figure 3.14. Surface generated by the fuzzy system to approximate the function $f(x, y) = \sin\left(\frac{\pi x}{10}\right)\sin\left(\frac{\pi y}{10}\right)$ (Courtesy of IEEE [27])

3.5.2 Modeling of a Three-Input Nonlinear Function

For this example the data were generated using the function

$$f(x, y, z) = (1 + x^{0.5} + y^{-1} + z^{-1.5})^2 \tag{3.27}$$

The training set for this example is composed of 216 random points from the input range $[1, 6] \times [1, 6] \times [1, 6]$ while the validation set has 125 random points from the input range $[1.5, 5.5] \times [1.5, 5.5] \times [1.5, 5.5]$. As a performance index, we used the average percentage error (APE):

$$APE = \frac{1}{P} \sum_{i=1}^{P} \frac{|T(i) - O(i)|}{|T(i)|} \times 100\% \tag{3.28}$$

where $T(i)$ is the desired output and $O(i)$ is the predicted output. This performance index allows us to compare the present result with previous works. First a mountain-clustering procedure was used and 11 clusters were found; further refinement was obtained by using fuzzy C-means clustering algorithm. In Figure 3.15 the projected membership functions can be observed. After reduction using FuZion with a minimum distance factor of 15% of the size of the universe of discourse, the membership functions shown in Figure 3.16 were obtained.

Figure 3.17 shows the singleton consequences and the consequences after FuZion. Table 3.1 shows the comparative results with previous work. From the results it can be observed that the numeric performance is very similar to other proposed methods. It is important to note the generalization capabilities of the system generated by the AFRELI method. Observe the small degradation of the performance when the training and the validation set are used. All the other techniques show significant degradation (*i.e.*, ANFIS almost 2 orders of magnitude). This example shows that the method provides an acceptable numerical performance with the advantage that the interpretability is guaranteed.

Table 3.1. Performance Comparison with Previous Work

Model	APE_{TRN}	APE_{VAL}	Param. num.	Size train. set	Size valid. set
AFRELI	1.002 %	1.091 %	80	216	125
ANFIS	0.043 %	1.066 %	50	216	125
GMDH	4.7 %	5.7 %	-	20	20
Fuzzy model 1	1.5 %	2.1 %	22	20	20
Fuzzy model 2	0.59 %	3.4 %	32	20	20

The results from previous works were taken from [19]

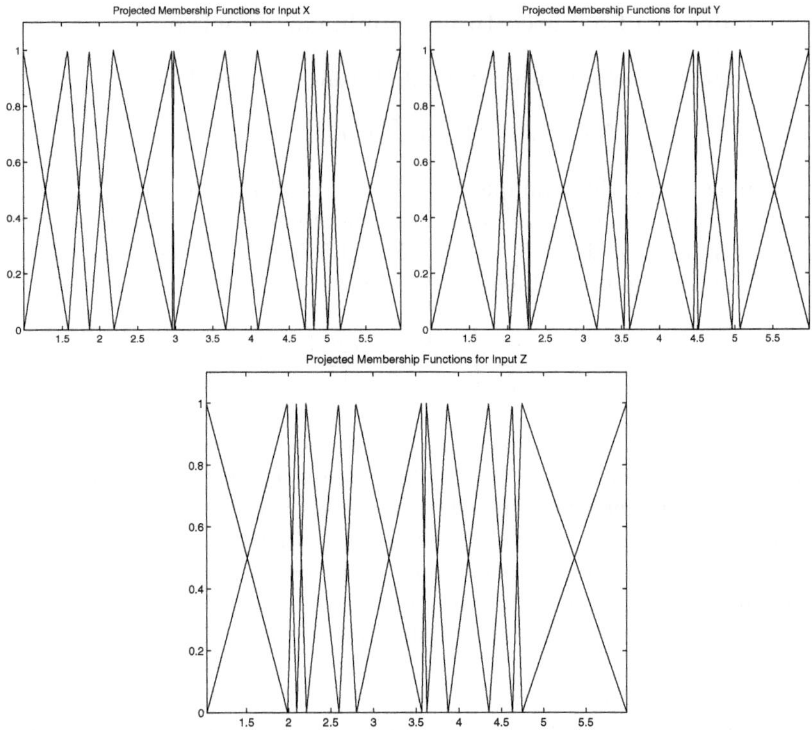

Figure 3.15. Projected membership functions to approximate the function $f(x, y, z) = (1 + x^{0.5} + y^{-1} + z^{-1.5})^2$.

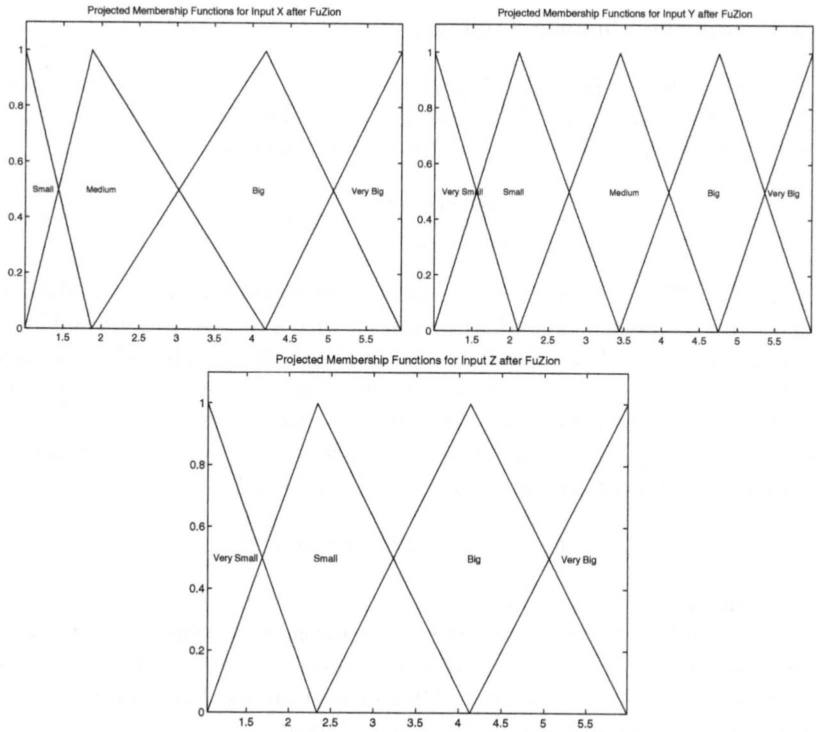

Figure 3.16. Membership functions after FuZion in the approximation of the function $f(x, y, z) = (1 + x^{0.5} + y^{-1} + z^{-1.5})^2$

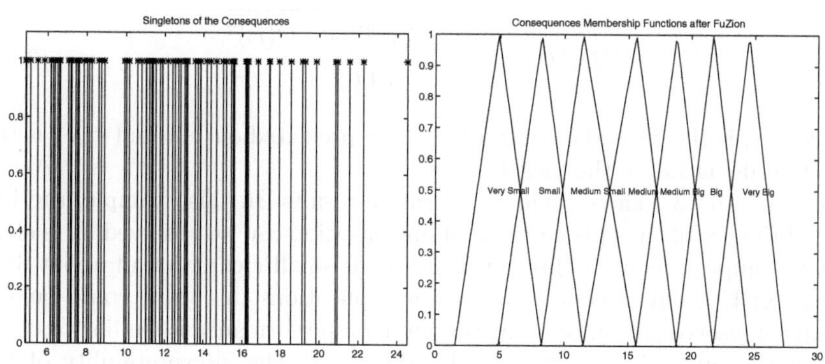

Figure 3.17. Consequences of the rules in the approximation of the function $f(x, y, z) = (1 + x^{0.5} + y^{-1} + z^{-1.5})^2$.(a) Singletons (b) Membership functions with linguistic meaning

3.5.3 Predicting Chaotic Time Series

This example shows the capability of the algorithm to capture the dynamics governing the Mackey–Glass chaotic time series. These time series were generated using the following delay differential equation:

$$\dot{x}(t) = \frac{0.2x(t-\tau)}{1+x^{10}(t-\tau)} - 0.1x(t) \tag{3.29}$$

where $\tau = 17$. The numerical solution of this differential equation was obtained using the fourth-order Runge–Kutta method, with a time step of 0.1 and initial condition $x(0) = 1.2$. The simulation was run for 2000 seconds and the samples were taken each second. To train and test the fuzzy system, 1000 points were extracted, $t = 118$ to 1117. The first 500 points were used as the training set and the remaining as the validation set. First, a six-step-ahead predictor is constructed using past outputs as inputs of the model:

$$[x(t-18)\, x(t-12)\, x(t-6)\, x(t)] \tag{3.30}$$

and the output will be $x(t+6)$.

After applying the mountain-clustering method, 57 clusters were found. Some refinement on the position of the clusters was obtained by using the Fuzzy C-Means clustering method. After projection and FuZion the membership functions shown in Figure 3.18 were obtained.

To allow a comparison with previous works, the prediction error was evaluated using the so called nondimensional error index (NDEI) defined as

$$NDEI = \frac{\sqrt{\frac{1}{N}\sum_{i=1}^{N}(T(i)-O(i))^2}}{\sigma(T)} \tag{3.31}$$

where $T(i)$ is the desired output, $O(i)$ is the predicted output and $\sigma(T)$ is the standard deviation of the target series.

Tables 3.2 and 3.3 show some comparative results. In this example, the impact of the use of the optional step of optimization can be observed. It is clear that the improvement of this optional step is small (reduction of about 30% on the NDEI), but of course on certain applications this value could be significant. Observe once more that the numeric performance is similar to other techniques but a significant value is added with the interpretability of the obtained rule base.

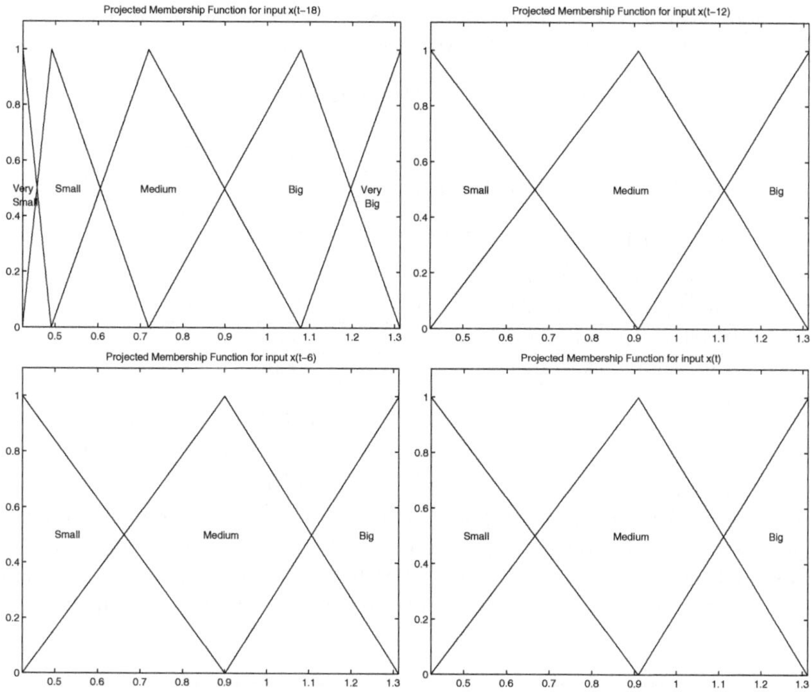

Figure 3.18. Mackey–Glass chaotic time series. Membership functions after projection and FuZion (Courtesy of IEEE [27])

Table 3.2. Mackey–Glass Chaotic Time Series. Performance for Prediction Six Steps Ahead

Method	Training cases	Nondimensional error index
AFRELI	500	0.0493
AFRELI (with optional step)	500	0.0324
ANFIS	500	0.007
AR model	500	0.19
Cascaded-correlation NN	500	0.06
Back-propagation MLP	500	0.02
sixth-order polynomial	500	0.04
Linear predictive method	2000	0.55

The results from previous works were taken from [19].

Table 3.3. Mackey–Glass chaotic time series. Performance for Prediction 84 Steps Ahead (the first seven rows) and 85 (the last four rows)

Method	Training cases	Nondimensional error index
AFRELI	500	0.1544
AFRELI (with optional step)	500	0.1040
ANFIS	500	0.036
AR model	500	0.39
Cascaded-correlation NN	500	0.32
Back-propagation MLP	500	0.05
Sixth-order polynomial	500	0.85
Linear predictive method	2000	0.60
LRF	500	0.10–0.25
LRF	10000	0.025–0.05
MRH	500	0.05
MRH	10000	0.02

Results for the first seven methods are obtained by simulation of the model obtained for prediction six steps ahead. Results for localized receptive fields (LRFs) and multiresolution hierarchies (MRHs) are for neurons trained to predict 85 steps ahead. The results from previous works were taken from [19].

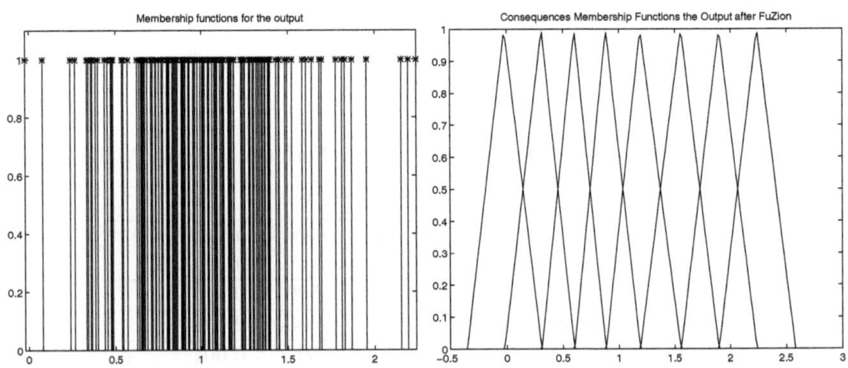

Figure 3.19. Mackey–Glass chaotic time series. Consequences of the rules: (a)Singletons (b) Membership functions with linguistic meaning (Courtesy of IEEE [27])

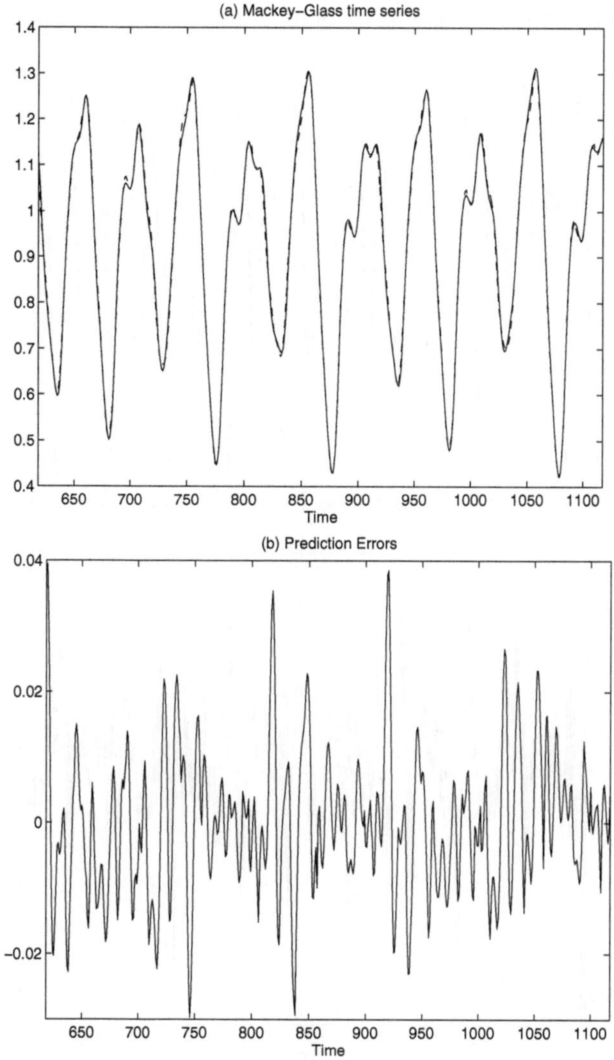

Figure 3.20. Mackey–Glass chaotic time series approximation. (a) Mackey–Glass time series (solid line) from $t = 618$ to 1117 and six-steps-ahead prediction (dashed line) (b) Prediction errors (Courtesy of IEEE [27])

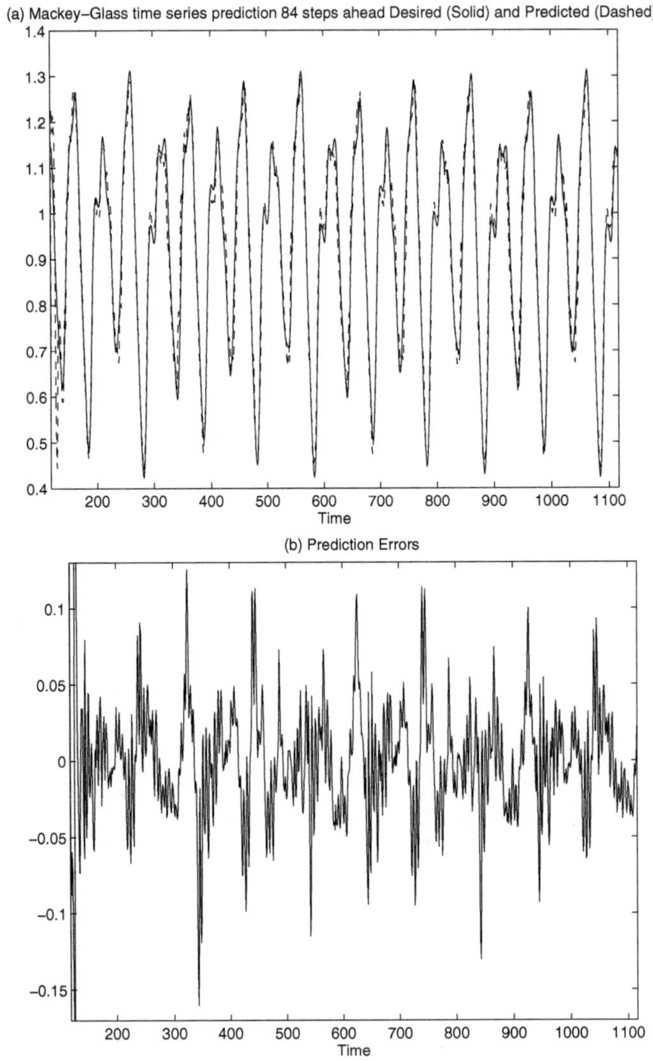

Figure 3.21. Mackey–Glass chaotic time series approximation. (a) Mackey–Glass time series (solid line) from $t = 118$ to 1117 and 84-steps-ahead prediction (dashed line) (b) Prediction errors (Courtesy of IEEE [27])

3.5.4 Modeling of the Quality Properties on a High-Density Polyethylene (HDPE) Reactor

In this example the purpose is to predict the density of the polyethylene produced in a gas phase HDPE reactor. For this purpose three signals are collected and preprocessed to eliminate dynamic information. Finally, 254 samples were selected and from this set two subsets were chosen, one for training (178 samples) and one for validation (76 samples). The input signals are C_4/C_2 ratio, H_2/C_2 ratio and product outflow (see Figure 3.22), and the output signal is the density of the polyethylene (see Figure 3.23).

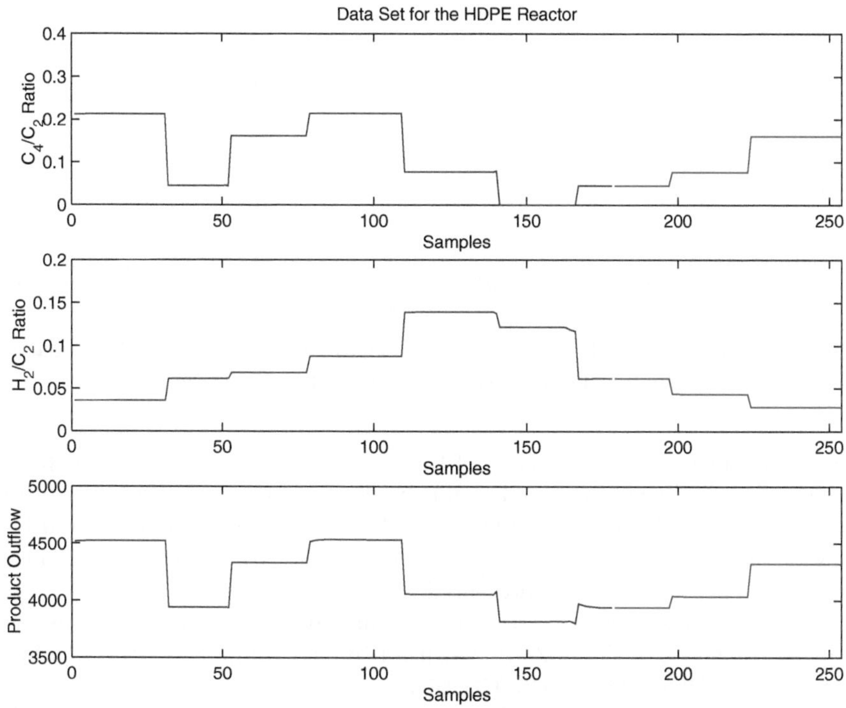

Figure 3.22. Modeling of an HDPE Reactor. Signals used to predict the density in the HDPE reactor (Courtesy of Springer-Verlag [25])

The training set was clustered using mountain clustering ([12]) with a grid of five divisions per dimension. From this procedure, 6 clusters were selected as the most important candidates, and they were refined using fuzzy C-means ([2]) (**step 2**). These clusters were projected into the input space and the membership functions were constructed (**steps 3, 4, 5, 6**). Figure 3.24 shows the projected membership functions. The FuZion algorithm was applied to

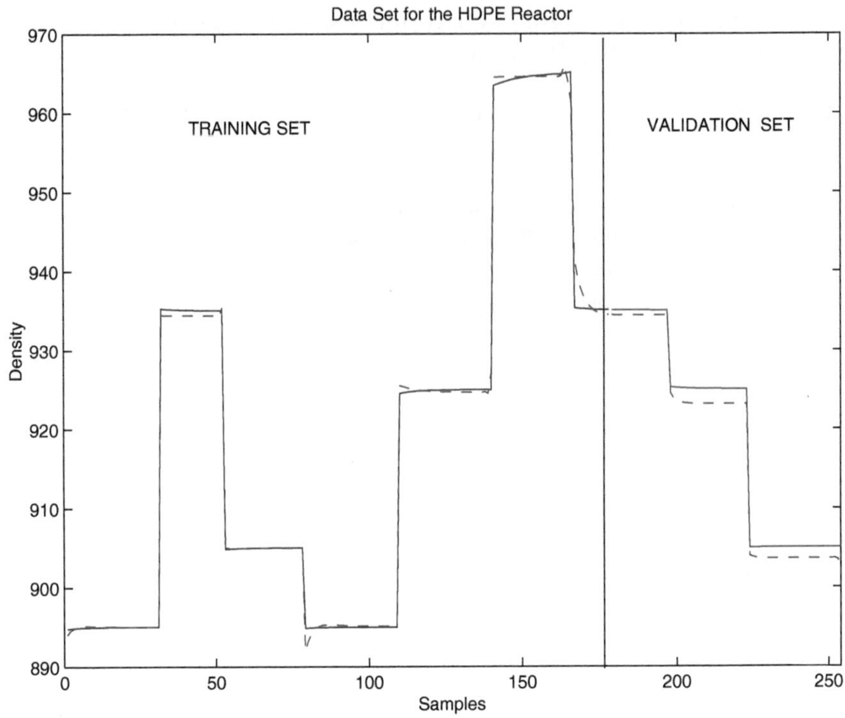

Figure 3.23. Modeling of an HDPE reactor. Data for density in the HDPE reactor Original data (-) Predicted data (- -) (Courtesy of Springer-Verlag [25])

the input domains with M equal to 10% of the universe of discourse (**steps 7, 8**). Figure 3.25 shows the membership functions after the FuZion algorithm was applied. A rule base of 60 rules ($4 \times 5 \times 3$) was constructed and the consequences of the rules were calculated (**steps 9, 10**). In Figure 3.26(a) the singleton consequences are represented.

The singletons were converted into triangular membership functions and reduced by means of the FuZion algorithm to only six membership functions [see Figure 3.26(b)] (**steps 12, 13, 14**).

The new membership functions are associated with the rules (**steps 15, 16**). Some of the obtained rules are

- **IF** C4-C2 Ratio is Small AND H2-C2 Ratio is Very Small AND Prod.Outflow is Small **THEN** Density is Medium Low with strength 0.9 AND Low with strength 0.1
- **IF** C4-C2 Ratio is Small AND H2-C2 Ratio is Very Small AND Prod.Outflow is Large **THEN** Density is Very High with strength 0.99 AND High with strength 0.01
- **IF** C4-C2 Ratio is Small AND H2-C2 Ratio is Very Large AND Prod.Outflow is Small **THEN** Density is Medium Low with strength 0.89 AND Medium High with strength 0.11

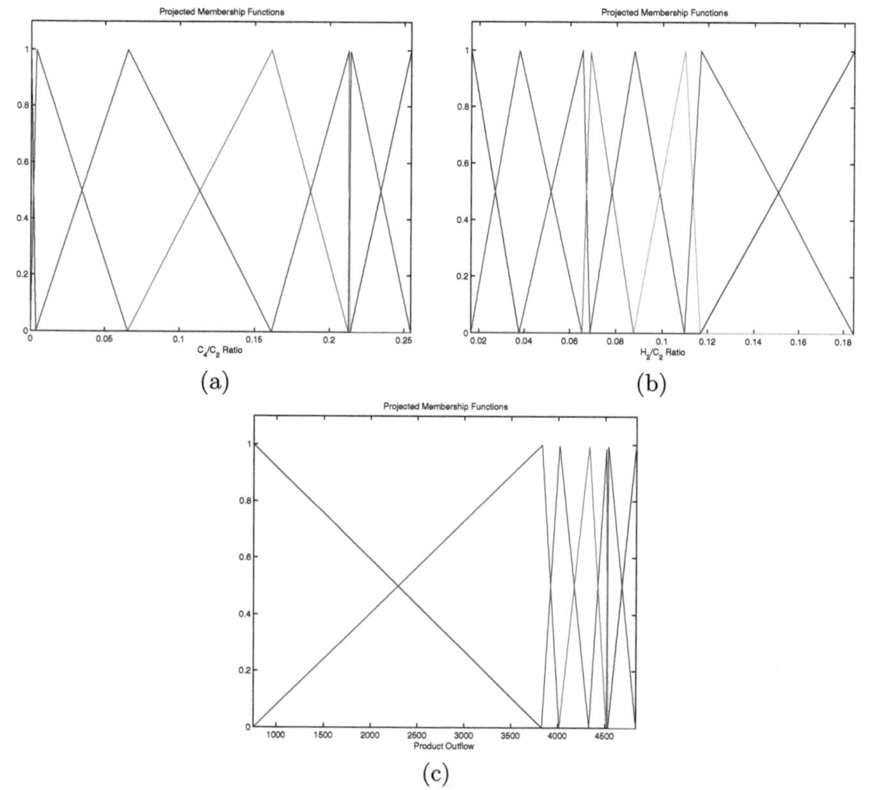

Figure 3.24. Modeling of an HDPE Reactor. Membership functions after cluster projection. (a) C_4/C_2 ratio (b) H_2/C_2 ratio (c) Product outflow (Courtesy of Springer-Verlag [25])

- **IF** C4-C2 Ratio is Small AND H2-C2 Ratio is Very Large AND Prod.Outflow is Large **THEN** Density is Very High with strength 0.99 AND High with strength 0.01
- **IF** C4-C2 Ratio is Very Large AND H2-C2 Ratio is Very Small AND Prod.Outflow is Small **THEN** Density is Very Low with strength 0.6 AND Low with strength 0.4
- **IF** C4-C2 Ratio is Very Large AND H2-C2 Ratio is Very Small AND Prod.Outflow is Large **THEN** Density is High with strength 0.95 AND Very High with strength 0.05
- **IF** C4-C2 Ratio is Very Large AND H2-C2 Ratio is Very Large AND Prod.Outflow is Small **THEN** Density is Very Low with strength 1
- **IF** C4-C2 Ratio is Very Large AND H2-C2 Ratio is Very Large AND Prod.Outflow is Large **THEN** Density is Medium High with strength 0.92 AND Medium Low with strength 0.08

Finally, Figure 3.23 shows the prediction of the fuzzy model, observe in Figure 3.22 that the conditions of the validation set are different to the ones

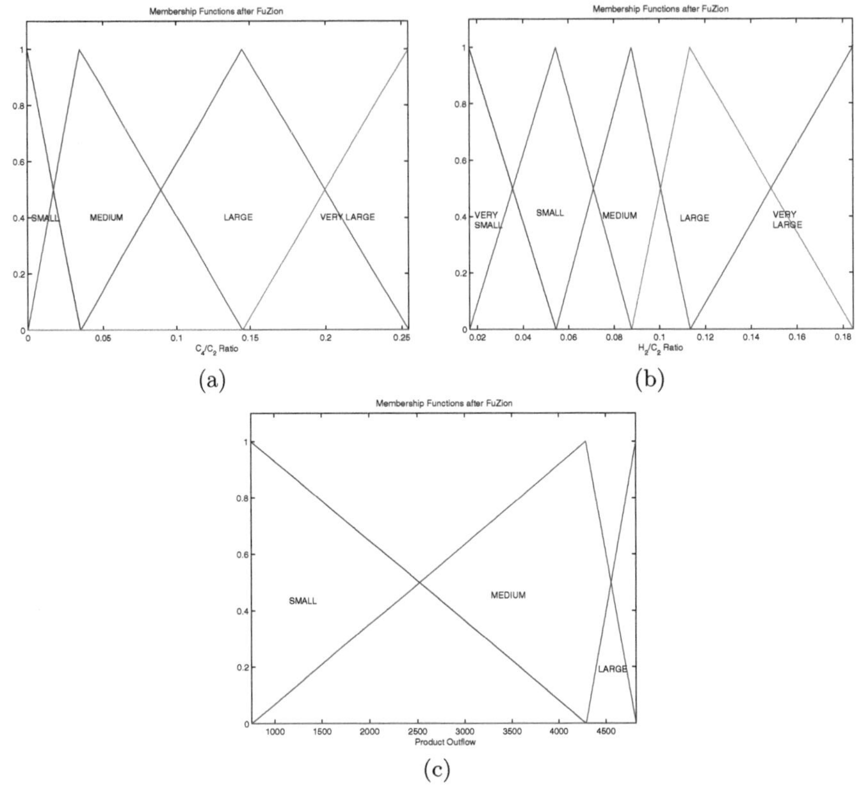

Figure 3.25. Modeling of an HDPE Reactor. Membership functions after FuZion. (a) C_4/C_2 ratio, (b) H_2/C_2 ratio, (c) Product outflow (Courtesy of Springer-Verlag [25])

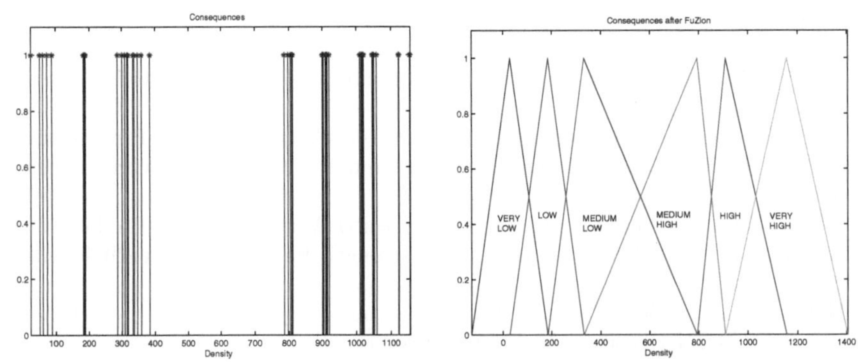

Figure 3.26. Modeling of an HDPE Reactor. Output membership functions. (a) Singletons (b) After FuZion (Courtesy of Springer-Verlag [25])

presented in the training set. However, the prediction is still very good for the validation set.

3.6 Complexity of the AFRELI Algorithm

The AFRELI algorithm is an algorithm based on clustering methods to locate the rules. The clusters are projected into the input space and converted into fuzzy sets. The FuZion algorithm reduces the large number of projected sets on each input so that the linguistic integrity is preserved. With the reduced group of fuzzy sets a combinatorial number of IF–THEN rules is generated using only the AND operation in the antecedents. The rule base will cover every possible case in the compact set defined by the universes of discourse of the input domains. This is an important advantage because the system will be able to make accurate predictions values even if there were no similar values in the data set used to construct the model. The generalization of this method for the case of rules badly excited will be as good as the generalization given by a multilinear model of dimension N where N is the number of inputs.

However, such a large number of rules present some problems: (1) a large set of rules with a large set of antecedents is difficult to understand and analyze; (2) a large set of rules demands large memory storage; for instance, a rule base created for a system with five inputs and six membership functions on each input will demand approximately 30 Kbytes of memory, and a similar system with 10 inputs will demand 230 Mbytes of memory. On the other hand, the evaluation time will not grow that fast; for the first system only $2^5 = 32$ rules need to be evaluated, for the second system $2^{10} = 1024$ rules.

Summary:
The exponential growth of the number of rules with respect to the increase in the number of inputs affects mainly the storage demanded by the model obtained using the AFRELI method. The performance of rule evaluation, however, remains limited since it grows only in powers of 2.

3.7 Conclusions

The AFRELI algorithm has been created to further exploit the comparative advantage of fuzzy systems by securing their linguistic interpretability. The AFRELI algorithm in combination with the FuZion algorithm guarantees a good trade-off between numerical accuracy and interpretability. The method exploits some successful elements proposed in other methods to reduce the complexity of the model construction.

The algorithm generates automatically the fuzzy sets and the interactive labeling process (with intervention of the designer) guarantees an agreement between the fuzzy set and the assigned label (semantic agreement).

The method generates a rule base covering all the possible cases, this guarantees the completeness of the rule base, but the associated drawback is the exponential growth of the rule base as the number of inputs increases. However, this is only a storage problem because the description of the fuzzy sets guarantees that only 2^N rules (N number of inputs) are activated on each inference. This makes the inference process fast because only a limited number of rules are evaluated.

The numerical accuracy of the algorithm is directly related with the choices of the M parameter governing the FuZion algorithm, and the choices in the clustering algorithm. When the number of inputs is large, the clustering will be limited to use the Fuzzy C-means algorithm with an overestimated number of clusters. The poor performance of the mountain clustering method with large dimensions motivated this modification.

Some improvements of the numerical performance of the model can be obtained by making a fine-tuning of the parameters of the antecedents by means of constrained gradient descent techniques.

4

Nonlinear Identification Using Fuzzy Models

This chapter is oriented to the study of system identification using fuzzy models. The chapter presents the main aspects of system identification using fuzzy models such as the structure of the fuzzy systems, experiment design, regressors selection, structure selection, parameter calculation and validation. The chapter includes the analysis and derivation of the gradients for dynamic calculation of the parameters of dynamic fuzzy models. This result is condensed at the end of the book in Appendix C.

It is important to remark that fuzzy systems are one of several possibilities in the area of nonlinear system identification. Neural networks, Volterra series, wavelets and other universal approximators represent other possibilities in nonlinear system identification. The advantages of the use of fuzzy systems is their capacity to interact and to extract linguistic information from input–output data and to describe the dynamics of the system in local regions described by the rules. These features are very valuable and make fuzzy models different from other traditional black-box techniques. The capacity to handle linguistic information adds an extra dimension to the identification and modeling because the validation process will be based not only on quantitative criteria but also on qualitative criteria such as whether or not the extracted rules "make sense." In this way, expert knowledge and empirical knowledge (which is normally not represented as quantitative information) can be exploited. Extra knowledge is gained when users of the models observe the extracted rules and realize what kind of reasoning process they apply on certain decisions. This is typically the case when the objective is to make the identification of a group of human experts. Information is collected about their decision under given circumstances. The human experts know how to take a decision, but they fail to explain the reasoning mechanism that they are using. Once the knowledge is extracted in the form of a fuzzy model with a rule base, they realize the reasoning mechanism that they are using. One of the main failures when human experts try to generate rules is to neglect the dynamic information, such as trends and other dynamic effects, which are tightly linked to the decision process. This capacity to extract knowledge does not come for free and the price paid for this property is a reduction in accuracy and the so-called *curse of dimensionality.* The curse of dimensionality arises because the number of rules grows exponentially when the number of inputs to the model is increased.

Modeling and function approximation using fuzzy models is a topic that has been studied extensively in recent years [7] [9] [8] [15] [10] [14], but very few authors had studied the whole problem in a formal way; most of them focus on training algorithms and function approximation. This chapter tries to fill this gap by studying most of the issues involved on system identification.

Summary:
Identification using fuzzy models offers a new dimension to the subject of nonlinear system identification. These dynamic fuzzy models can offer linguistic and numerical information together with "local" descriptions of the system behavior.

4.1 System Identification

System identification is a technique to build mathematical models of dynamic systems based on input–output data. The output of the dynamic system at time t is $y(t)$ and the input $u(t)$. The "data set" will be described as

$$Z^t = \{y(1), u(1), \dots, y(t), u(t)\} \tag{4.1}$$

A model of the dynamic system can be constructed as a mapping from past data Z^{t-1} to the next output $y(t)$. This model is known as the predictor model and is represented by

$$\hat{y}(t) = f(Z^{t-1}) \tag{4.2}$$

where $\hat{y}(t)$ represents the estimated output. The essence of identification using fuzzy systems is to try to represent the function f by means of a fuzzy model. It is important to see the fuzzy system as a parameterizable mapping,

$$\hat{y}(t|\theta) = f(Z^{t-1}|\theta) \tag{4.3}$$

where θ is the vector of parameters to be chosen (position and shape of the membership functions, consequences of the rules, *etc.*). The choice of these parameters is guided by the information embedded in the data. The structure of (4.2) is a very general structure and it has the drawback that the data set is continuously increasing. For this reason, it is better to use a vector $\varphi(t)$ of fixed dimension. So the general model is now formulated as

$$\hat{y}(t|\theta) = f(\varphi(t)|\theta) \tag{4.4}$$

The vector φ is known as *the regression vector* and its elements *regressors.*

$$\varphi(t) = [y(t-1), \dots, y(t-n), u(t-1), \dots, u(t-m)] \tag{4.5}$$

Using this parameterization, the problem can be decomposed in three subproblems:

1. How to choose the regressors in $\varphi(t)$ from the set of past inputs and outputs
2. How to find the structure of the Fuzzy System $f(.,.)$
3. How to find the parameters θ

In the following sections these topics will be addressed.

Summary:
System identification using fuzzy models can be formulated in the "classical" framework of system identification where the structure definition and the parameter estimation are the subproblems that must be solved in order to obtain a model.

4.2 Basic Structure of the Fuzzy System

Fuzzy systems are suitable for identification, from the mathematical point of view, because these structures are "universal approximators," as shown in a previous chapter. Using this way of reasoning, there is a guarantee that the nonlinear system identification problem can be approached using fuzzy systems. There are many universal approximators; RBF, MLP, wavelets, Fourier series, Volterra kernels, *etc.* are just few to mention. The use of fuzzy systems for nonlinear identification is not motivated only by their approximation capabilities but also by their capacity to extract linguistic information in the form of IF–THEN rules which typically describe compact sets.

The structure selection for fuzzy system demands the selection of a large set of diverse type of parameters: shapes of the membership functions, *AND* and *OR* operations, implications, defuzzification methods, consequence type (Mamdani or Takagi–Sugeno), *etc.*

Since some of the parameter adjustment methods are based on gradient descent methods, it is preferable to use operations with continuous derivatives. This fact motivated the selection of *product* and *bounded sum* as *AND* and *OR* operators because their derivatives are continuous. The shape of the membership functions of the consequences will be singletons or linear combinations of the inputs (Takagi–Sugeno models). It is important to remember that the singletons can be converted into triangular membership functions and reduced using the rule description presented in the previous chapter in the AFRELI algorithm.

The choice of the type of consequence will determine the linguistic properties of the model: the use of singletons gives more linguistic meaning to the rules, but the use of Takagi–Sugeno models can improve the approximation properties, especially in the case of piecewise linear functions.

The shape of the membership functions for the fuzzification process will be preferred to be triangular or polynomial for their local and linguistic properties (see previous chapter), but Gaussian membership functions can also be used if minor attention is devoted to interpretability and locality of the rules.

Additional constraints are needed if the model should have some linguistic meaning. Some of them are

- The number of membership functions in every universe of discourse for the inputs should be limited to as much as 7 ± 2 and at least 2.
- The overlap among neighboring membership functions should be enforced to be 0.5.
- The distance between the "center" of the membership functions should guarantee a minimum amount of coverage over the universe of discourse.

In this section we will use the parameterization of the fuzzy system presented in previous sections:

$$f(\mathbf{x}) = \frac{\sum_{l=1}^{L} \bar{y}^l \mu_l(\mathbf{x})}{\sum_{l=1}^{L} \mu_l(\mathbf{x})} \tag{4.6}$$

where $\mathbf{x}$ represents the input vector. In the present case the input vector will be the regressors $\varphi(t)$, and $\mu_l(\mathbf{x})$ represent the membership function of the rule l constructed during the inference process. Finally, $\bar{y}^l$ is the singleton consequence of the l rule.

For the case of Takagi–Sugeno models the parameterization will be,

$$f(\mathbf{x}) = \frac{\sum_{l=1}^{L} A^l \mathbf{x} \mu_l(\mathbf{x})}{\sum_{l=1}^{L} \mu_l(\mathbf{x})} \tag{4.7}$$

where A^l is a row vector. In this case, the consequences are linear combinations of the inputs or other variables.

These initial choices are complemented with other choices that can be guided by I/O data using clustering and projections methods, as shown in Chapter 2.

These structures as they are described will be used to identify multiple-input–single-output (MISO) systems. Most of the comments expressed here are also valid for multiple-input–multiple-output (MIMO) systems. However, the authors consider that the problem of nonlinear MIMO identification problems only makes sense, from the practical point of view, if the output variables are of the same type (*e.g.* all temperatures or all displacements) and with similar orders of magnitude. Otherwise, the formulation of a cost function that balances the approximation of different signals with different magnitude is very difficult. In addition, MIMO parameterizations where some outputs are weakly related with some inputs tend to generate overfitting and bad generalizations. In these cases, it is better to use multiple MISO models (one per output) where only strongly related inputs are used to construct the model. Because there is only one output, no compromise among outputs should be taken into account to formulate the cost function for the approximation.

Summary:
The structure of fuzzy systems for identification can be fixed by using different criteria: Continuity of the function, linguistic interpretability and locality are just some worthy of mention.

4.3 Experiment Design for System Identification

The main objective of the experiment design is to extract as much information as possible from the dynamic system by means of a “good” input signal. What we understand for a “good” input signal is a signal that exposes the most important features of the system. Typically, a good input signal for the identification of nonlinear systems is a signal that should be rich on amplitudes and frequency, is limited in duration and, according to the application of the model, must excite those characteristics that are relevant for the desired application.

The problem of optimal experiment design can be formulated as the minimization in the uncertainty of the estimated parameters in the model. This statement includes a fundamental assumption implying that the structure of the model is already known. This leads to the consequence that every *optimal experiment design* for the identification of a nonlinear system should be an iterative process.

In this iterative process, an initial signal is used to make a rough identification of the structure of the model (regressors and noise model). Once this structure is selected, an initial description of the model is created, and with this description an optimization can be performed.

In heuristic terms, a good excitation signal for a fuzzy system is a signal that will excite the complete rule base. The signal should expose all possible cases presented in the rule base and, if the rule base is complete (all possible combinations of antecedents), the excitation signal will cover all possible situations in a compact set. It is important to remark that the excitation of the rule implies not only the excitation of the consequence but also the excitation of the membership functions of the antecedents associated with the rule.

In general, the probability distribution of the parameter estimates can be characterized by their bias $B(\hat{\theta})$ and covariance $\text{cov}(\hat{\theta})$. If the estimator is unbiased, the bias will approach zero asymptotically as the number of training point grows and the covariance will be bounded [32]. The estimator in this case will be unbiased if the correct structure (regressors, membership functions and inputs) is selected.

Assuming the description of the fuzzy system given in Equation (4.6), the fuzzy system can be described as

$$\hat{y}(t) = \sum_{l=1}^{L} \bar{y}^l w_l(\varphi(t)) \tag{4.8}$$

with

$$w_l(\varphi(t)) = \frac{\mu_l(\varphi(t))}{\sum_{l=1}^{L} \mu_l(\varphi(t))} \tag{4.9}$$

such that

$$y(t) = \bar{Y}^T W(t) + e(t) \tag{4.10}$$

where $\bar{Y} = \{\bar{y}^1, \bar{y}^2, \ldots, \bar{y}^L\}^T$, $W(t) = \{w_1(\varphi(t)), w_2(\varphi(t)), \ldots, w_L(\varphi(t))\}^T$ and $e(t)$ is the prediction error that is assumed to have zero mean and variance σ^2. If the vector of the consequence values is estimated using least squares, the *covariance matrix* of errors on the estimates of $\bar{Y}$ will be given by

$$\text{cov}\bar{Y} = \sigma^2 \left[\sum_{t=1}^{N} W(t)W(t)^T \right]^{-1} \tag{4.11}$$

Now the problem is to minimize (or maximize) some measure of the *covariance matrix* or its inverse, the *information matrix* ($M_{\bar{Y}} = [\text{cov}\bar{Y}]^{-1}$). Formally, the problem can be written as

$$\min_{u(1),\ldots,u(N)} \phi(M_{\bar{Y}}) \tag{4.12}$$

where ϕ is a cost function (typically a norm) calculated over the *information matrix* and $u(1), \ldots, u(N)$ is a sequence of N inputs.

According to the purpose of the model, the cost function $\phi(.)$ can be defined in different ways including even criteria such as *control relevance* of the inputs [33]. This optimization problem is nonlinear and nonconvex, so there is no guarantee that the obtained solution is the global minimum. However, a local minimum can be an acceptable solution.

When no prior knowledge about the structure is present, using a combined sequence of random movements can create an excitation signal. The experience of the authors shows that a good excitation can be obtained by constructing a signal composed by a "slow" random signal with some discrete values selected in the input range and a random wide band signal (fast signal) with an amplitude that covers the gap between the discrete values of the "slow signal." The reasoning behind the formulation of this kind of signal is that the slow signal will drive the plant to different operating points and the fast signal will guarantee enough excitation around the operating points. It is important to remark that the fast signal will be limited in amplitude and frequency such that saturation and slew rate limitations in the actuators will be respected. Figure 4.1 shows an example of such a signal. Combinations of

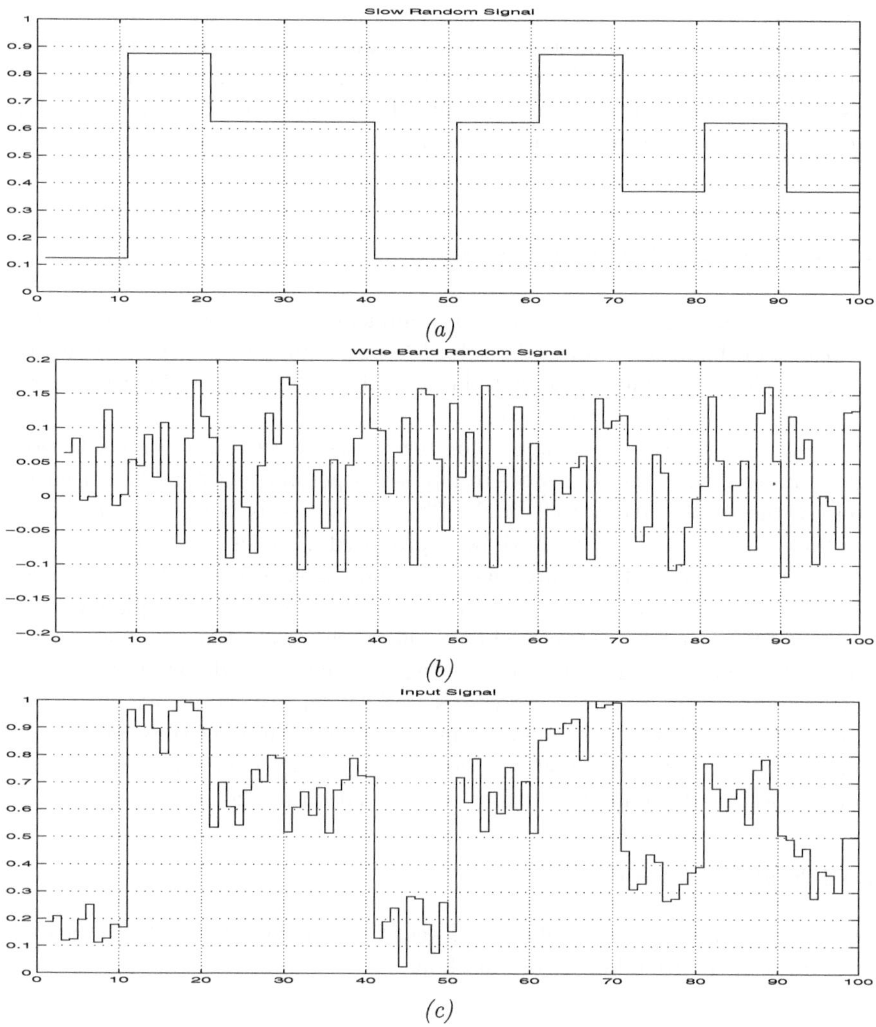

Figure 4.1. Heuristically designed excitation signal with four discrete levels between 0 and 1. (a) Slow random signal with 4 discrete levels (b) Wide band signal (c) Combined signals

this type of signals can be used for the identification of multivariable systems. Other excitation signals are the multisine signals with variable frequency and the swept sinus with random frequencies. These signals are frequently used in the identification of mechanical systems.

Summary:
The identification of nonlinear system demands a good experiment design such that the sequence of input signals excites all the rules in the rule base as well as all the important dynamics. The design of an optimal input sequence is conditioned by the knowledge of the model structure.

4.4 Choosing the Regressors

The problem of regressors selection for nonlinear modeling is a complex combinatorial problem that cannot be solved in polynomial time. The complexity of the problem is, in fact, exponential $O(2^n)$: if there are n possible regressors, 2^n possibilities must be evaluated.

Figure 4.2 shows the sequence for regressors selection. First a selection algorithm to pickup the regressors from a set of possible candidates is needed, because the exhaustive search of the 2^n possibilities is not practical. Some shortcuts to guide the search have been proposed. Among these methods it is worth mentioning the tree search methods and the genetic selection methods. The scheme shows that once a set of regressors is postulated the quality must be evaluated. To evaluate the quality of the regressors two types of strategies can be distinguished: model-based strategies, where a model should be built to evaluate the quality of the regressors, and model-free strategies, where the quality of the regressors is evaluated using only the input–output data. Next sections will explore these issues in detail. Section 4.4.1 will show the search methods and Section 4.4.2 will show the evaluation methods.

4.4.1 Search Methods

This section describes two classes of search methods, the heuristic search methods and the pseudo-random methods.

Heuristic Search

In the class of heuristic search methods, a tree selection method is used. In this method, two approaches can be used: one approach constructs a model adding new inputs one by one. The other method makes a model with all possible inputs and regressors are dismissed according to the impact in the performance evaluation (those with small impact are removed first). For fuzzy systems, the first one is preferred because the size of the models made for the evaluation is smaller (in terms of number of inputs). The method works as follows: first a set of all possible models using one regressor is proposed (n models where n is the number of regressor candidates). The quality of the models is evaluated using any of the model-based or model-free methods shown in Section 4.4.2. The regressor with the best performance index is selected and a new set of

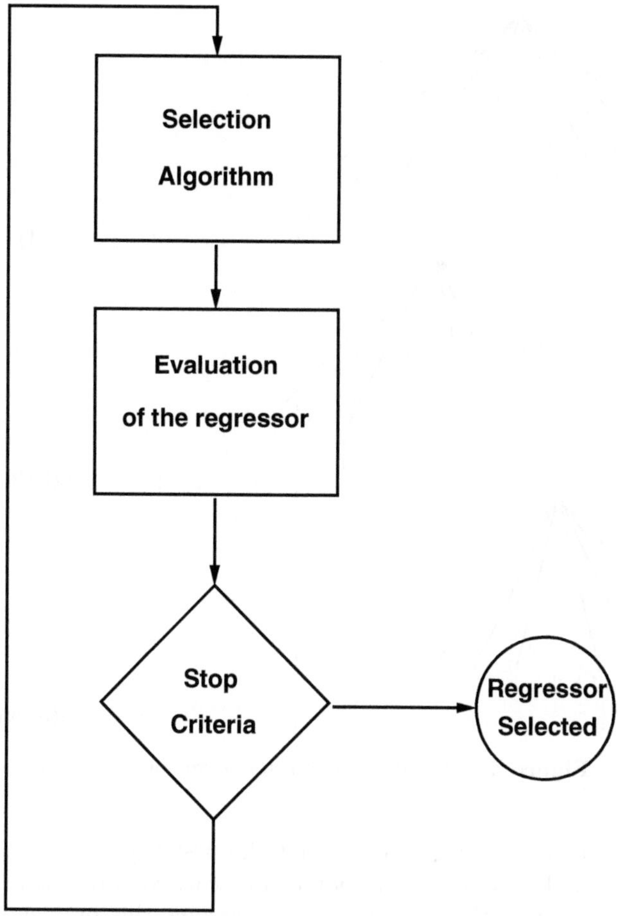

Figure 4.2. Sequence for regressors selection

models with two regressors is proposed. One of the two regressors will be always the one selected in the previous step; in this way $n-1$ possible models will be proposed and evaluated. Again the best regressors will be selected and the procedure will be repeated until a prescribed number of inputs is reached or no improvement in the performance index is observed. Figure 4.3 shows a graphical representation of an example of regressor selection using this method. The complexity of this method is polynomial ($O(n^2)$). At the worst case, this algorithm reduces the number of evaluated possibilities to

$$\sum_{i=1}^{n} i = \frac{n^2+n}{2} \tag{4.13}$$

where n is the number of regressor candidates.

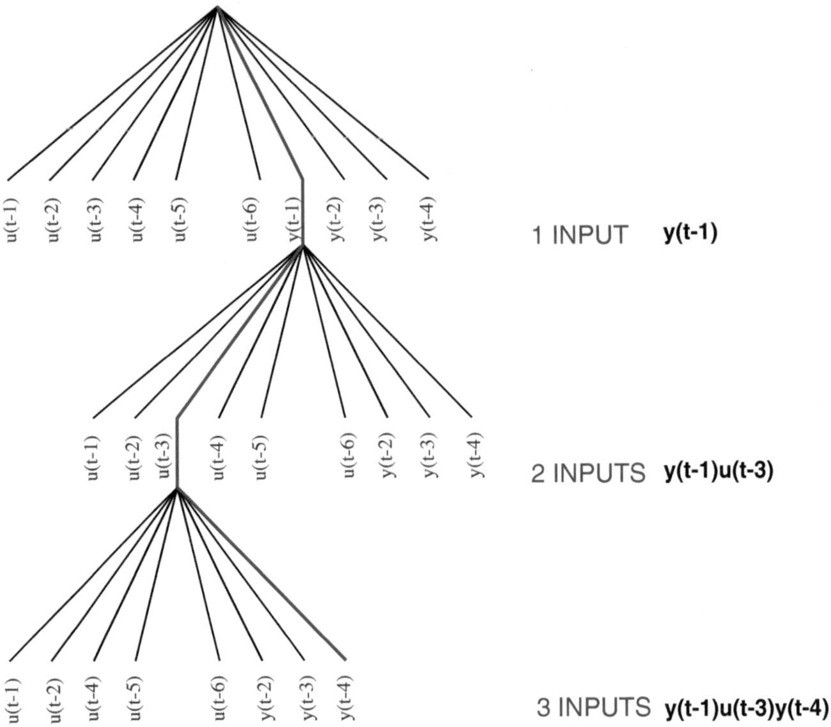

Figure 4.3. Heuristic search for regressors selection

This tree search does not guarantee that the best regressors are selected. It ignores the fact that some information can be carried only when two or more signals are present at the same time. To overcome this problem, the search can evaluate the addition of two or more inputs at the same time.

Pseudo-Random Methods

Genetic algorithms are included among the pseudo-random methods. The selection of the candidate regressors to be evaluated is given by the genetic algorithm. It was explained in Chapter 2 that there is a high probability that after a given number of iterations the optimal solution is found. The advantage of using the genetic algorithm is that the codification of the problem is straightforward and each bit of the string will be used to represent an active (bit equal "1") or an inactive (bit equal "0") regressor. Figure 4.4 shows the codification for the selection of the regressors among 10 candidates and the effect of the *crossover* operator of the genetic algorithm. The number of evaluations for this method is at the worst:

$$\text{Size of the population} \times \text{Number of generations} \tag{4.14}$$

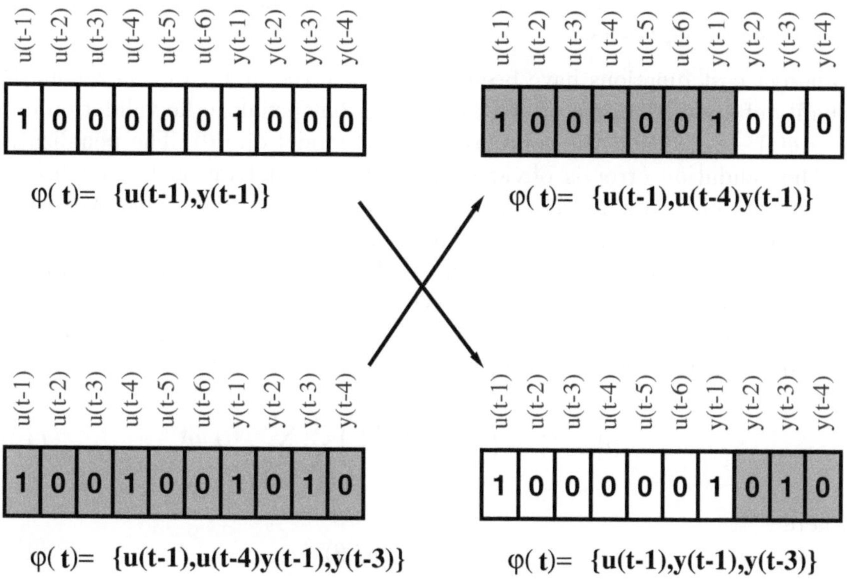

Figure 4.4. Regressors selection using genetic algorithms. Effect of the crossover operator

Summary:
A rigorous selection of the regressors for a model demands the evaluation of 2^n possible solutions. This exponential complexity of the problems motivates the use of search methods. The most important are the heuristic search and the pseudo-random search.

4.4.2 Regressors Evaluation

Model-Based Evaluation

Model-based evaluation evaluates the prediction error of the model constructed with the suggested set of regressors. Special attention must be taken during the construction of the model to avoid problems such as lack of excitation, overparameterization and large training time which could jeopardize the evaluation of the set of regressors. An acceptable solution to avoid these problems is the use of the *smallest fuzzy model* described in Section 2.5.1. This model is evaluated very fast (only a least-squares solution is needed) and there is a guarantee of enough excitation and no chance of overparameterization. In addition, the model is more complex than a linear model; in fact, the model is multilinear. The evaluation of a complete model with more than two membership functions per input is not recommended for two reasons:

- Computational cost could be very large since the regressor selection demands many model evaluations.
- Lack of excitation and overparameterization could spoil the evaluation of the set of regressors.

Different cost functions have been presented in the literature to evaluate the quality of a model. Some of these cost functions penalize the validation error $\varepsilon = y(t) - \hat{y}(t|\theta)$ and the complexity of the model $\dim(\theta)$. The penalization of the validation error is obvious; however, the motivation to penalize the complexity of the model should be explained as an effort to avoid overparameterization. The fact that these two measures have different units requires the use of weights to combine them in a cost function. These weights are in most of the cases arbitrary. Among these types of criteria are Akaike's information criteria (AIC) and Rissanen's minimum description length (MDL) [34]. The cost function for Akaike's information criterion is given by

$$V_N(\theta, Z^N) = \left(1 + \frac{2\dim\theta}{N}\right)\frac{1}{N}\sum_{t=1}^{N}\varepsilon^2(t,\theta) \tag{4.15}$$

where

$$\varepsilon = y(t) - \hat{y}(t|\theta) \tag{4.16}$$

and $\dim\theta$ represents the length of the vector of the parameters, and N is the size of the validation set.

Rissanen's minimum description length criterion is very similar to the AIC and its cost function is described as

$$V_N(\theta, Z^N) = \left(1 + \frac{2\log N\dim\theta}{N}\right)\frac{1}{N}\sum_{t=1}^{N}\varepsilon^2(t,\theta) \tag{4.17}$$

The arbitrary weighting between the objectives makes the cost functions very "subjective." Therefore, it is very difficult to say which cost function is more suitable.

Other criteria are practically motivated and they take into account only the validation error and try to avoid the overparameterization by other means. Most of these criteria try to minimize the impact of the data set Z^N used for the evaluation. One example of this type of methods is the so-called **Regularity Criterion**.

The application of this method to fuzzy modeling was suggested by Sugeno and Yasukawa [11]. The regularity criterion was first proposed in other modeling technique known as GMDH (group method of data handling). The cost function for this criterion is defined as follows:

$$RC = \left[\sum_{i=1}^{k_A}(y_i^A - \hat{y}_i^{AB})^2/k_A + \sum_{i=1}^{k_B}(y_i^B - \hat{y}_i^{BA})^2/k_B\right]/2 \tag{4.18}$$

where

- k_A and k_B, number of data of the groups A and B
- y_i^A and y_i^B, outputs of the data groups A and B
- $\hat{y}^{AB}$, model output for the group A input estimated using the model constructed with the data set B
- $\hat{y}^{BA}$, model output for the group B input estimated using the model constructed with the data set A

This cost function evaluates the prediction error in a cross-validation scheme. Figure 4.5 shows a representation of the cross-validation scheme. Cross-validation aims to prevent overparameterization and an overtraining.

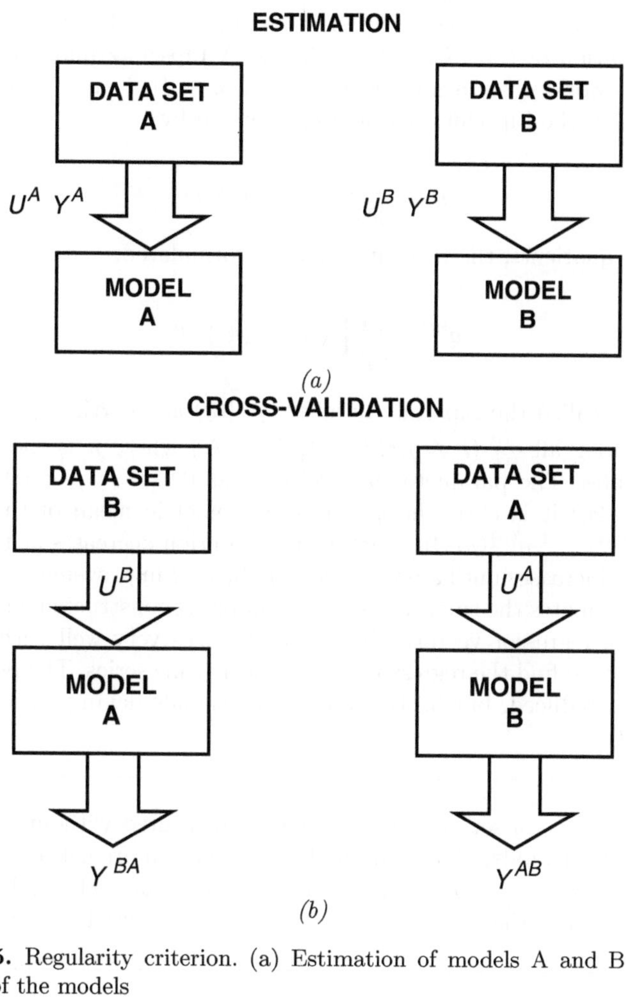

Figure 4.5. Regularity criterion. (a) Estimation of models A and B (b) Cross-validation of the models

Summary:
The cost function to measure the quality of a model punishes the distance of the predictions of the model to the original data and introduce a punishment in the complexity of the models to avoid overparameterization. Other methods avoid the overparameterization by means of cross-validation.

Model-Free Evaluation

The main drawback of the model-based techniques is that the evaluation can be influenced by estimation errors or, in some cases, the calculation of the parameters of the model can be very time-consuming. A good alternative is the model-free test proposed by He and Asada in [35]. The method is based in the evaluation of the *Lipschitz quotients.* A Lipschitz quotient is defined in this framework: given a nonlinear function $y = f(\mathbf{x})$ and N input output pairs $(y_i, (x)_i)$, the Lipschitz quotient q_{ij} is given by

$$q_{ij} = \frac{|y_i - y_j|}{|\mathbf{x}_i - \mathbf{x}_j|}, \quad (i \neq j) \tag{4.19}$$

Using these quotients, the following index is formulated:

$$\bar{q}^{(n)} = (\prod_{k=1}^{p} \sqrt{n} q^{(n)}(k))^{1/p} \tag{4.20}$$

where $\bar{q}^{(n)}$ is called the Lipschitz number, $q^{(n)}(k)$ is the kth largest Lipschitz quotient among all $q_{ij}^{(n)} (i \neq j; i, j = 1, 2, \ldots, N)$ where n is the number of input variables. The parameter p is selected as $0.01N < p < 0.02N$. From Equation (4.20) it is clear the $\bar{q}^{(n)}$ is the geometric mean of the sequence $q^{(n)}(1), q^{(n)}(2), \ldots, q^{(n)}(p)$. In practice, the criterion decreases as the number of regressors increases until a point where no further improvement is obtained. The index evaluates the smoothness of the mapping constructed with the data points in the regressor vector. This method works very well with noise-free data; it can even find the regressors for a chaotic time series. The performance with noise is reduced, but it still works with signals having an S/N ratio of around 8 dB.

Summary:
The performance of set of regressors can be measured without constructing any explicit model. These methods of evaluation are called model-free methods. *Lipschitz quotients* are a good example of such methods. The method estimates the smoothness of function constructed with a given set of regressors.

4.5 Choosing the Structure

Following the nomenclature for linear models the following structures has been defined for nonlinear systems [36] [37] as follows:

- NFIR models, in this case the vector of regressors is composed only with past inputs $\varphi = [u(t-k), \ldots, u(t-n)]$.
- NARX models, in this case the vector of regressors has past inputs and outputs $\varphi = [u(t-k), \ldots, u(t-n), y(t-k), \ldots, y(t-m)]$.
- NOE models, the vector of regressors has past inputs and past estimated outputs: $\varphi = [u(t-k), \ldots, u(t-n), \hat{y}(t-k|\theta), \ldots, \hat{y}(t-m|\theta)]$.
- NARMAX models, which use as regressors past inputs, past outputs and estimation errors $\varphi = [u(t-k), \ldots, u(t-n), y(t-k), \ldots, y(t-m), \varepsilon(t-k|\theta), \ldots, \varepsilon(t-l|\theta)]$.
- NBJ models, in this case the regressors are past inputs, estimation errors using past outputs and estimation errors using past estimated outputs ($\varepsilon_u(t-k|\theta)$): $\varphi = [u(t-k), \ldots, u(t-n), y(t-k), \ldots, y(t-m|\theta), \varepsilon(t-k|\theta), \ldots, \varepsilon(t-l|\theta), \varepsilon_u(t-k|\theta), \ldots, \varepsilon_u(t-p|\theta)]$.
- State-space models, for the deterministic case the model will be

$$\hat{x}(t+1) = f(\hat{x}(t), u(t)) \tag{4.21}$$

$$\hat{y}(t) = g(\hat{x}(t), u(t)) \tag{4.22}$$

  for the stochastic case the model will be

$$\hat{x}(t+1) = f(\hat{x}(t), u(t), \varepsilon(t)) \tag{4.23}$$

$$\hat{y}(t) = g(\hat{x}(t), u(t)) \tag{4.24}$$

It is clear that there is a wide set of possible structures. The application, the information available and the complexity of the model condition the selection of one type of structure. Some remarks about the models and their applications are reviewed in the following lines.

The NFIR model type is simple and has some interesting properties with respect to the stability (in fact, it is always stable), but its main drawback is that it needs a significant number of past inputs to capture simple dynamics. This large number of past inputs can make the rule base of the fuzzy model very big, thus making the model inefficient. In general, it is not a practical structure for fuzzy models.

The NARX model type is the most used structure since it is very easy to estimate due to its nonrecursive structure. For this reason, the NARX models are used during the phase of regressor selection where the computation of several models requires the use of simple models. It is interesting for applications where short-term forecasting is needed (one or few steps ahead).

The NARMAX, NOE, NBJ and the state-space stochastic model types are recurrent models where the estimated output is fed back. This characteristic makes the estimation of the model more complicated. The advantage of the use of these models is that the information provided by the simulation model is corrected via a noise model providing a synchronization mechanism with the real process.

NOE models are preferred for simulation purposes because no information from the "real system" is needed to operate a simulation, since past real outputs are not included in the model.

All these model structures can be used either with Mamdani models- or with Takagi–Sugeno models. For the Mamdani models the regressors are processed by the inference system, making in some cases the construction of models with many regressors very difficult. In these cases it is preferable to use Takagi–Sugeno models with rules where the consequences are linear models. These linear models use the above-mentioned structures and they are "scheduled" by the inference mechanism, which is driven by either states, inputs or outputs. The selection of the variable used by the inference mechanism can be done with the methods mentioned above, although in most of the cases the variable is selected using prior knowledge about the dynamic system.

Summary:
There are multiple structures to construct dynamic fuzzy models. The selection of one structure is guided by the features of the dynamic system as well as the applications of model.

4.6 Calculating the Parameters

The parameter adjustment techniques for dynamic systems are very similar to the ones used for function approximation, as explained in Chapter 2. Clustering techniques and projection can be used to estimate initial distributions of the membership functions. This initial estimation must be done using the NARX structure, because the construction of an NARX model is equivalent to the construction of a model for static function approximation. The initialization of the consequences must use the method explained in Section 2.5.

Once the structure of the model has been selected, the NARX model can be used to initialize the other structures; further improvement must be achieved using gradient descent techniques.

The classical cost function to be minimized is the quadratic cost function, defined as:

$$V_N(\theta) = \frac{1}{2N} \sum_{t=1}^{N} |y(t) - \hat{y}(t|\theta)|^2 \tag{4.25}$$

where θ is a vector containing the parameters of the membership functions and the position of the singletons of the consequences or the parameters of the vectors $\mathbf{A}^l$ in the Takagi–Sugeno models. The iterative scheme to update the parameters is defined as

$$\hat{\theta}^{(i+1)} = \hat{\theta}^{(i)} - \mu_i R_i^{-1} \hat{g}_i \tag{4.26}$$

where μ_i is the step size, $\hat{g}_i$ is an estimate of the gradient $V'_N(\hat{\theta}^{(i)})$ and R_i is a matrix that modifies the search direction. *online* (*recursive*) and *offline* (*batch*) methods are available [38]. Depending on the R_i the methods will be

- *Gradient direction* $R_i = I$
- *Gauss–Newton direction*

$$R_i = H_i = \frac{1}{N} \sum_{t=1}^{N} \psi(t, \hat{\theta}^{(i)}) \psi^T(t, \hat{\theta}^{(i)})$$

  where

$$\psi(t, \theta) = \frac{\partial}{\partial \theta} \hat{y}(t|\theta)$$

- *Levenberg–Marquard direction*

$$R_i = H_i + \delta I$$

- *Conjugate gradient direction*

$$R_i = V''_N(\hat{\theta}^{(i)})$$

It is very important to remark that the gradient of some of the parameters involved in the model must be generated dynamically due to the recursive structure of the model. The following example illustrates the complexity of the problem.

Example 4.1. Given a nonlinear dynamic system defined as $\hat{y}(t) = f(\hat{y}(t - k), \theta)$ where $\hat{y}(t)$ is the output of the model, $\hat{y}(t)$ delayed k units of time and $\theta \in \Re$ is a parameter of the model. The gradient of $\hat{y}(t)$ with respect to θ is given by the expression

$$\frac{\partial \hat{y}(t)}{\partial \theta} = \frac{\partial f(\hat{y}(t-k), \theta)}{\partial \theta} + \frac{\partial f(\hat{y}(t-k), \theta)}{\partial \hat{y}(t-k)} \frac{\partial \hat{y}(t-k)}{\partial \theta} \tag{4.27}$$

Replacing $\partial \hat{y}(t)/\partial \theta$ by $g(t)$, the expression will look like

$$g(t) = \frac{\partial f(\hat{y}(t-k), \theta)}{\partial \theta} + \frac{\partial f(\hat{y}(t-k), \theta)}{\partial \hat{y}(t-k)} g(t-k) \tag{4.28}$$

The expression clearly shows that the gradient $\partial \hat{y}(t)/\partial \theta$ is given by a dynamic system.

Some constraints can be imposed on these models to enforce certain properties (*i.e.*, limited gains, monotonicity, *etc.*). In general, the use of the gradients will improve the performance of the optimization algorithms used to tune the models. Gradient expressions for different types of fuzzy systems are presented in Appendix C.

Summary:
Gradient-based methods are the most reliable method to calculate the parameters of an identified model. The use of such methods demands the calculation of the gradient, which is in many cases (for fed-back variables) the simulation of a dynamic system.

4.7 Validation

For nonlinear systems, validation methods are restricted to criteria in time but not in frequency. The most typical validation test is the prediction error [32] $\varepsilon(t) = y(t) - \hat{y}(t|\theta)$. Once this test is acceptable, further tests can be introduced. The most important is the *residual validation test.* This test is based in correlation analysis of the residuals. The test allows us to detect information that can still be modeled and has not been captured by the model. The test also detects unmodeled dynamics or bias in the estimation.

Billings *et al.* [39] have shown that the residual validation test applied for linear systems,

$$\left.\begin{aligned} \phi_{\varepsilon\varepsilon}(\tau) &= \delta(\tau) \\ \phi_{\varepsilon u}(\tau) &= 0 \qquad \forall\tau \end{aligned}\right\} \tag{4.29}$$

(where ϕ stands for the normalized cross-correlation) was not enough to detect biases or unmodeled dynamics. The reason is that the test can only detect the residuals that are uncorrelated with themselves and with the inputs. The test simply establishes that no more "linear" relations can be found in the current data.

An extension of this test to nonlinear systems includes the correlation test between powers of the inputs and the residuals [39], for instance:

$$\left.\begin{aligned} \phi_{\varepsilon\varepsilon}(\tau) &= \delta(\tau) \\ \phi_{\varepsilon^2 u^2}(\tau) &= 0 \\ \phi_{\varepsilon u}(\tau) &= 0 \\ \phi_{\varepsilon u^2}(\tau) &= 0 \\ \phi_{\varepsilon\varepsilon u}(\tau) &= 0 \qquad \forall\tau \end{aligned}\right\} \tag{4.30}$$

These results can be used in a constructive way. For instance, if the residuals are correlated with an input to the second power, the model can be increased to include directly this new input ($u^2(t)$) or the number of membership functions used in the input $u(t)$ can be increased.

As a final remark, it is important to remember that fuzzy models have an additional validation mechanism by comparing the linguistic rules with the knowledge of an expert. In dynamic systems, this is not always a simple task because of the tendency of human beings to represent expert knowledge in a static form.

Summary:
Validity of the models can be assessed by evaluating the prediction error or by a correlation analysis of the residuals. The classical correlation analysis can be extended to nonlinear systems.

4.8 Example: Identification of the Box and Jenkins Gas Furnace Data Set

In this section, some of the techniques discussed in this chapter are applied to the data of the example given by Box and Jenkins [1]. The process is a gas furnace with a single input $u(t)$ (gas flow) and a single output $y(t)$ (CO_2 concentration). The data set contains 296 data points; here only the last 290 points are used.

For the selection of the regressors the *heuristic search* method explained in Section 4.4.1 was applied. In order to evaluate the "quality" of the regressors we used the *Lipschitz quotients* and the *regularity criterion* (see Section 4.4.2). The set of initial candidates for regressors were $\{u(t-1), u(t-2), u(t-3), u(t-4), u(t-5), u(t-6), y(t-1), y(t-2), y(t-3), y(t-4)\}$.

Figure 4.6 shows the results of the evaluation of the regressors using *Lipschitz quotients* (with $p = 6$) method combined with the input selection using the *heuristic search.* The bar graph shows that among the models built with one input the regressor selected was $y(t-1)$. Among the models with two inputs, $y(t-1)$ and $y(t-2)$ were selected. Finally the selection process was stopped with three inputs $y(t-1), y(t-2)$ and $u(t-3)$.

This example also evaluated the input selection using the *heuristic search* and evaluating the sets of regressors by means of the *regularity criterion* generating the results shown in Figure 4.7. The *regularity criterion index* obtained for a model with one regressor makes us select $y(t-1)$. Among the models with two regressors $y(t-1), u(t-3)$ were the best and for the model with three regressors $y(t-1), u(t-3), y(t-4)$. The constructed models were the "smallest" fuzzy models described in Section 2.5.1.

The methods selected different regressors. The selections using *Lipschitz quotients* proposes $y(y-1), y(t-2), u(t-3)$ and the selection using the *regularity criterion* $y(t-1), u(t-3), y(t-4)$. The reason for this discrepancy is that the quality of the regressors is measured with a completely different criterion. The *Lipschitz quotients* method rewards the smoothness of the candidate function and the *regularity criterion* rewards models with small prediction error.

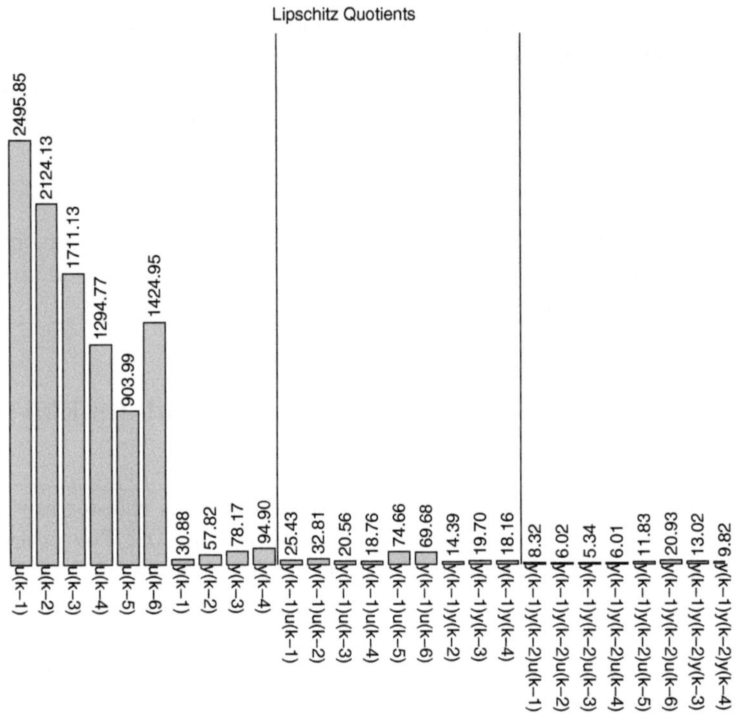

Figure 4.6. Input selection using Lipschitz quotients and heuristic search

Once the regressors are selected, the problem is to find the structure of the fuzzy system (number of rules, membership functions, consequences, *etc.*). In order to extract the model structure, clustering, projection and FuZion were used (see Section- 3.3). In this example the *mountain-clustering* method (see Section B.3) was applied with $\alpha = 4$ and $\beta = 4$. For both sets of regressors 9 clusters were found and projected into the input space (see Figures 4.8 and 4.9). The membership functions were reduced using the FuZion algorithm (see Section 3.4) with $M = 10\%$ of the universe of discourse of the variable. The model using the regressors obtained using the *Lipschitz quotients* $(y(t-1), y(t-2), u(t-3))$ used 4 triangular membership functions on each input, as shown in Figure 4.10. The total number of rules for this model is 64 $(4 \times 4 \times 4)$. The model using the regressors obtained using the *regularity criterion* $(y(t-1), y(t-4), u(t-3))$ used 4 triangular membership functions on the inputs $y(t-1)$ and $y(t-3)$ and 5 triangular membership functions on the input $y(t-4)$, as shown in Figure 4.11. The total number of rules for this model is 80 $(4 \times 4 \times 5)$.

With each set of regressors two models were constructed: one NARX and one NOE. The consequences of the NARX model were calculated using the method

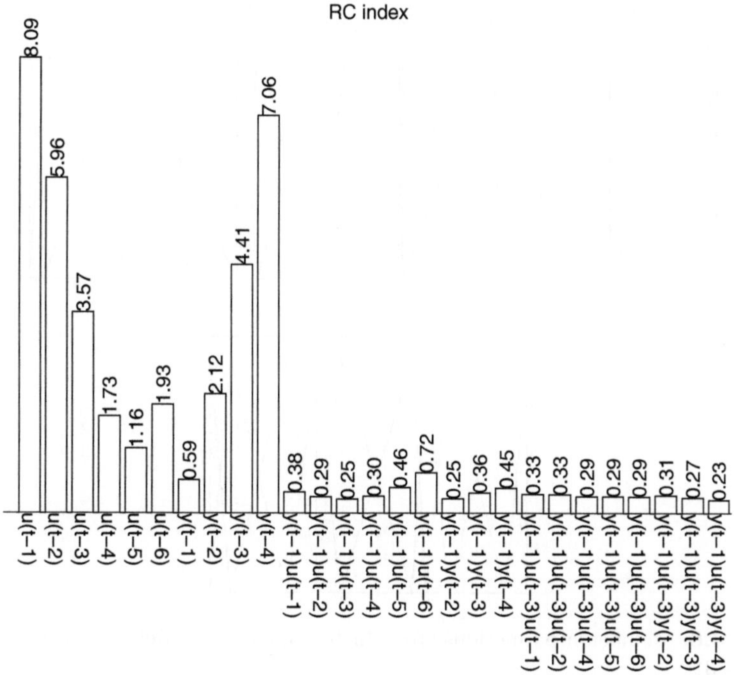

Figure 4.7. Input selection using regularity criterion and heuristic search

described in Section 2.5. The NOE model was calculated by initializing the system with the NARX solution and by running an optimization over the consequences, using the prediction error as a cost function. The results of the obtained models can be seen in Table 4.1 and in Figures- 4.12, 4.13, 4.14 and- 4.15.

Table 4.1. Example: Performance Comparison Between the Models

Model description	Number of rules	Pred. error whole data	Pred.error validation data
NARX $[y(t-1), y(t-2), u(t-3)]$	64	0.3189	0.5175
NOE $[y(t-1), y(t-2), u(t-3)]$	64	0.8128	1.3812
NARX $[y(t-1), y(t-4), u(t-3)]$	80	0.0648	0.1837
NOE $[y(t-1), y(t-4), u(t-3)]$	80	0.0646	0.1741

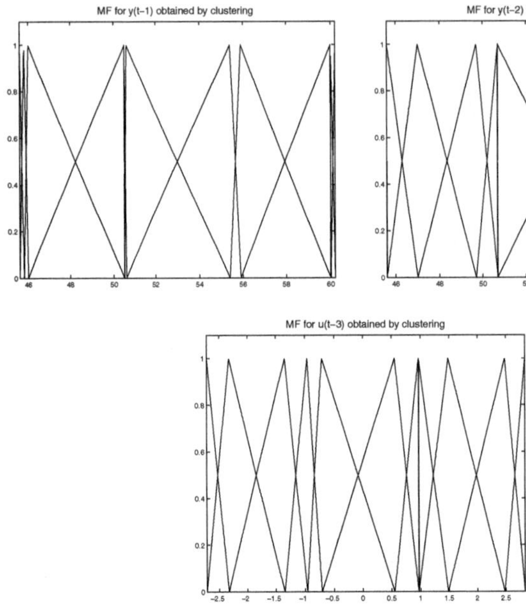

Figure 4.8. Membership functions after clustering for the model using $y(t-1), y(t-2), u(t-3)$

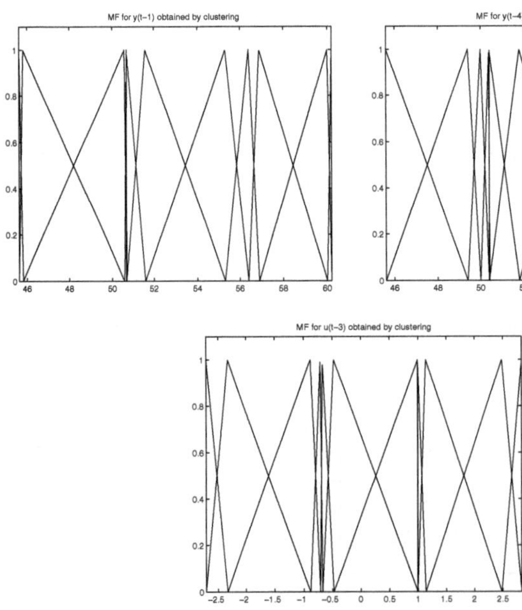

Figure 4.9. Membership functions after clustering for the model using $y(t-1), y(t-4), u(t-3)$

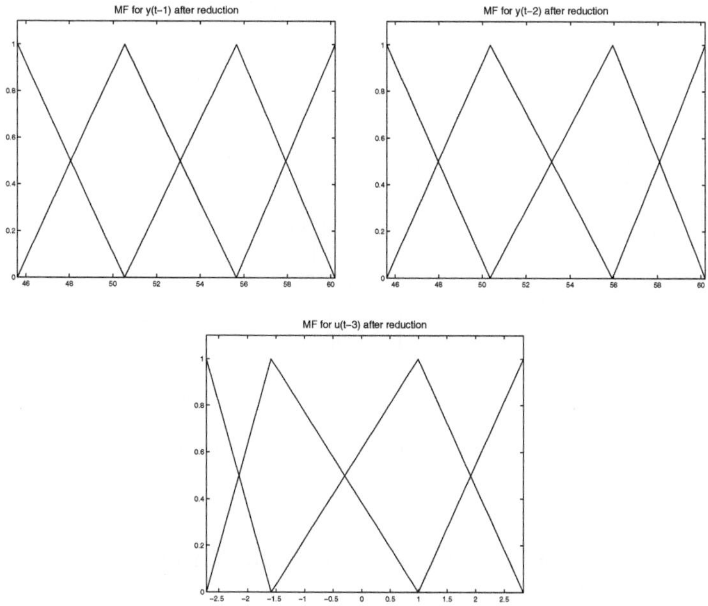

Figure 4.10. Membership functions after clustering and FuZion for the model using $y(t-1), y(t-2), u(t-3)$

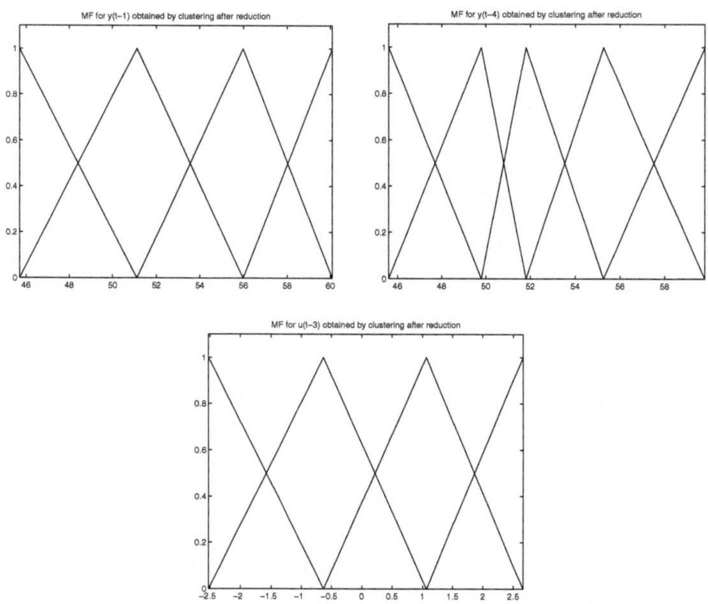

Figure 4.11. Membership functions after clustering and FuZion for the model using $y(t-1), y(t-4), u(t-3)$

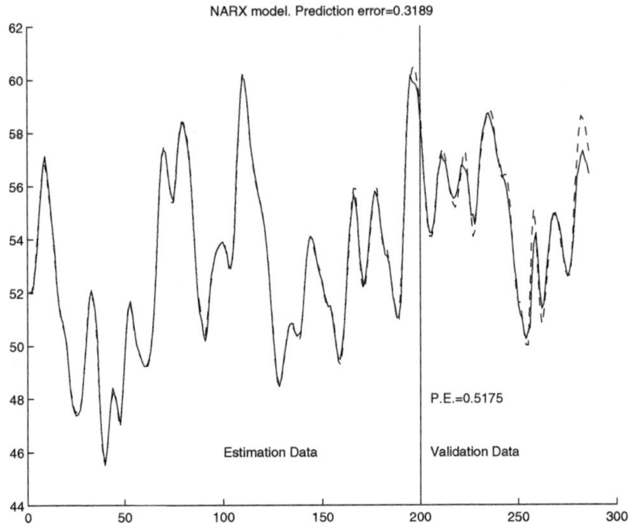

Figure 4.12. Prediction error one step ahead for the NARX model with regressors $y(t-1), y(t-2), u(t-3)$

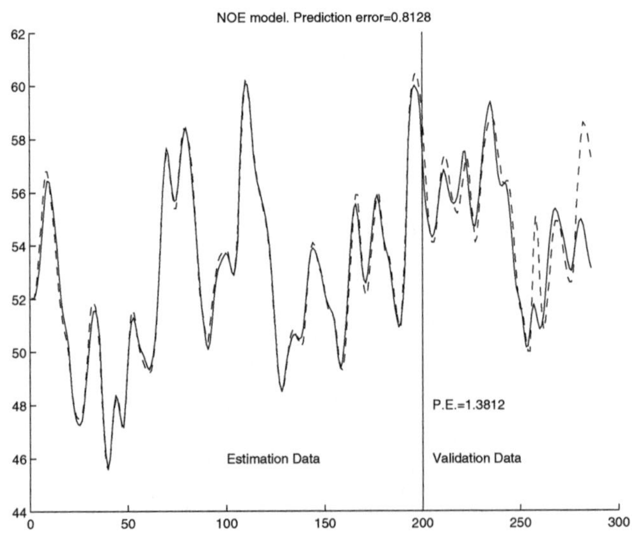

Figure 4.13. Prediction error one step ahead for the NOE model with regressors $y(t-1), y(t-2), u(t-3)$

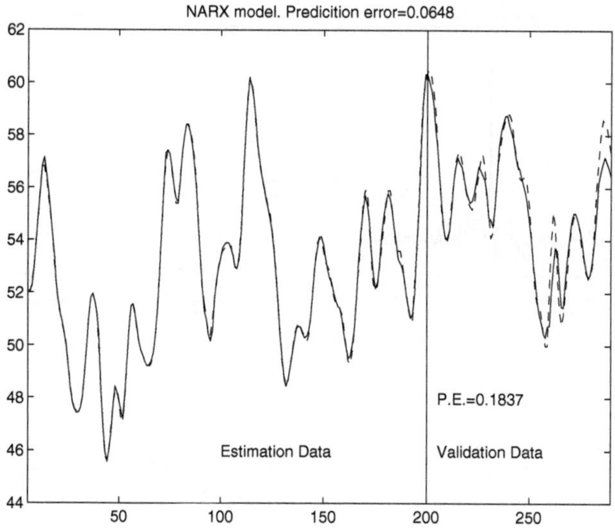

Figure 4.14. Prediction error one step ahead for the NARX model with regressors $y(t-1), y(t-4), u(t-3)$

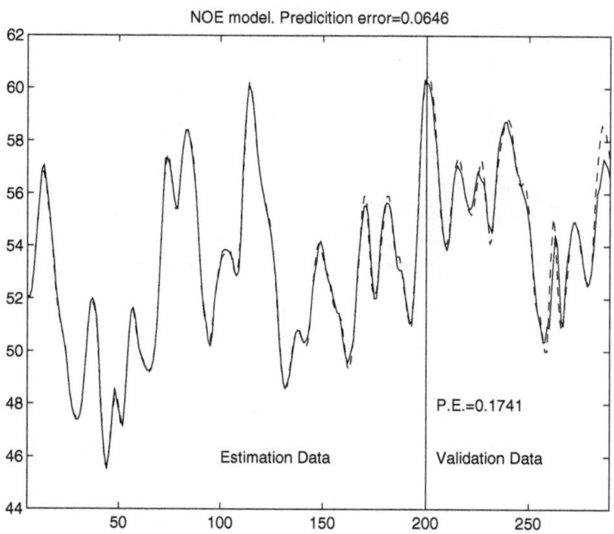

Figure 4.15. Prediction error one step ahead for the NOE model with regressors $y(t-1), y(t-4), u(t-3)$

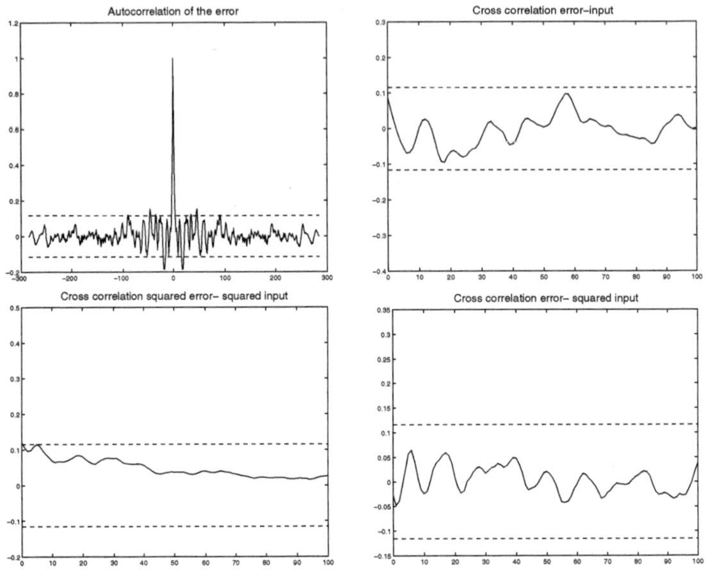

Figure 4.16. Correlation analysis for the NARX model with regressors $y(t-1), y(t-2), u(t-3)$

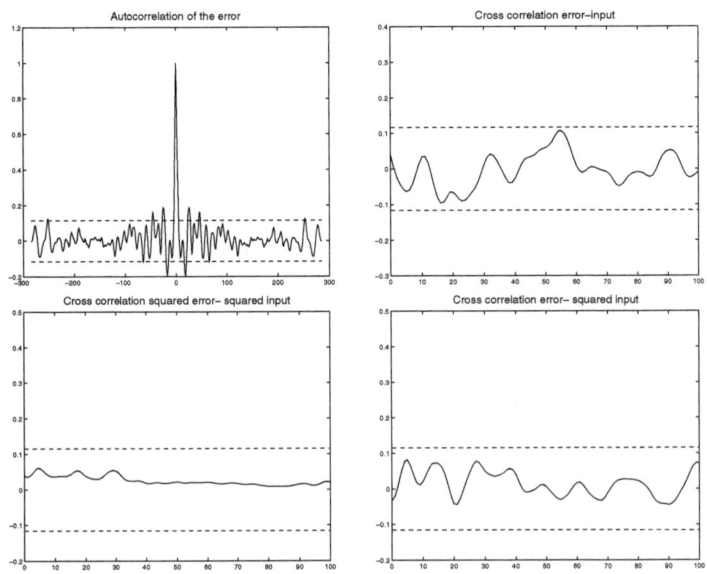

Figure 4.17. Correlation analysis for the NOE model with regressors $y(t-1), y(t-2), u(t-3)$

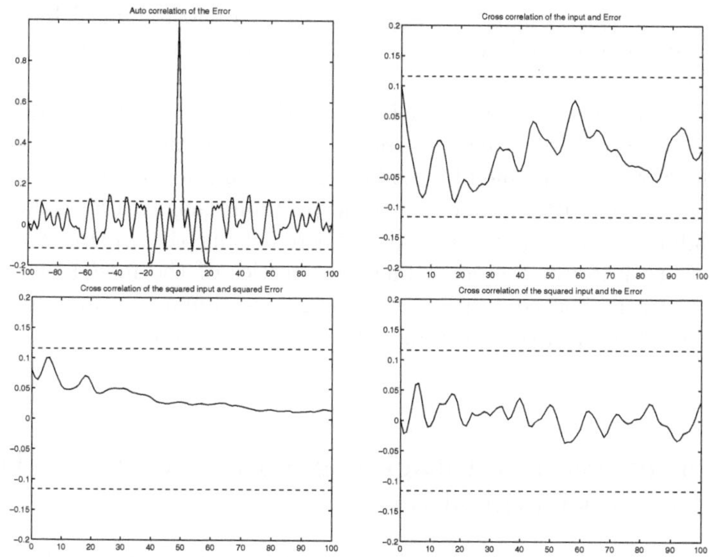

Figure 4.18. Correlation analysis for the NARX model with regressors $y(t-1), y(t-4), u(t-3)$

A final validation is obtained by using the correlation analysis. This is shown in the Figures 4.16 and 4.17, which correspond to the NARX and the NOE

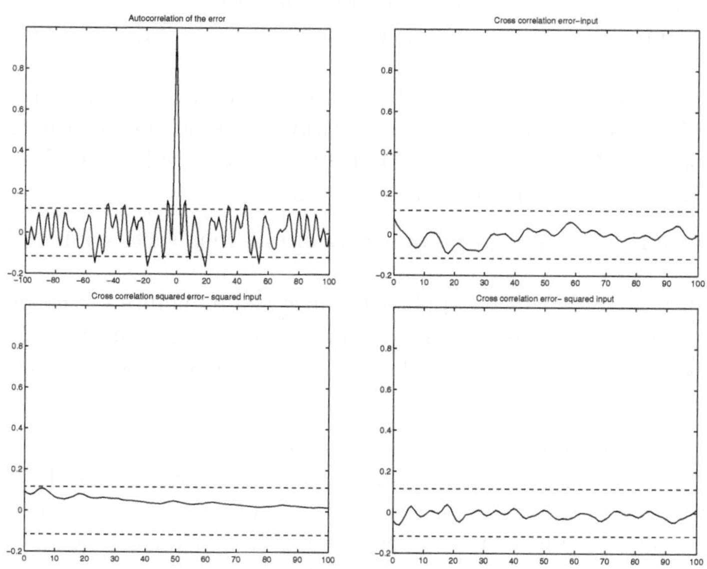

Figure 4.19. Correlation analysis for the NOE model with regressors $y(t-1), y(t-4), u(t-3)$

models for the regressors $y(t-1), y(t-2), u(t-3)$ and Figures 4.18 and 4.19, which correspond to the NARX and the NOE models for the regressors $y(t-1), y(t-4), u(t-3)$.

From the pictures and the table it is clear that the models constructed with the regressors obtained using the *regularity criterion* have smaller prediction error. This is easily explained by the fact that the input selection based on the *regularity criterion* attempts to find the set of input that generates the smaller cross-validation error. The regressors selected using the *Lipschitz quotients* are chosen to improve the smoothness properties of the model but not the prediction error. The correlation analysis shows similar features for all the models, being acceptable according with this validation criterion.

4.9 Identification of Takagi–Sugeno Fuzzy Models Using Local Linear Identification

Classical control techniques are based in the design of a feedback compensator (controller) based on a "local" linear model. Understand by "local" a given neighborhood where the dynamics of the system can be described by a linear perturbation model.

The ever-increasing demands in the performance of control systems motivated the use of different linear controllers for different *localities*. In this way the control system could guarantee the performance under different operating conditions. So far each local model has been treated as disconnected from other models on its neighborhood, under the classical model and control scheduling theory.

Takagi–Sugeno fuzzy models provide an automatic mechanism to combine local models and calculate smooth transitions among the localities generating a global nonlinear dynamic system described by rules with local linear dynamics. Such models offer many advantages:

- Simple analysis is possible, because each rule represents a linear dynamic system.
- Stability analysis can be applied because the dynamics of the plant will always be described by a convex combination of local linear models.
- Simple local linear models can be obtained and extended every time a new operating point is reached.

The identification problem demands the selection not only of the regressors but also of the scheduling variables. The scheduling variables are the variables that govern the changes of dynamical regime. The scheduling variables can be outputs, inputs or states. In practice, to guarantee a smooth behavior of the model the scheduling variables are selected to be slowly varying.

The combination of the local models demands consistency among them. By consistency is understood the use of the same regressors in all the models or the same state-space representation in case the local model are obtained in state-space form.

One way to guarantee the consistency of rules using state-space models is to identify all the local models with the same order and convert all of them to the so-called observer canonical form (see Chapter 2 in [40]). In this way all the states of the local model will be consistent and their evolution will be perfectly synchronic.

The identification of Takagi–Sugeno fuzzy models offers many advantages for practitioners since it can generate a smooth transition from linear to nonlinear models. Current linear models can be used as the initial description of the consequences of the rules and gradually as the process is moved to new operating points the new information can be use to update the local rules.

Summary:
The identification of Takagi–Sugeno fuzzy models offers many advantages for practitioners specially in process controls. The fact that the changes from one operating point to another is slow favors the use of this since it can generate a smooth transition from linear to nonlinear models. Current linear models can be used as the initial description of the local models.

4.10 Conclusions

Dynamic fuzzy models can be constructed and validated using classical system identification theory. This chapter presented the main problems faced along the construction of dynamic fuzzy models using system identification, regressors selection, experiment design, structure, parameter adjusting and validation.

The problem of regressor selection faces the so-called curse of dimensionality (exponential growth of the possible solutions). The chapter presented some efficient methods based on heuristic search and genetic algorithms to trade off the complexity of the calculation with the accuracy of the solution. Such methods improve the search of the parameter by means of some educated decision.

Once a set of regressors is selected, the optimization of the parameters of the models is a task that involves the use of gradient descent techniques. The gradients are generated by a dynamic system whose states are the gradients of some of the parameters. The chapter included the derivation of the dynamic systems that generates such gradients for the most common membership functions (see Appendix C).

Validation is a very important phase that defines the criteria to compare, accept or reject certain model. The validation methods discussed in this chapter included the prediction error and a nonlinear correlation analysis.

Finally, the chapter was closed with a brief explanation of an empirical method for system identification based on Takagi–Sugeno models . However simple, the methods is powerful, intuitive and reliable. It does not demand the use of complex system identification tools, since it can construct models using only linear identification techniques and very simple similarity transformations to guarantee the consistency of the models.

Part II

Fuzzy Control

5

Fuzzy Control

The use of fuzzy logic and fuzzy systems for control has promised the development of powerful control strategies. These expectations can be explained by the linguistic representation of the control actions and the flexible nonlinearities that can be constructed with such systems. On the other hand, some limitations to the analysis of these control systems arise from the complex mathematical description of the nonlinearities.

Fuzzy controllers can be constructed in many different ways but it is possible to establish a classification between model-based designed controllers and model-free. Model based controllers usually demand a "complete" description of the plant dynamics. The model-free strategies are called model-free because they are not based on complete mathematical models; however, they are not completely model-free. They are based on information extracted from simple experiments (relay experiments, *etc.*) or a heuristic model present in the designer's mind.

This chapter presents an overview of "classical" methods to build fuzzy controllers. The chapter includes some novel results and applications using these "classical" techniques. Appendixes D and E are a complement to this chapter. The content of Appendix D is a proof of the theorem stating that any linear controller with bounded states and inputs can be made into an exact equivalent fuzzy controller. The content of Appendix E shows an application of these techniques to a system used in the automotive industry.

Summary:
According to the information used to construct fuzzy controllers, they can be classified as model-free and model-based control systems.

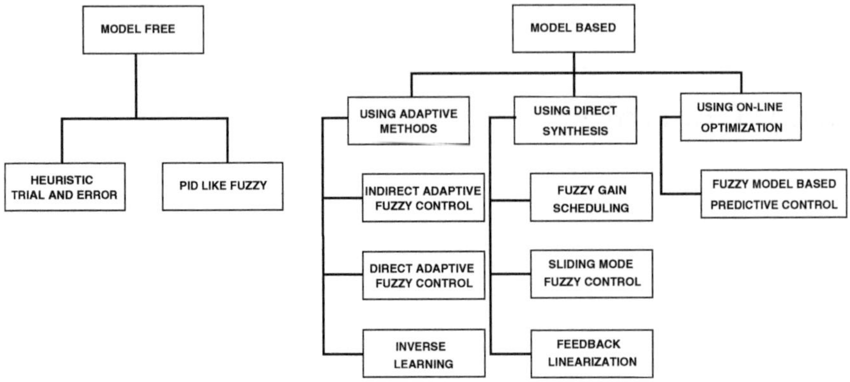

Figure 5.1. Classification of the fuzzy controllers

5.1 Model-Free Fuzzy Control

5.1.1 Heuristic Trial-and-Error Design

This methodology is probably the first technique ever used to design fuzzy controllers. The technique uses the experience cumulated over years of manual control. The typical approach to construct these controllers has been the formulation of rule bases using the information provided by the operator's manual. Most of the time, these controllers are used at high level as a kind of supervisory control where issues such as stability are not critical. The stabilization of the plant is a task accomplished by low-level controllers. In many cases, the fuzzy controller is not used directly in the system in automatic mode, but it is used as a support system for the operator. Successful applications of this technique have been reported in the areas of cement kiln control [41], boiler startup sequences, washing powder production [42], waste incineration and waste water treatment [43].

5.1.2 Design of PID-like Fuzzy Controllers

This PID (proportional integral and derivative)-like fuzzy controller has been included in this model-free class because a fuzzy controller can be designed using the same experiments designed to tune linear PID controllers in a model-free basis or using simple models (step response models). The main idea behind this approach is that any PID with bounded input and output can be reproduced exactly by a fuzzy system (see theorem proof in Appendix D). The design method proceeds as follows:

- Tune a PID controller using any of the traditional methods (Ziegler Nichols or Kappa–Tau from Aström and Hägglund [44].
- Construct a fuzzy controller equivalent to the tuned PID.

- Do further tuning of the fuzzy controller using heuristics.

Perhaps the most popular presentation of the low-level (nonsupervisory) fuzzy controller is the so called e, Δe controller. This is a fuzzy controller with two inputs, e=*error* and Δe=*change of the error*, and one output, which is the control action u or Δu depending, on whether the controller acts as PD (proportional derivative) or PI (proportional integral). One interesting characteristic of this description of the controller is the fact that a direct analogy can be established with the classical PD and PI controllers. The fuzzy controller with a direct action will be analog to a PD controller, and the fuzzy controller with incremental action will be analog to a PI controller (see Figures 5.2 and 5.3).

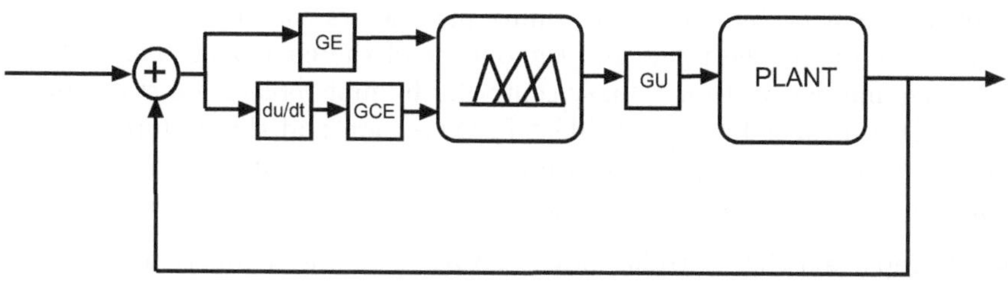

Figure 5.2. Fuzzy PD controller

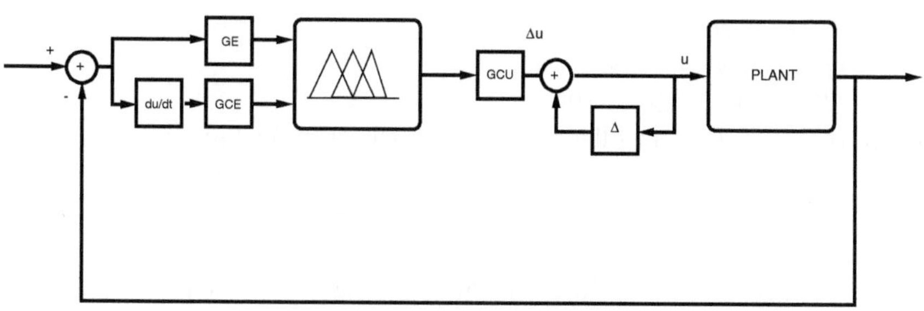

Figure 5.3. Fuzzy PI controller

Assuming five fuzzy sets (NL: Negative Large, NS: Negative Small, ZE: Zero, PS: Positive Small and PL: Positive Large) on each input of the fuzzy controller and seven fuzzy sets in the output (the previous five plus PM: Positive Medium, NM: Negative medium, PVL: Positive Very Large and NVL: Negative Very Large) a typical rule base is shown in Table 5.1.

The distribution of the membership functions in the inputs can be very uniform, as shown in Figure 5.4. Observe that the domain of the membership

Table 5.1. Rule Base for a PD- or PI-like Fuzzy Controller

e \ Δe	NL	NS	ZE	PS	PL
NL	PVL	PL	PM	PS	ZE
NS	PL	PM	PS	ZE	NS
ZE	PM	PS	ZE	NS	NM
PS	PS	ZE	NS	NM	NL
PL	ZE	NS	NM	NL	NVL

functions is distributed in the interval –100 to 100. Also, the output membership functions have been scaled between –200 to 200 and are described using singletons. Using this rule base, triangular membership equally distributed and with 0.5 overlap, a fuzzy controller perfectly equivalent to a PI or PD controller can be built and the tuning parameters can be initially given by scaling factors at the input and the output of the controller. This will be a very safe method to tune a fuzzy controller by first replacing a stabilizing PI or a PD controller by its equivalent fuzzy version and then a further tuning of the membership functions can improve the performance of the controller. The procedure can be summarized as follows:

- To replace a PD controller described by the transfer function

$$C(s) = K_p(1 + \tau_d s) \tag{5.1}$$

  the scaling factors should fulfill the following requirements:
  - $GE \times GU = K_p$
  - $GCE \times GU = K_p \tau_d$
  - $GE \times \max |e| \leq 100$
  - $GCE \times \max |\Delta e| \leq 100$
- To replace a PI controller described by the transfer function

$$C(s) = K_p(1 + \frac{1}{\tau_i s}) \tag{5.2}$$

  the scaling factors should fulfill the following requirements:
  - $GCE \times GCU = K_p$
  - $GE \times GCU = \frac{K_p}{\tau_i}$
  - $GE \times \max |e| \leq 100$
  - $GCE \times \max |\Delta e| \leq 100$

Observe that for both cases the first two conditions guarantee the correct mapping of the gains of the controller. Meanwhile the last two conditions avoid the saturation. In this way the obtained fuzzy controller will behave exactly as the replaced PD or PI controller. The main guidelines of this tuning method were mentioned in [45].

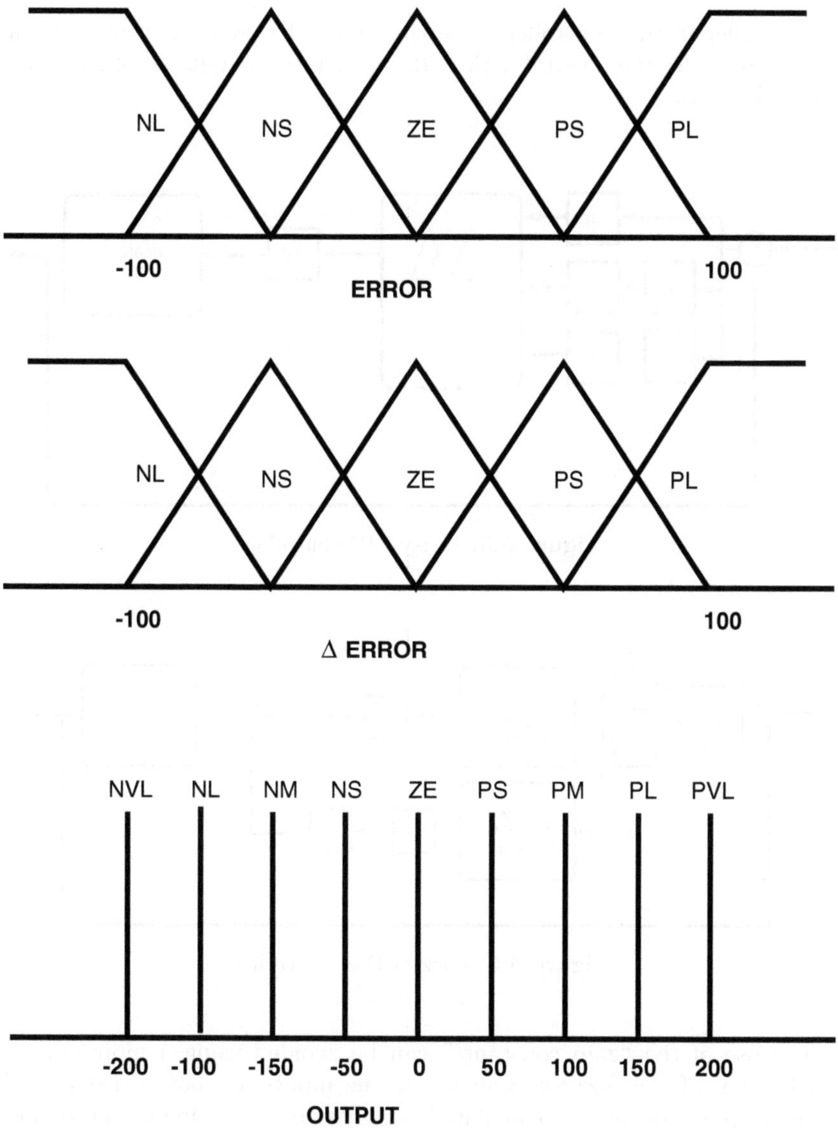

Figure 5.4. Fuzzy system with linear relations

The design of a fuzzy-PID controller can be achieved in different ways. One way is to build a fuzzy controller with three inputs: the error (proportional action), the delta error (derivative action) and the sum of the error (integral action) (see Figure 5.5). The inconvenience for such a controller is that the number of rules will grow and instead of 25 rules (for systems with five membership functions on each input), this controller will have 125 rules, making the tuning task very difficult. A more efficient solution is to divide the controller in two controllers, one that is the PD equivalent and another one that provides the integral action. It will reduce the number of rules to 30 (see Figure 5.6).

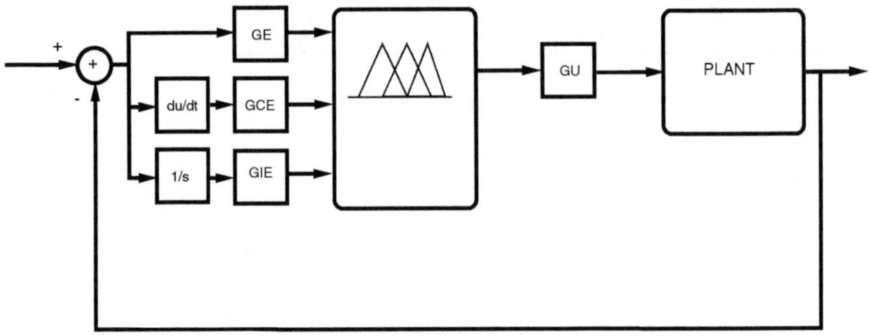

Figure 5.5. Fuzzy PID controller

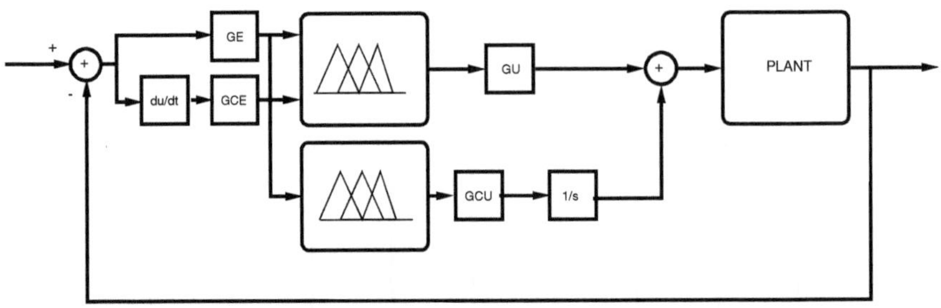

Figure 5.6. Fuzzy PD+I controller

The use of the "gain constants" can be avoided using a more complex initialization of the fuzzy system where the inputs are not scaled and the singletons of the output are calculated using its direct relation with the modal values of the membership functions in the antecedents, in a similar way as the rules are initialized in the training algorithm presented in Section 2.5.1.

Further improvement will be done by trial and error, if there is no model of the plant, or via optimization, if a model is available. Using this initialization, the controller will guarantee a performance at least as good as the one provided by the linear PID controller.

Some of the typical actions in the trial-and-error procedure are to increase the impact of the proportional action when the error is very large and slightly reduce the integral action. These two actions tend to reduce the settling time and the overshoot.

Summary:
Fuzzy controllers can be built as "exact" copies of PID controllers. This initial representation guarantees a smooth migration from linear PID to nonlinear PID controllers. Further improvement of the performance can be achieved by tuning the new degrees of freedom provided by the fuzzy controller.

5.2 Model Based Fuzzy Control

This section presents three types of methods to construct model-based fuzzy controllers. These are considered as the most representative methods. In practice, these strategies appear mixed up to some level.

- **Using adaptive methods** This design technique is an optimization-based technique that uses the model of the plant and intensive simulations to optimize the parameters of the fuzzy controller. In most of the cases the optimizations are performed offline, but in cases like the direct adaptive control the optimization can be done online.
- **Using direct synthesis** This design technique uses either the information given by the parameters of the model or the model itself to construct the controller.
- **Using online optimization** This technique uses a fuzzy model to predict the future behavior of the plant in a receding horizon and with this information calculates the future movements of the control actions using optimization.

5.2.1 Using Adaptive Methods

Inverse Learning

The fundament of this kind of controllers is the construction of an inverse model of the plant such that the controller generates an input to drive the state of the plant from the current state x_k to a desired state x^d_{k+n} [19]. For the application of this technique, it is assumed that the states of the plant are measurable. The dynamics of the plant are assumed to be discrete or at least sampled and represented by the function

$$x_{k+1} = f(x_k, u_k) \tag{5.3}$$

where k represents the discrete time, x_k is the state and u_k is the input of the plant. The state of the plant for the time $k + N$ is given by

$$x_{k+N} = \underbrace{f(f(f(x_k, u_k), u_{k+1}), \ldots, u_{k+N-1})}_{N\text{times}} \tag{5.4}$$

It is equivalent to say

$$x_{k+N} = F(x_k, U) \tag{5.5}$$

where F is a function representing the multiple composition of the function $f(.,.)$ and U is a vector with the input sequence $u_k, \ldots, u_{k+N-1}$. With this description, and assuming the invertibility of the function F, an inverse map of the plant can be constructed as

$$U = G(x_k, x_{k+N}) \tag{5.6}$$

This function will generate the control sequence U to move the plant from the current state x_k to the state x_{k+N} in N steps. The existence of this inverse map is equivalent to the controllability condition for linear systems [46].

The existence of the map G does not guarantee the existence of an analytical closed form. The fuzzy system $\hat{G}$ is used to approximate this map. The dimension of the map for a system of order n with one input will be $2n$ inputs and N outputs. It is clear that even for systems of low order the number of inputs will make the fuzzy system very big (with a large number of rules).

The use of the inverse model for control works as follows: the reference is given as the future state x^d_{k+N} and with the current state the inverse model $\hat{G}$ generates a vector U. The first entry of the vector is implemented and the function is evaluated once more at the next sampling time. If the reference is not known in advance, the future reference will be replaced by the current reference, generating a system that behaves as a pure delay system.

For practical purposes it is better to use an inverse model that generates only one value for the input sequence $N = 1$. Figure 5.7 shows the construction of the controller.

This control technique is limited by the condition of invertibility of the plant and the fact that the minimization of the norm $||u_k - \hat{u}_k||$ does not guarantee the minimization of $||x_k - x^d_k||$. For practical purposes this method is very limited because it demands full access to the states of the system. Moreover, the tuning demands that a large number of possible transitions from x_k to x_{k+1} are tested to guarantee full coverage of the operating range of the controller. This problem is even worse when the order of the system is large. Already a fifth-order model can generate some serious problems, because the controller will have $2 \times 5 = 10$ inputs and at least 1024 rules. In summary, this strategy can be applied only to low-order (first- or second-order) invertible systems with full access to the state variables of the system. The performance of the controller will be limited by the characteristics of the experiments used during the training phase. This control strategy can be seen during the learning phase as an identification experiment. Other authors proposed the use of adaptive schemes where the inverse model is continuously updated online while the process is in operation [47].

Summary:
Fuzzy inference systems can be used to map the inverse dynamics of a plant and by these means achieve simple nonlinear control of a plant. The performance of the controller will be limited by the bandwidth on the input signals used along the training phase.

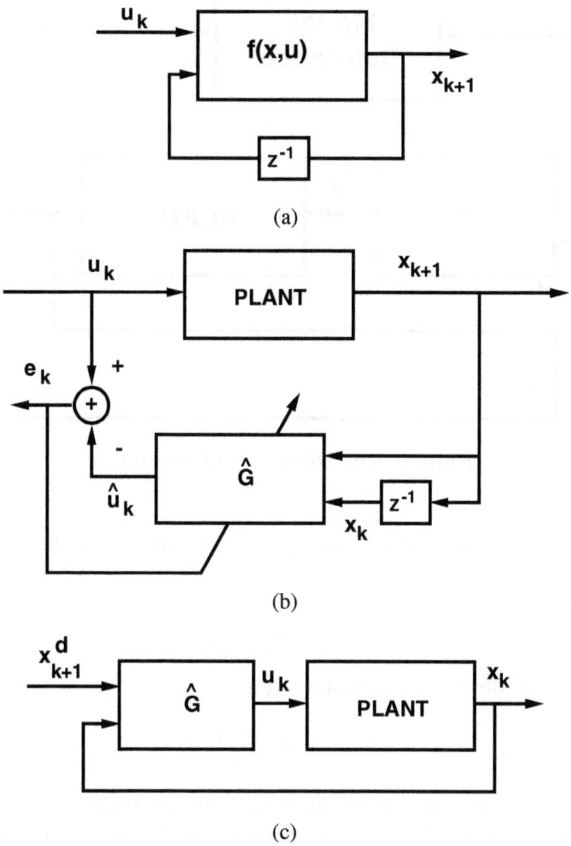

Figure 5.7. Scheme for design of an inverse controller (a) Plant (b) Learning Phase (c) Scheme of operation

Specialized Learning or Direct Adaptive Fuzzy Control

This technique adjusts the parameters of the controller according to some performance measurement [48]. In this case, the controller is represented by a fuzzy system. Due to the multiple possible schemes the discussion will be restricted to two cases: *state feedback* and *output feedback*.

The performance measure can be made against some performance specification (settling time, maximum overshoot, raise time) or a reference model (normally a linear one) that already includes the desired specifications. The general scheme is shown in Figure 5.8.

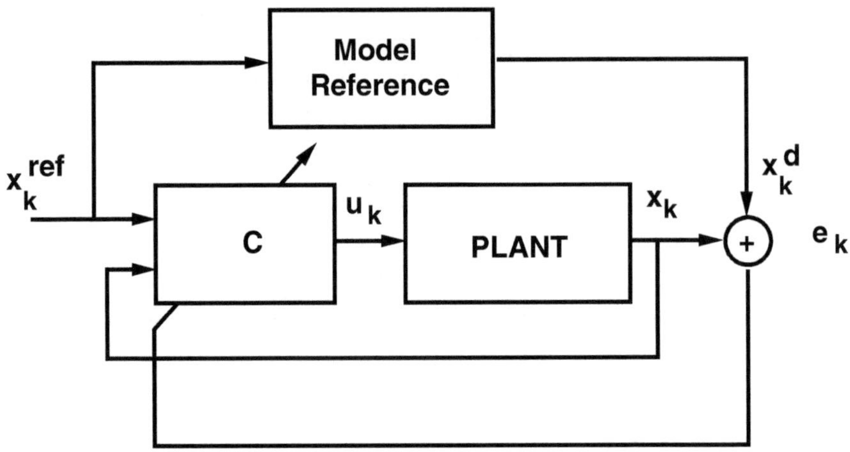

Figure 5.8. Direct adaptive control

Assuming a controller using *state feedback* the formulation of the method is as follows:
Given a plant model

$$x_{k+1} = f(x_k, u_k) \tag{5.7}$$

the objective is to design a control system

$$u_k = g(x_k, x_k^{ref}, \theta) \tag{5.8}$$

this controller is a static nonlinear map with $2n$ inputs and a parameterization is given by the vector θ. The nonlinear map will be constructed using a fuzzy system. The goal is to obtain a set of parameters θ such that the closed loop dynamics should mimic the dynamics of the *reference model* depicted in Equation (5.9).

$$x_{k+1}^d = \bar{f}(x_k^d, x_k^{ref}) \tag{5.9}$$

where x_k^d is the output of the reference model and x_k^{ref} is the reference. The design of this controller can be formulated by the optimization problem

$$\min_\theta J = \min_\theta \sum_{k=1}^{N} ||x_k^d - x_k|| \tag{5.10}$$

the closed-loop expression for the system is

$$x_{k+1} = f(x_k, g(x_k, x_k^{ref}, \theta)) \tag{5.11}$$

then the minimization can be written as

$$\min_{\theta} \sum_{k=0}^{N-1} ||\bar{f}(x_k^d, x_k^{ref}) - f(x_k, g(x_k, x_k^{ref}, \theta))||^2 \tag{5.12}$$

If a gradient descent technique is going to be applied, the adjustments will be directed by the gradient

$$\frac{\partial J}{\partial \theta} = 2 \sum_{k=0}^{N-1} -(x_{k+1}^d - x_{k+1}) \frac{\partial x_{k+1}}{\partial \theta} \tag{5.13}$$

where the gradient $\partial x_{k+1}/\partial \theta$ is generated by the dynamic system

$$\begin{aligned}\frac{\partial x_{k+1}}{\partial \theta} = &\left[\frac{\partial f(x_k, u_k)}{\partial x_k} + \frac{\partial f(x_k, u_k)}{\partial u_k} \frac{\partial g(x_k, x_k^{ref})}{\partial x_k} \right] \frac{\partial x_k}{\partial \theta} \\ &+ \frac{\partial f(x_k, u_k)}{\partial u_k} \frac{\partial g(x_k, x_k^{ref})}{\partial \theta}\end{aligned} \tag{5.14}$$

It is clear that in order to apply this method a good model of the plant $f(.,.)$ is needed to derive the expressions $\partial f(x_k, u_k)/\partial x_k$, $\partial f(x_k, u_k)/\partial u_k$ analytically or by means of numerical methods.

Now the analysis will be oriented to the case of output feedback, where the model of the plant will be described by

$$\begin{aligned} x_{k+1} &= f(x_k, u_k) \\ y_k &= g(x_k) \end{aligned} \tag{5.15}$$

and the controller will be a dynamic fuzzy model represented by the equation:

$$\begin{aligned} z_{k+1} &= h(z_k, y_k, y_k^{ref} | \alpha) \\ u_k &= d(z_k, y_k, y_k^{ref} | \beta) \end{aligned} \tag{5.16}$$

where z_k are the states of the controller, y_k^{ref} is the reference input, α and β are the parameters of the functions $h(.,.,.)$ and $d(.,.,.)$. The closed-loop system will be expressed as the dynamic system

$$\begin{aligned} s_{k+1} &= m(s_k, y_k^{ref} | \theta) \\ u_k &= p(s_k, y_k^{ref} | \beta) \\ y_k &= q(s_k) \end{aligned} \tag{5.17}$$

where $s_k = \{x_k, z_k\}^T$, $\theta = \{\alpha, \beta\}^T$, $d(z_k, y_k, y_k^{ref} | \beta) = p(s_k, y_k^{ref} | \beta)$ and $q(s_k) = g(x_k)$. Assuming a cost function

$$\min_\theta J = \min_\theta \sum_{k=1}^{N} ||y_k^{ref} - y_k|| + \lambda u_k^T u_k \tag{5.18}$$

the derivative will be given by

$$\frac{\partial J}{\partial \theta} = 2\sum_{k=0}^{N-1} -(y_k^{ref} - y_k)\frac{\partial y_k}{\partial \theta} + \lambda u_k \frac{\partial u_k}{\partial \theta} \tag{5.19}$$

with the terms $\frac{\partial y_k}{\partial \theta}$ and $\frac{\partial u_k}{\partial \theta}$ generated by the dynamic system:

$$\begin{aligned}\frac{\partial s_{k+1}}{\partial \theta} &= \frac{\partial m}{\partial s_k}\frac{\partial s_k}{\partial \theta} + \frac{\partial m}{\partial \theta}\\ \frac{\partial y_k}{\partial \theta} &= \frac{\partial q}{\partial s_k}\frac{\partial s_k}{\partial \theta}\\ \frac{\partial u_k}{\partial \theta} &= \frac{\partial p}{\partial s_k}\frac{\partial s_k}{\partial \theta} + \frac{\partial p}{\partial \theta}\end{aligned} \tag{5.20}$$

According to the parameters to be adjusted (membership functions, consequences or scaling values) the computation of the gradients will be more or less complex. Figures 5.9 and 5.10 show the tuning of the values for a PID controller. The optimization procedure should evaluate a cost function that can include the optimization for disturbance rejection or for tracking a set-point or a compromise between these two objectives.

Different parameters can be tuned using this method, but according to the impact into the performance the priority will be to tune the scaling factors if there are any, then tune the consequences of the rules of the fuzzy system and finally if a very fine tuning is required, the parameters of the membership functions of the antecedents of the rules.

This method does not demand full access to the states and an output feedback strategy can be implemented if the system is both controllable and observable.

Summary:
The use of specialized learning or direct adaptive fuzzy control provides the means to construct optimal fuzzy controllers. The scheme demands a good model of the plant dynamics, such that the desired optimality can be achieved.

5.2.2 Using Direct Synthesis

Feedback Linearization

This methodology is applied to nonlinear systems of order n of the form

$$\begin{aligned}x^{(n)} &= f(x, \dot{x}, \ldots, x^{(n-1)}) + g(x, \dot{x}, \ldots, x^{(n-1)})u\\ y &= x\end{aligned} \tag{5.21}$$

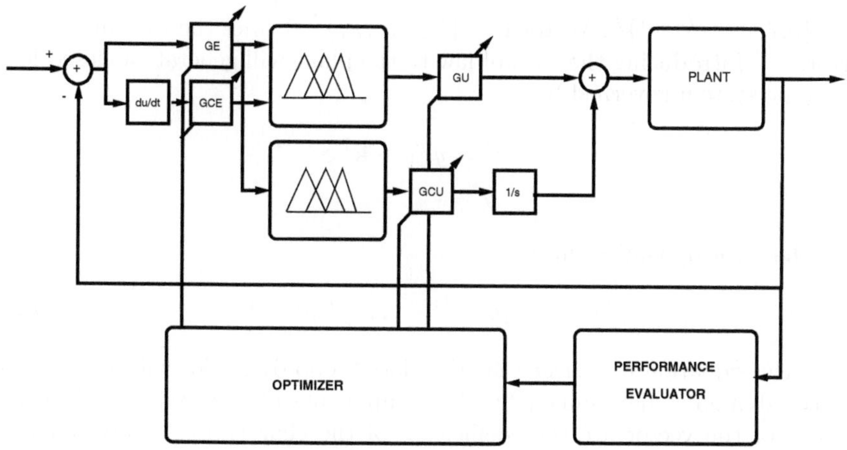

Figure 5.9. Tuning of the gain constants using nonlinear optimization

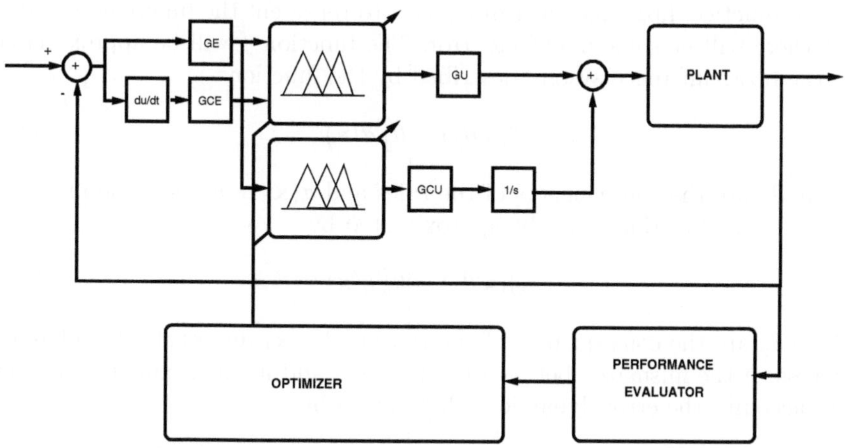

Figure 5.10. Tuning of the internal parameter of the fuzzy controller (membership functions, consequences, *etc.*)

these systems are known as nonlinear affine systems. A very interesting feature of these type of nonlinear systems is that the knowledge of the functions $f(x)$ and $g(x)$ can be used in a straight manner to design a control signal u so that the nonlinearity is cancelled and the controller can be designed using linear techniques such as pole placement [49][50][17].

The control law is described by this equation,

$$u = \frac{1}{g(\mathbf{x})}(-f(\mathbf{x}) + y_{ref}^{(n)} + \mathbf{k}^T \mathbf{e}) \tag{5.22}$$

Once this controller is applied to the plant, then the control error will be defined as $e = y_{ref} - y$, and the vector of state errors will be defined as

$\mathbf{e} = \{e, \dot{e}, \ldots, e^{(n-1)}\}^T$. Vector $\mathbf{k} = \{k_n, \ldots, k_1\}^T$ defines the dynamics of the error. Introducing the control law to the plant will generate a closed-loop dynamic system governed by

$$\begin{aligned} x^{(n)} &= y_{ref}^{(n)} + \mathbf{k}^T \mathbf{e} \\ y &= x \end{aligned} \tag{5.23}$$

and the error dynamics will be

$$e^{(n)} + k_1 e^{(n-1)} + \ldots + k_n e = 0 \tag{5.24}$$

From this equation it is clear that the closed-loop dynamics will be governed by the dynamics determined by the components of the vector $\mathbf{k}$ since the entries of the vector are the coefficients of the characteristic polynomial of the closed-loop system. A proper selection of the elements of $\mathbf{k}$ guarantees the stability and the convergence of y toward y_{ref}.

In practice, fuzzy models can be used to represent the functions f and g, but there will be some modeling error. The function f will be approximated by the fuzzy inference system described by the function

$$\hat{f}(\mathbf{x}|\theta_f) = \theta_f^T \sigma(\mathbf{x}) \tag{5.25}$$

where θ_f are the consequences of the rules and $\sigma(\mathbf{x})$ represents the inference process. The function g will be approximated by

$$\hat{g}(\mathbf{x}|\theta_g) = \theta_g^T \eta(\mathbf{x}) \tag{5.26}$$

where θ_g are the consequences of the rules and $\gamma(\mathbf{x})$ represents the inference process. If the mismatch between the function and its approximator is taken into account, the error dynamics will be given by

$$e^{(n)} = -\mathbf{k}^T \mathbf{e} + [\hat{f}(\mathbf{x}|\theta_f) - f(\mathbf{x})] + [\hat{g}(\mathbf{x}|\theta_g) - g(\mathbf{x})]u^* \tag{5.27}$$

where

$$u^* = \frac{1}{\hat{g}(\mathbf{x}|\theta_g)}(-\hat{f}(\mathbf{x}|\theta_f) + y_{ref}^{(n)} + \mathbf{k}^T \mathbf{e}) \tag{5.28}$$

The error dynamics can be written in matrix form using

$$\Lambda = \begin{bmatrix} 0 & 1 & 0 & 0 & \ldots & 0 & 0 \\ 0 & 0 & 1 & 0 & \ldots & 0 & 0 \\ \ldots & \ldots & \ldots & \ldots & \ldots & \ldots & \ldots \\ 0 & 0 & 0 & 0 & \ldots & 0 & 1 \\ -k_n & -k_{n-1} & \ldots & \ldots & \ldots & \ldots & -k_1 \end{bmatrix}, \quad \mathbf{b} = \begin{bmatrix} 0 \\ \vdots \\ 0 \\ 1 \end{bmatrix} \tag{5.29}$$

generating the following differential equation to describe the error dynamics:

$$\dot{\mathbf{e}} = \Lambda \mathbf{e} + \mathbf{b}[\hat{f}(\mathbf{x}|\theta_f) - f(\mathbf{x})] + \mathbf{b}[\hat{g}(\mathbf{x}|\theta_g) - g(\mathbf{x})]u^* \tag{5.30}$$

The minimum approximation error is defined as

$$w = [\hat{f}(\mathbf{x}|\theta_f^*) - f(\mathbf{x})] + [\hat{g}(\mathbf{x}|\theta_g^*) - g(\mathbf{x})]u^* \tag{5.31}$$

where θ_f^* and θ_g^* represent the optimal approximation values of θ_f and θ_g.

Using this representation, the error dynamics can be written as

$$\dot{\mathbf{e}} = \Lambda\mathbf{e} + \mathbf{b}\{[\hat{f}(\mathbf{x}|\theta_f) - \hat{f}(\mathbf{x}|\theta_f^*)] + [\hat{g}(\mathbf{x}|\theta_g) - \hat{g}(\mathbf{x}|\theta_g^*)]u^* + w\} \tag{5.32}$$

Replacing the expressions for $\hat{f}(\mathbf{x}|\theta_f)$ and $\hat{g}(\mathbf{x}|\theta_g)$ given in Equations (5.25) and (5.26), the error dynamics become

$$\dot{\mathbf{e}} = \Lambda\mathbf{e} + \mathbf{b}[(\theta_f - \theta_f^*)^T\sigma(\mathbf{x}) + (\theta_g - \theta_g^*)^T\eta(\mathbf{x})u^* + w] \tag{5.33}$$

The task is to define a stable adaptation law for θ_f and θ_g so that the tracking and the parameter errors are minimized as time evolves. Wang [17] proposed the following procedure based on the following Liapunov function:

$$V = \frac{1}{2}\mathbf{e}^T P\mathbf{e} + \frac{1}{2\gamma_1}(\theta_f - \theta_f^*)^T(\theta_f - \theta_f^*) + \frac{1}{2\gamma_2}(\theta_g - \theta_g^*)^T(\theta_g - \theta_g^*) \tag{5.34}$$

where γ_1 and γ_2 are positive constants and P is a positive definite matrix satisfying the Lyapunov equation:

$$\Lambda^T P + P\Lambda = -Q \tag{5.35}$$

where Q is an arbitrary positive definite matrix. The derivative of the Lyapunov function is given by

$$\begin{aligned}\dot{V} = &-\frac{1}{2}\mathbf{e}^T P\mathbf{e} + \mathbf{e}^T P\mathbf{b}w + \frac{1}{\gamma_1}(\theta_f - \theta_f^*)^T[\dot{\theta}_f + \gamma_1\mathbf{e}^T P\mathbf{b}\sigma(\mathbf{x})]\\ &+ \frac{1}{\gamma_2}(\theta_g - \theta_g^*)^T[\dot{\theta}_g + \gamma_g\mathbf{e}^T P\mathbf{b}\eta(\mathbf{x})u^*]\end{aligned} \tag{5.36}$$

The minimization of the tracking error $\mathbf{e}$ and the parameter error is equivalent to the minimization of the Lyapunov function V. The adaptation law should guarantee that $\dot{V}$ is negative. To guarantee the negativeness of $\dot{V}$ let us take a look at the terms of the expression: the term $-\frac{1}{2}\mathbf{e}^T P\mathbf{e}$ is always negative, since P is positive definite, the term $\mathbf{e}^T P\mathbf{b}w$ can be positive or negative, but if the initial modeling task is well done w must be small and therefore $\mathbf{e}^T P\mathbf{b}w << \frac{1}{2}\mathbf{e}^T P\mathbf{e}$, a condition that still will guarantee a negative value for $\dot{V}$. Finally, by forcing the last two terms to be equal to zero, we obtain the following adaptation law:

$$\dot{\theta}_f = -\gamma_1\mathbf{e}^T P\mathbf{b}\sigma(\mathbf{x}) \tag{5.37}$$

$$\dot{\theta}_g = -\gamma_2\mathbf{e}^T P\mathbf{b}\eta(\mathbf{x})u^* \tag{5.38}$$

Figure- 5.11 shows a scheme of the proposed control system.

In summary, the sequence of design is defined in the following steps:

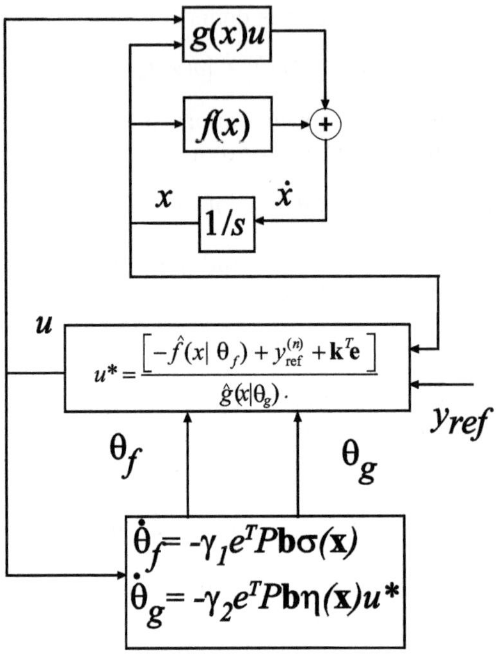

Figure 5.11. Scheme for adaptive feedback linearization

1. Design an initial model of the plant $\hat{f}(\mathbf{x}|\theta_f), \hat{g}(\mathbf{x}|\theta_g)$ using the identification techniques depicted in previous section.
2. Design the vector $\mathbf{k}$ according to the desired behavior.
3. Connect the system in closed loop and use the updating rule to update online the consequence values θ_f, θ_g of the rules.

The main advantage of this control strategy is that it makes possible to construct a controller directly using the model and the desired linear behavior. The main disadvantage is that it is only applicable to a limited set of continuous-time nonlinear systems and since the controller is implemented in discrete time, care must be taken during the implementation phase. Additionally, the scheme does not assume the presence of disturbances and relies on the adaptation of the parameters θ_f and θ_g to compensate the effect of the disturbances. The stability of the system is guaranteed, but the tracking error will be proportional to the mismatch between the plant and the model.

Summary:
Fuzzy models can be used for the synthesis of feedback linearizing controllers. The technique offers the possibility to construct nonlinear controllers using the model of the system and the desired closed-loop dynamics.

Sliding Mode Fuzzy Control

This control strategy keeps some similarities with the feedback linearization technique; but its analysis include disturbances, making the formulation more realistic. The objective is to "force" the system to behave like a linear system with dynamics described by the so-called sliding surface [51].

The control law is designed to steer the plant toward the "sliding surface." Once the system reaches the sliding surface, the controller guarantees that the closed-loop system behaves according to the dynamics of the sliding surface.

This control design methodology can be applied to plants of the form

$$x^{(n)} = f(\mathbf{x}) + g(\mathbf{x})u + \tilde{d} \tag{5.39}$$

where $\mathbf{x}(t) = \{x, \dot{x}, \ldots, x^{(n-1)}\}^T$ is the state vector and $\tilde{d}(t)$ are disturbances with known upper bounds, and $f(\mathbf{x})$ and $g(\mathbf{x})$ are nonlinear functions describing the dynamics of the plant. Additionally the set ν_u is defined to represent the spectrum of unmodeled frequencies. The desired trajectory is given by $\mathbf{x}_{ref}(t)$ and the tracking error by

$$\mathbf{e}(t) = \mathbf{x}(t) - \mathbf{x}_{ref}(t) = \{e, \dot{e}, \ldots, e^{(n-1)}\}^T \tag{5.40}$$

The "sliding surface" is determined as

$$s(\mathbf{x}, t) = (\frac{d}{dt} + \lambda)^{n-1} e = \sum_{k=0}^{n-1} \binom{n-1}{k} \lambda^k e^{(n-1-k)} = 0 \tag{5.41}$$

The stability of the "sliding surface" is guaranteed if $\lambda > 0$, which guarantees that the linear dynamics of the "sliding surface" are governed by poles placed in the negative real axis. The control problem requires the design of a control law to guarantee that the vector $\mathbf{e}(t)$ remains in the sliding surface $s(\mathbf{x}, t) = 0$ for $t \geq 0$. To construct this control law a Lyapunov function is proposed:

$$V = \frac{1}{2} s^2 \tag{5.42}$$

Observe that $V(0) = 0$ and $V > 0$ for $s > 0$. A sufficient condition for the stability of (5.39) is

$$\dot{V} = \frac{1}{2} \frac{d}{dt}(s^2) = s\dot{s} \leq -\eta |s| \tag{5.43}$$

with $\eta > 0$.

From (5.43) the *reaching condition* can be derived:

$$\dot{s}\, \mathrm{sgn}(s) \leq -\eta \tag{5.44}$$

The first parameter that must be selected during the design phase is the parameter λ. Assuming $\nu_{u\min}$ as the lower bound of the unmodeled frequencies, then λ must be selected as

$$\lambda << \nu_{u\min} \tag{5.45}$$

Calculating the time derivative of s, we obtain

$$\dot{s} = x^{(n)} - x_{ref}^{(n)} + \sum_{k=1}^{n-1} \binom{n-1}{k} \lambda^k e^{(n-k)} \tag{5.46}$$

Replacing this expression and the plant model in the *reaching condition* (5.44), we obtain

$$\mathrm{sgn}(s) \left(f(\mathbf{x}) + g(\mathbf{x})u + \tilde{d} - x_{ref}^{(n)} + \sum_{k=1}^{n-1} \binom{n-1}{k} \lambda^k e^{(n-k)} \right) \leq -\eta \tag{5.47}$$

The sliding mode control law is defined as follows:

$$u = \hat{g}(\mathbf{x})^{-1}(\tilde{u} - \hat{f(\mathbf{x})}) \tag{5.48}$$

$$\tilde{u} = G(\hat{u} - K(\mathbf{x}, t)\mathrm{sgn}(s)) \tag{5.49}$$

$$\hat{u} = x_{ref}^{(n)} - \sum_{k=1}^{n-1} \binom{n-1}{k} \lambda^k e^{(n-k)} \tag{5.50}$$

where $K(\mathbf{x}, t) > 0$ and $\hat{f}(\mathbf{x})$ and $\hat{g}(\mathbf{x})$ are estimates of the functions $f(\mathbf{x})$ and $g(\mathbf{x})$. Observe that, so far, with the exceptions of the terms G and $K(\mathbf{x}, t)\mathrm{sgn}(s)$, the control strategy resembles the feedback linearization control strategy.

The term G is determined using the following bounds:

$$0 \leq \beta^{\min} \leq g(\mathbf{x})\hat{g}(\mathbf{x})^{-1} \leq \beta^{\max} \tag{5.51}$$

as

$$G = (\beta^{\min}\beta^{\max})^{-1/2} \tag{5.52}$$

Also a parameter β will be defined as

$$\beta = \left(\frac{\beta^{\max}}{\beta^{\min}} \right)^{1/2} \tag{5.53}$$

The goal of the design now is to find a $K(\mathbf{x}, t)$ satisfying the reaching condition. Replacing the control law in the reaching condition,

$$\mathrm{sgn}(s)(\Delta f(\mathbf{x}) + (g(\mathbf{x})\hat{g}(\mathbf{x})^{-1}G - 1)\hat{u} + \tilde{d} - g(\mathbf{x})\hat{g}(\mathbf{x})^{-1}GK(\mathbf{x}, t)\mathrm{sgn}(s)) \leq -\eta \tag{5.54}$$

where $\Delta f(\mathbf{x}) = f(\mathbf{x}) - g(\mathbf{x})\hat{g}(\mathbf{x})^{-1}\hat{f}(\mathbf{x})$. It is equivalent to

$$(\Delta f(\mathbf{x}) + (g(\mathbf{x})\hat{g}(\mathbf{x})^{-1}G - 1)\hat{u} + \tilde{d})\mathrm{sgn}(s) - g(\mathbf{x})\hat{g}(\mathbf{x})^{-1}GK(\mathbf{x}, t) \leq -\eta \tag{5.55}$$

This inequality is always true if

$$g(\mathbf{x})\hat{g}(\mathbf{x})^{-1}GK(\mathbf{x},t) \geq |\Delta f(\mathbf{x}) + (g(\mathbf{x})\hat{g}(\mathbf{x})^{-1}G - 1)\hat{u} + \tilde{d}| + \eta \tag{5.56}$$

A stronger condition is given by

$$g(\mathbf{x})\hat{g}(\mathbf{x})^{-1}GK(\mathbf{x},t) \geq |\Delta f(\mathbf{x})| + |(g(\mathbf{x})\hat{g}(\mathbf{x})^{-1}G - 1)||\hat{u}| + |\tilde{d}| + \eta \tag{5.57}$$

$g(\mathbf{x})\hat{g}(\mathbf{x})^{-1}$ can be substituted by its lower bound $\beta^{\min}$ and because $\beta^{-1} = \beta^{\min}G$ the condition for $K(\mathbf{x},t)$ is given by

$$K(\mathbf{x},t) \geq \beta(|\Delta f| + (1 - \beta^{-1})|\hat{u}| + |\tilde{d}| + \eta) \tag{5.58}$$

Taking the upper bounds for

$$|\Delta f(\mathbf{x})| < \tilde{F}$$
$$|\hat{u}| < \tilde{U}$$
$$|\tilde{d}| < D$$

then the condition for the sliding surface $s = 0$ to be a global domain of attraction is satisfied if

$$K(\mathbf{x},t) \geq \beta(\tilde{F} + (1 - \beta^{-1})\tilde{U} + D + \eta). \tag{5.59}$$

The steps to design the controller are as follows:

1. Select λ from the lower bound of the unmodeled frequencies $\nu_{\min}$.
2. Derive $\hat{u}$ and its upper bound $\tilde{U}$.
3. Find estimates $\hat{f}(\mathbf{x})$ and $\hat{g}(\mathbf{x})$ for $f(\mathbf{x})$ and $g(\mathbf{x})$ using a fuzzy inference system.
4. Derive G from (5.52).
5. Determine η and the upper bounds $\tilde{F}$ for $|\Delta f(\mathbf{x})|$ and D for $|\tilde{d}|$.
6. Compute $K(\mathbf{x},t)$

An important advantage of this technique compared with the feedback linearization is that this technique takes into account disturbances and modeling mismatches. However, a significant disadvantage is that the control actions are very strong (with a large amplitude and abrupt changes). Nevertheless the impact of this drawback can be reduced by the introduction of the so-called boundary layer (BL) near the sliding surface $s = 0$. The introduction of this boundary layer will affect the tracking accuracy. The width of the BL is defined as 2Φ, $|s|$ is the distance of the state $\mathbf{e}$ to the sliding surface. A state $\mathbf{e}$ is located inside the BL if $|s| \leq \Phi$ and is outside if $|s| > \Phi$. The BL is introduced in the control law by replacing $\text{sgn}(s)$ by $\text{sat}(s/\Phi)$,

$$u = \hat{g}(\mathbf{x})^{-1}(\tilde{u} - \hat{f}(\mathbf{x}))$$
$$\tilde{u} = G[(\hat{u} - K(\mathbf{x},t).\text{sat}(s/\Phi)] \tag{5.60}$$

where the function $\text{sat}(z)$ is defined as

$$\text{sat}(z) = \begin{cases} z & \text{if } |z| < 1 \\ \text{sgn}(z) & \text{if } |z| \geq 1 \end{cases} \tag{5.61}$$

For practical purposes $K(\mathbf{x}, t)$ is reduced to a constant value.

Palm *et al.* [52] propose replacing the term $K(\mathbf{x}, t)\text{sat}(s/\Phi)$ with a fuzzy nonlinearity $K_{fuzz}(\mathbf{x}, t, s, \Phi)$ showing some improvements in the performance and reducing the chattering effect. In the same document some examples show the introduction of an integral action in this type of structure.

Figure 5.12(a) shows a fuzzy system with one input s. The fuzzy system represents the term $K(\mathbf{x}, t).\text{sat}(s/\Phi)$ when $K(\mathbf{x}, t)$ is a constant. Observe that the location of the membership functions will affect the shape of the function. In Figure 5.12(b), it is clear that the separation of the membership functions from the center will cause the curve to flatten in the middle. This will make the control action react slowly for small errors or disturbances. The third plot shows a system where the slope is steep in the middle, generating strong control actions even for small errors.

In summary, fuzzy systems can be used in sliding mode control to generate the estimates of the model ($\hat{f}$ and $\hat{g}$) and the function $K_{fuzz}(\mathbf{x}, t, s, \Phi)$, which forces the error vector $\mathbf{e}$ to remain in the sliding surface.

Summary:
The design of sliding mode controllers can be achieved by means of the combination of a fuzzy model of the plant and a nonlinear monotonic function used to fine-tune the controller and reduce the aggressiveness of "classical sliding mode controllers."

Fuzzy Gain Scheduling

Gain scheduling has been used extensively on industrial applications including aircraft control. The advantage of this technique is the use of linear techniques to design controllers for nonlinear systems by applying linearization around the different operating points. In this way several linear controllers are designed and "connected" to the plant according to the current operating point. This design method is also known as the *the paradigm of parallel distributed compensation.* Fuzzy systems offer to this techniques the advantage of integrating in one mechanism the detection of the operating point and the interpolation among the different operating points, providing a smooth nonlinear control law. It has been proved that the fuzzy controllers can be seen as smooth gain scheduling controllers [53].

To apply this compensation technique, the model of the plant is assumed to be given by a Takagi–Sugeno fuzzy system with rules of the form

Rule i: IF $x_1(k)$ is μ_{i1} AND … AND $x_n(k)$ is μ_{in}
THEN $x(k+1) = A_i x(k) + B_i u(k)$

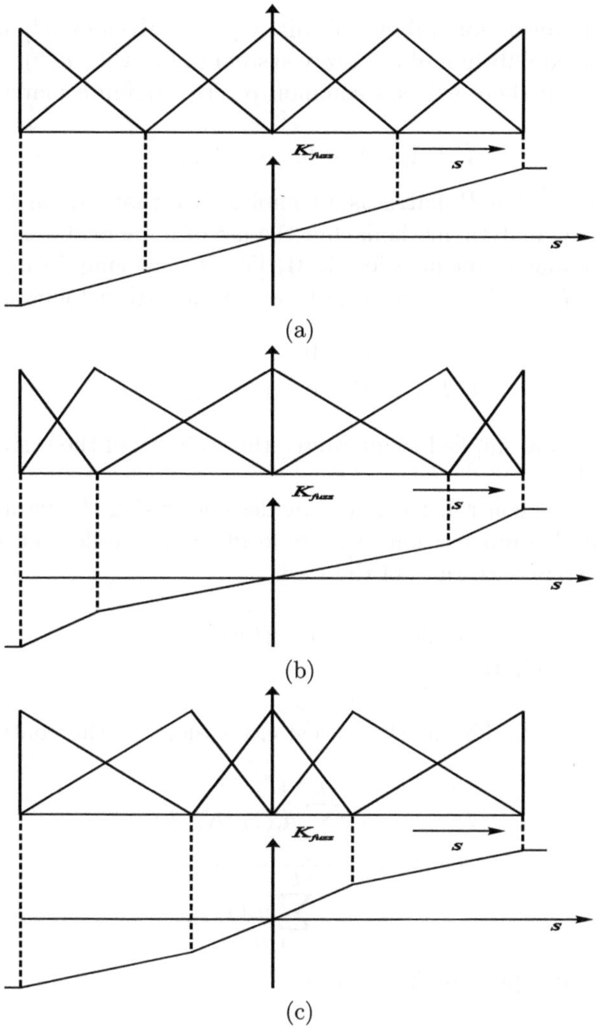

Figure 5.12. Relation between the position of the membership functions and the shape of $K_{fuzz}(\mathbf{x}, t, s, \Phi)$

where $x(k) = \{x_1(k), \ldots, x_n(k)\}^T$, $u(k) = \{u_1(k), \ldots, u_{Ni}(k)\}^T$ and $i = 1, \ldots, L$ with L is the number of rules. The dynamic system defined by the fuzzy system is given by

$$x(k+1) = \frac{\sum_{i=1}^{L} \mu_i(x_k)[A_i x(k) + B_i u(k)]}{\sum_{i=1}^{L} \mu_i(x_k)} \tag{5.62}$$

A sufficient condition for stability of this system is given in [54] and is defined as follows: the equilibrium of the fuzzy system (5.62) with $u(k) = 0$ is asymptotically stable if there exists a common positive definite matrix P, so that

$$A_i^T P A_i - P < 0, \quad i = 1, 2, \ldots, L \tag{5.63}$$

The search for such a P matrix is a complex task that cannot be solved efficiently by using analytic methods, but the use of numerical tools to solve this type of linear matrix inequalities (LMI) [55][56] has simplified the problem. The search of P can be described in terms of an LMI as follows:

$$\begin{aligned} P &> 0 \\ A_i^T P A_i - P &< 0, \quad i = 1, 2, \ldots, L. \end{aligned}$$

This technique was applied to guarantee the stability of the system described in Appendix E.

A state feedback controller can also be designed using the paradigm of parallel distributed compensation. A state feedback controller can be described by a fuzzy system with rules of the form

Rule i: IF $x_1(k)$ is μ_{i1} AND ... AND $x_n(k)$ is μ_{in}
THEN $u(k) = -K_i x(t)$

where $i = 1, \ldots, L$. The mathematical expression for the controller will be given by

$$u(k) = \frac{-\sum_{i=1}^{L} \mu_i(x_k) K x(k)}{\sum_{i=1}^{L} \mu_i(x_k)} \tag{5.64}$$

The closed-loop expression is given by

$$x(k+1) = \frac{\sum_{i=1}^{L}\sum_{j=1}^{L} \mu_i(x_k)\mu_j(x_k)[A_i - B_i K_j]x(k)}{\sum_{i=1}^{L}\sum_{j=1}^{L} \mu_i(x_k)\mu_j(x_k)} \tag{5.65}$$

So the equilibrium of the closed-loop system (5.65) is asymptotically stable if there exists a common positive definite P matrix such that

$$[A_i - B_i K_j]^T P [A_i - B_i K_j] - P < 0 \tag{5.66}$$

for $\mu_i . \mu_j \neq 0$.

The state feedback matrices K_j can be calculated by solving the following LMIs (see proof, [57]):

$$\begin{pmatrix} Q & A_iQ + B_iY_i \\ QA_i^T + Y_i^TB_i^T & Q \end{pmatrix} > 0 \qquad (5.67)$$

$$\begin{pmatrix} Q & \frac{A_iQ+B_iY_j+A_jQ+B_jY_i}{2} \\ \frac{QA_i^T+Y_j^TB_i^T+QA_j^T+Y_i^TB_j^T}{2} & Q \end{pmatrix} > 0, \quad i < j \quad (5.68)$$

and the fuzzy state feedback matrices are constructed as $K_i = Y_iQ^{-1}$, $i = 1, 2, \ldots, L$. The design methods using LMIs can be extended also to guarantee some decaying rate and even to account for uncertainties (for further details see [57]).

Summary:
Takagi–Sugeno fuzzy controllers can be constructed for systems described by local linear models. Such controllers can have guaranteed stability and performance. The design methodology demands the solution of linear matrix inequalities using efficient numerical methods.

Example Control of a Helicopter Laboratory Process

This example[1] presents the comparative results of a fuzzy controller against some classical controllers (an LQG and a PID) tested on a didactic system that imitates some of the dynamics of a helicopter.

The system is part of a laboratory of the CUAO University in Cali, Colombia. As mentioned before the system with two degrees of freedom imitates some of the dynamics present in a helicopter (see Figure 5.13). The objective of the control system is to position the arm at a certain elevation angle and certain azimuth angle by using the main rotor and the side rotor.

The main nonlinearities in this system are in the elevation angle; the azimuth angle behaves almost linearly and in fact it has integral dynamics. The model used to build the controllers was obtained using subspace identification. The linear controllers were tuned with the model obtained around zero angle of elevation. For the fuzzy controller the variable used for the inference system was the elevation angle. The model constructed was a Takagi–Sugeno fuzzy model with three rules: one for elevation angle around -30°, one for elevation angle around 0° and one for elevation angle around 30°.

The three controllers were constructed: the LQR, the PIDs and the fuzzy controller. The fuzzy controller was constructed using three linear LQG controllers, one for each rule. The global stability of the system was tested by solving the feasibility stability problem shown in previous section. The three controllers deliver a similar performance, as shown in Figures 5.14, 5.15 and 5.16. The only remarkable difference is the "nervous" behavior of the control action of the PID controller.

However, when the system is perturbed by putting a counterweight into

[1] The results presented in this example were obtained by one of the master students under the supervision of one of the authors [58].

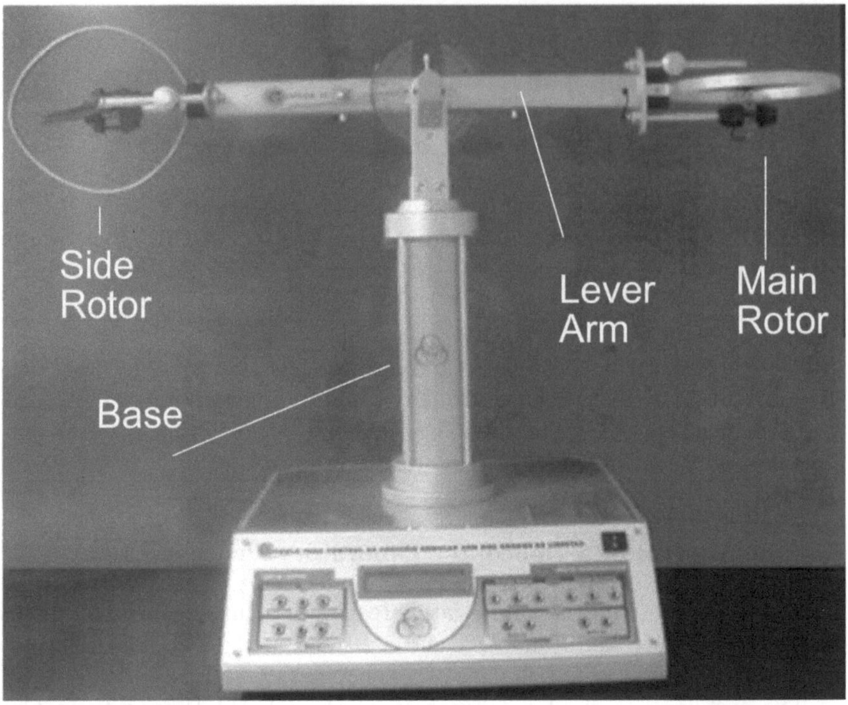

Figure 5.13. Helicopter laboratory process. COPTER II. Photo: Courtesy Oscar Mauricio Agudelo

the lever arm, the performance of the PID and the LQR degraded significantly. Observe the coupling of the PID and the oscillations of the LQR (see Figures 5.17 and 5.18). The fuzzy controller preserves its good behavior despite the disturbance (see Figure 5.19). One can conclude from this experiment that the additional information included in the fuzzy model and in the fuzzy controller not only generates a good performance but also improves the robustness of the controllers.

5.3 Conclusions and Future Perspectives

This chapter has presented "classical" ways to design fuzzy controllers. The description has covered the widely used model-free strategies, characterized by their simplicity. Heuristic methods are popular for first-order systems and supervisory systems because they can provide an acceptable performance using the expert knowledge of the operators.

PID-like fuzzy controllers represent an interesting alternative to improve the performance of existing PID controllers with a small additional effort of tuning.

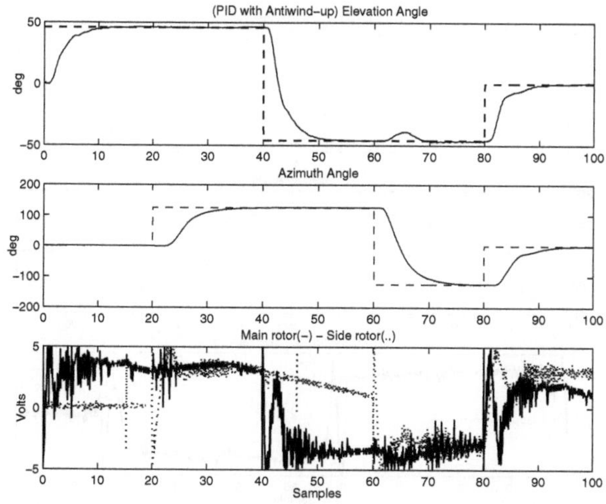

Figure 5.14. PID controllers applied to the helicopter laboratory process in nominal conditions. In the first two plots: (- -) reference trajectory (–) helicopter's trajectory. In the third plot the control voltages: main rotor (–) and side rotor (..)

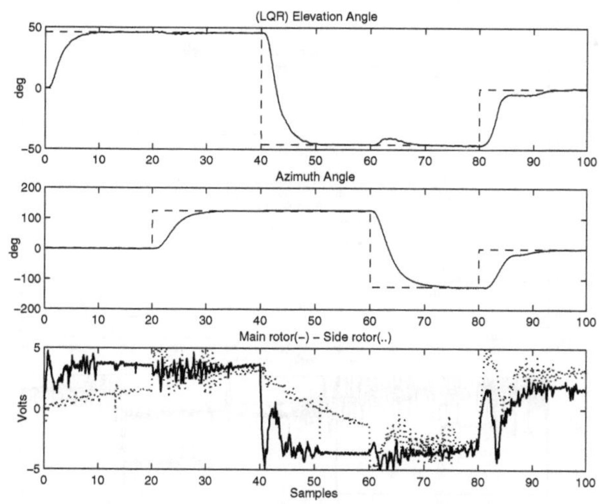

Figure 5.15. LQR controller applied to the helicopter laboratory process in nominal conditions. In the first two plots: (- -) reference trajectory (–) helicopter's trajectory. In the third plot the control voltages: main rotor (–) and side rotor (..)

Model based control techniques using adaptive methods have been described. These techniques require the use of a "good" plant model, and the robustness of the obtained controller is not guaranteed. Also, model-based control techniques using direct synthesis have been presented. The feedback

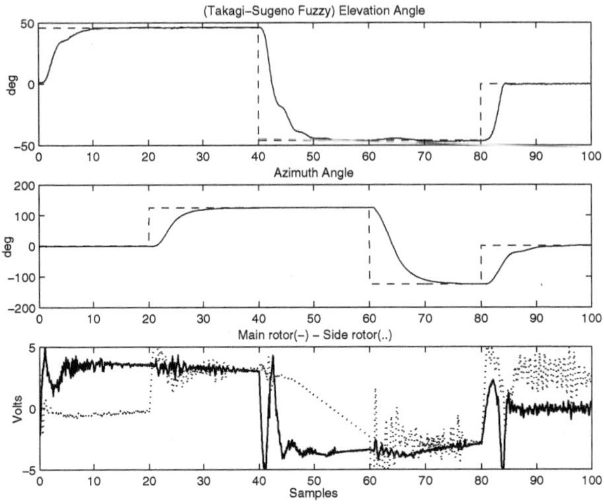

Figure 5.16. Takagi–Sugeno fuzzy controller applied to the helicopter laboratory process in nominal conditions. In the first two plots: (- -) reference trajectory (–) helicopter's trajectory. In the third plot the control voltages: main rotor (–) and side rotor (..)

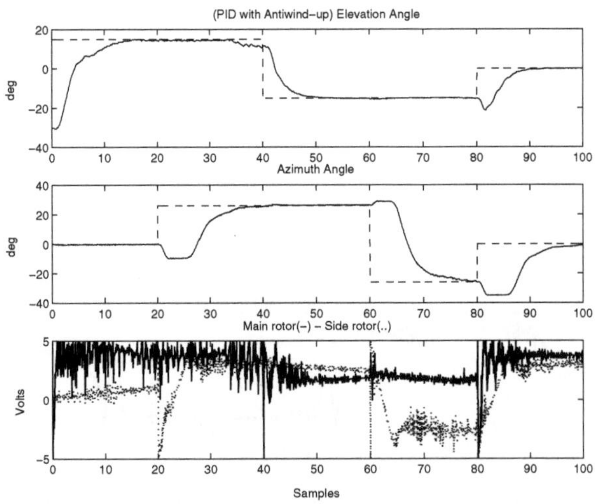

Figure 5.17. PID controllers applied to the helicopter laboratory process in disturbed conditions. In the first two plots: (- -) reference trajectory (–) helicopter's trajectory. In the third plot the control voltages: main rotor (–) and side rotor (..)

linearization technique is a technique limited to a certain class of nonlinear systems (affine nonlinear systems). This technique has been criticized for its lack of robustness. To overcome this drawback a stable adaptive control strategy has been presented: the sliding mode fuzzy control strategy. This strategy

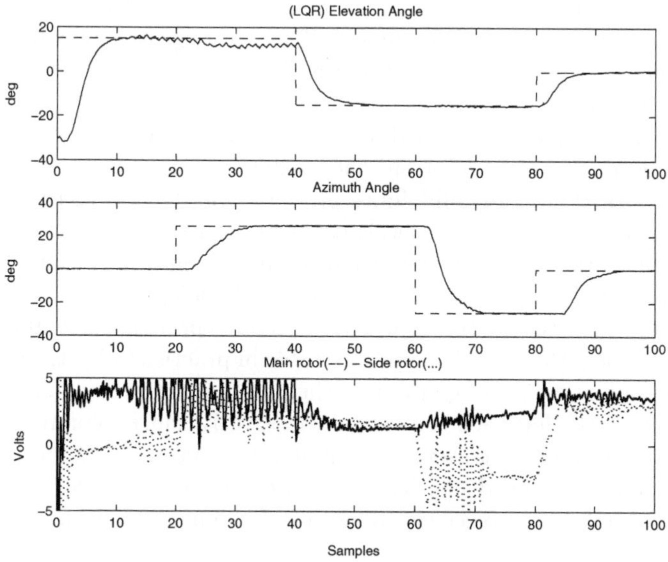

Figure 5.18. LQR controller applied to the helicopter laboratory process in disturbed conditions. In the first two plots: (- -) reference trajectory (–) helicopter's trajectory. In the third plot the control voltages: main rotor (–) and side rotor (..)

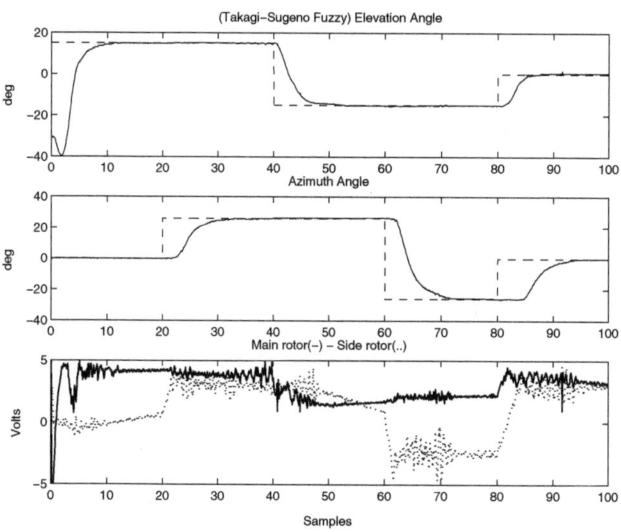

Figure 5.19. Takagi–Sugeno fuzzy controller applied to the helicopter laboratory process in disturbed conditions. In the first two plots: (- -) reference trajectory (–) helicopter's trajectory. In the third plot the control voltages: main rotor (–) and side rotor (..)

also demands a special structure of the nonlinear model. The strategy requires only a rough description of the model. This strategy is very robust and the performance can be improved by the use of fuzzy systems in the design of the switching function.

Finally, the fuzzy gain scheduling technique was presented. This design technique is quite "elegant" from the analytical point of view. The use of advanced algorithms for semidefinite programming facilitates the solution of the LMIs generated by the synthesis problem. More advanced synthesis techniques include the H^∞ criterion for disturbance rejection, robust synthesis and observer design by applying the same methodologies.

Stability analysis is still an open problem, especially for empirically designed controllers that are shown to be stable in practice (after trial and error) but no mathematical proof guarantees their stability. The design of fuzzy controllers using LMIs has been considered conservative since a common P matrix has to be found for a large set of dynamics. New alternatives have been proposed by Johansson *et al.*[59]. The relaxation of the problem is obtained by means of formulating a piecewise quadratic Lyapunov function.

6

Predictive Control Based on Fuzzy Models

This chapter and the next one are devoted to the presentation of a control technique that will exploit the capacity of fuzzy systems to represent nonlinear plants. The author considers the use of fuzzy models for predictive control a step further in the implementation of modern control techniques in the process industry. Predictive control (MPC, model-based predictive control) based on linear models is a mature control technique with multiple applications in the process industry. The next natural step in this area is the development of predictive control based on nonlinear models. The use of controllers that take into account the nonlinearities of the plant implies an improvement in the performance of the plant by reducing the impact of the disturbances and by improving the tracking capabilities of the control system.

The inclusion of nonlinear information demands the use of a parametric representation for the model of the plant. These chapters assume that the representation of the plant is given in terms of a fuzzy model. The use of fuzzy models together with the concept of predictive control is a promising technique because both techniques can be explained in simple terms to operators and commissioning engineers. Simplicity is paramount for the success of a control technique in the industrial environment. In fact, some of these techniques are already embedded in leading-edge products for process control.

The use of nonlinear models increases the complexity of the problem and demands more information from the plant. Gains, raise times and in general data at different operating points are part of the information needed to construct a nonlinear model of the plant. In other cases rigorous physical models demand the knowledge of the physical properties of the materials and equipment involved in the process. One advantage of the use of fuzzy models is the fact that their complexity can be gradually increased as more information is gathered. This increase in complexity can be done automatically or manually by a careful commission of the new operating point.

From the computational point of view, nonlinear predictive control can be quite demanding. The quest for shortcuts and simplification is considered one of the main research topics in this subject. This chapter contains some ideas about ways to implement the simplification of the calculation by exploiting some features of the fuzzy models. The chapter starts by presenting a solution for the case of single-input–single-output (SISO) systems without constraints. The control technique demands the estimation of a step response. In this chapter we present a novel method to estimate such a step response to maximize the resemblance of the model and the plant. The method improves the estimation of the nonlinear information extracted from the model, improving the performance of the controller. In the second part of the chapter, the multivariable problem with constraints is studied. The method studied for SISO systems is extended for multivariable constrained systems. Thereafter two new methods are included where the structure of the Takagi–Sugeno fuzzy models is exploited to construct two types of predictive controllers. The chapter illustrates these control techniques with examples based on realistic simulations of three industrial systems: a chemical reactor, a steam generator and a polymerization reactor. The examples include analysis about the performance of the controllers.

6.1 The Predictive Control Strategy

The predictive control strategy is based on a receding horizon optimization, calculated online at each sampling time (see Figure 6.2). The algorithm can be described as follows:

1. Sample the output of the plant.
2. Use the model of the plant to predict its future behavior over a prediction horizon during N_p samples when a control action is applied along a control horizon during N_c. samples.
3. Calculate the optimal control sequence $\{u(k), \ldots, u(k+Nc)\}$ that minimizes

$$\min_{u(k),\ldots,u(k+Nc)} J(u(k), y(k), w(k)) \tag{6.1}$$

subject to

$$\begin{aligned} x(k+1) &= f(x(k), u(k)) \\ y(k) &= g(x(k), u(k)) \\ y_{min} &\leq y(k) \leq y_{max}, \ \forall k = 1, \ldots, Np \\ u_{min} &\leq u(k) \leq u_{max}, \ \forall k = 1, \ldots, Nu \\ \Delta y_{min} &\leq \Delta y(k) \leq \Delta y_{max}, \ \forall k = 1, \ldots, Np \\ &|\Delta u(k)| \leq \Delta u_{max}, \ \forall k = 1, \ldots, Nu \end{aligned}$$

where $J(.)$ is the "the cost function" and if it is quadratic it will be of the form

$$J(u(k), y(k), r(k)) = \sum_{t=k}^{k+Np} (w(t) - y(t))^T Q (w(t) - y(t))$$
$$+ \sum_{t=k}^{k+Nc} (u(t)^T R u(t) + \delta u(t)^t S \delta u(t)) \quad (6.2)$$

where $x(k)$ represents the states of the system, $u(k)$ the inputs, $y(k)$ the outputs, the functions $f(.,.)$ and $g(.,.)$ represent the dynamic model of the plant, $w(k)$ is the reference signal, $\delta u(t) = u(t) - u(t-1)$, Q is a positive definite matrix and R and S are positive semidefinite matrices.

4. Apply the input $u(k)$ and repeat the procedure at the next sampling time.

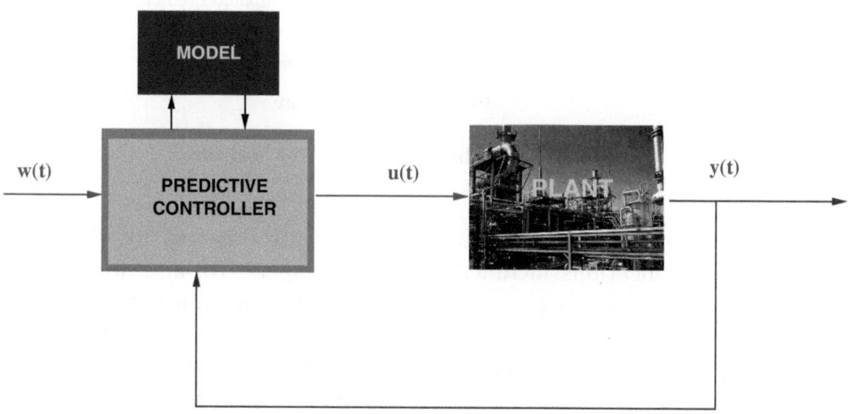

Figure 6.1. Block diagram of a predictive controller

The problem posed in Step 3 of the algorithm is a very complex optimization problem. This problem can be relaxed by assuming no constraints on the input nor on the outputs and a linear plant model. This problem has been extensively studied [60][61][62][63]. The most important characteristic of this relaxation (unconstrained linear predictive control) is the generation of a controller in a closed form, so that no optimization is solved online. A complete analysis of this strategy including stability and robustness issues has been presented in books such as [60][64].

The complete problem without relaxation has been analyzed in previous studies. [65] and [66] include some interesting studies. Probably the best overview published up today is [67]. Also interesting is [68]. The use of fuzzy techniques in predictive control was proposed for the first time by

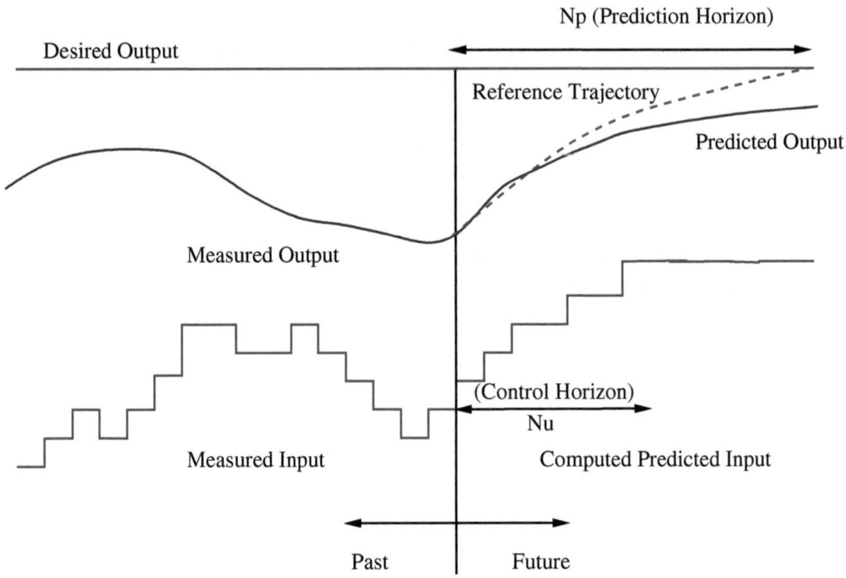

Figure 6.2. Prediction and control horizons

Yasunobu [69] on his implementation of the control system for the Sendai Train. Other publications includes conference papers [70] and doctoral theses [71] [72]. The analysis presented in this chapter assumes that the plant model included in the optimization (6.1) given by the nonlinear functions $f(.,.)$ and $g(.,.)$ will be represented by a fuzzy model. The control strategies presented in Sections 6.2 and 6.3.2 can be applied to any type of fuzzy model. However, the control strategies presented in Sections 6.3.3 and 6.3.4 demand the model to be a Takagi–Sugeno fuzzy model [73][74][75] [76].

Summary:
Predictive control is a model-based control strategy that calculates at each sampling time via optimization the optimal control action to maintain the output of the plant close to the desired reference.

6.2 Unconstrained Nonlinear Predictive Control

In this section, the study is focused on the optimization problem of the unconstrained nonlinear predictive control with quadratic cost. The section presents an approximate solution to the problem where the information given by a fuzzy model is used to solve the problem.

The problem can be written as

$$\min_{u(k),\ldots,u(k+Nc)} J(u(k),y(k),w(k)) = \sum_{t=k}^{k+Np} (w(t)-y(t))^T Q(w(t)-y(t)) \\ + \delta u(t)^t R \delta u(t)) \tag{6.3}$$

subject to

$$x(k+1) = f(x(k),u(k))$$
$$y(k) = g(x(k),u(k))$$

For the present case the plant model described by the functions $f(.,.)$ and $g(.,.)$ are parameterized by a fuzzy model. An alternative representation for the plant model is a NOE model represented as

$$y(t) = \hat{y}(t)+n(t) = f(\hat{y}(t-1),\ldots,\hat{y}(t-m),u(t-1),\ldots,u(t-n))+n(t) \tag{6.4}$$

where the function $f(.)$ corresponds to the fuzzy model and the noise model is given by

$$n(t) = \frac{C(z^{-1})}{D(z^{-1})}e(t) \tag{6.5}$$

where $e(t)$ is a *white noise* sequence. A typical choice of the noise model is

$$n(t) = \frac{C(z^{-1})}{D(z^{-1})}e(t) = \frac{e(t)}{\Delta} \tag{6.6}$$

where $\Delta = 1 - z^{-1}$. This choice of the noise model will guarantee a zero steady-state error for step disturbances and constant references (for details, see [61] [63] [62]).

An approximation of the predicted future output is given by

$$y(t+k|t) = y_{\text{forced}}(t+k|t) + y_{\text{free}}(t+k|t) \tag{6.7}$$

where $y_{\text{forced}}(t+k|t)$ depends only on the future increments on the input and $y_{\text{free}}(t+k|t)$ depends only on the past inputs and outputs. In the present case the y_{free} sequence is given by

$$y_{\text{free}}(t+k|t) = \\ f(\hat{y}(t+k-1),\ldots,\hat{y}(t+k-m),u(t+k-1),\ldots,u(t+k-n)) + n(t+k|t) \tag{6.8}$$

with $u(t) = u(t+1) = \ldots = u(t+k-1) = u(t-1)$, which is equivalent to simulate the system assuming all the future inputs constant and equal to the last input value applied to the plant $u(t-1)$, and the sequence of predicted noise $n(t+k|t)$ is calculated assuming that $e(t+k|t) = 0$ for $k > 1$ (white noise assumption). The predicted y_{forced} will be given by

$$y_{\text{forced}}(t+k|t) = \sum_{i=0}^{k-1} g_i \delta u(t+k-i-1|t) \tag{6.9}$$

where g_i are the step response coefficients of the plant, calculated on the present operating point by simulating the step response on the model.

It is important to remark that the representation given by Equation (6.7) is not a linear representation because the term y_{free} is generated via simulation using the nonlinear model and the coefficients g_i depend on the current operating point and the amplitude of the input signal.

This is an important contribution to the analysis of the problem because it simplifies the study and the solution without a significant degradation. Most of the nonlinear information is preserved, as shown in the simulation included in the example of Section 6.2.2.

The optimization problem can be formulated in matrix form. First, the predictor is constructed as

$$\mathbf{Y} = \mathbf{G}\Delta\mathbf{U} + \mathbf{Y}_{\text{free}} \tag{6.10}$$

where

$$\mathbf{Y} = \begin{bmatrix} y(t+1|t) \\ y(t+2|t) \\ \vdots \\ y(t+N_p|t) \end{bmatrix}, \qquad \mathbf{G} = \begin{bmatrix} g_0 & 0 & \cdots & 0 \\ g_1 & g_0 & \cdots & 0 \\ \vdots & \vdots & \ddots & \vdots \\ g_{N_u-1} & g_{N_u-2} & \cdots & g_0 \\ \vdots & \vdots & \vdots & \vdots \\ g_{N_p-1} & g_{N_p-2} & \cdots & g_{N_p-N_u-1} \end{bmatrix} \tag{6.11}$$

$$\Delta\mathbf{U} = \begin{bmatrix} \delta u(t|t) \\ \delta u(t+1|t) \\ \vdots \\ \delta u(t+N_u-1|t) \end{bmatrix}, \qquad \mathbf{Y}_{\text{free}} = \begin{bmatrix} y_{\text{free}}(t+1|t) \\ y_{\text{free}}(t+2|t) \\ \vdots \\ y_{\text{free}}(t+N_p|t) \end{bmatrix} \tag{6.12}$$

$$\begin{aligned} y_{\text{free}}(t+i|t) &= f(\hat{y}(t+i-1), \ldots, \hat{y}(t+i-m), \\ &\qquad \hat{u}(t+i-1), \ldots, \hat{u}(t+i-n)) + n(t+i|t) \\ \text{with: } \hat{u}(k) &= \begin{cases} u(t-1) \ \forall k > t-1 \\ u(k) \qquad \text{otherwise} \end{cases} \end{aligned} \tag{6.13}$$

and with the reference vector described as

$$\mathbf{W} = \begin{bmatrix} w(t) \\ w(t+1) \\ \vdots \\ w(t+N_p) \end{bmatrix} \tag{6.14}$$

The cost function can be written as

$$J = \mathbf{E}^T Q \mathbf{E} + \Delta \mathbf{U}^T R \Delta \mathbf{U} \tag{6.15}$$

$$\begin{aligned} J &= (\mathbf{W} - \mathbf{Y})^T Q (\mathbf{W} - \mathbf{Y}) + \Delta \mathbf{U}^T R \Delta \mathbf{U} \\ &= (\mathbf{W} - \mathbf{G}\Delta \mathbf{U} - \mathbf{Y}_{\text{free}})^T Q (\mathbf{W} - \mathbf{G}\Delta \mathbf{U} - \mathbf{Y}_{\text{free}}) + \Delta \mathbf{U}^T R \Delta \mathbf{U} \end{aligned} \tag{6.16}$$

The minimization of the function J can be obtained by calculating the input sequence $\Delta\mathbf{U}$ so that $\partial J/\partial \Delta \mathbf{U} = 0$:

$$\frac{\partial J}{\partial \Delta \mathbf{U}} = 2\mathbf{G}^T Q(\mathbf{Y}_{\text{free}} - \mathbf{W}) + 2(\mathbf{G}^T Q \mathbf{G} + R)\Delta \mathbf{U} = 0 \tag{6.17}$$

Then the optimal sequence $\Delta\mathbf{U}$ is

$$\Delta \mathbf{U} = (\mathbf{G}^T Q \mathbf{G} + R)^{-1} \mathbf{G}^T Q (\mathbf{W} - \mathbf{Y}_{\text{free}}) \tag{6.18}$$

The input applied to the plant at time t is

$$u(t) = u(t-1) + \delta u(t) \tag{6.19}$$

where $\delta u(t)$ is the first element of the vector $\Delta\mathbf{U}$.

Observe that the expression given by Equation (6.18) is the same expression obtained for the generalized predictive control (GPC) [61]. However, in the GPC formulation the components involved in the calculation of the formula (6.18) come from a linear model. In the present case the components introduced in the formula (6.18) are generated by the nonlinear model (a fuzzy model). A more rigorous formulation of (6.18) will be to represent the components as time-variant matrices, as they are shown in the expression (6.20). In this expression at each sampling time the vectors $\mathbf{G}(t), \mathbf{Y}_{\text{free}}(t), \mathbf{W}(t)$ are reconstructed. The vector $\mathbf{Y}_{\text{free}}(t)$ is obtained by simulating the fuzzy model with the current input $u(t)$; the matrix $\mathbf{G}(t)$ is also reconstructed at each sampling time by using a method described in the next section.

$$\Delta \mathbf{U}(t) = (\mathbf{G}(t)^T Q \mathbf{G}(t) + R)^{-1} \mathbf{G}(t)^T Q (\mathbf{W}(t) - \mathbf{Y}_{\text{free}}(t)) \tag{6.20}$$

6.2.1 Estimation of the Step Response to Construct $\mathbf{G}(t)$

The estimation of the step response is obtained by

$$g(k-1) = \frac{y_{\text{step}}(t+k|t) - y_{\text{free}}(t+k|t) - n(t+k|t)}{du(t)} \tag{6.21}$$

where $du(t)$ represents the step size and

$$\begin{aligned} y_{\text{step}}(t+k|t) &= f(\hat{y}(t+k-1), \ldots, \hat{y}(t+k-m), \\ &\qquad \hat{u}(t+k-1), \ldots, \hat{u}(t+k-n)) \\ \text{with: } \hat{u}(k) &= \begin{cases} u(k) & \forall k \leq t-1 \\ u(t-1) + du(t) & \forall k > t-1 \end{cases} \end{aligned} \tag{6.22}$$

The estimation of the step response is a very important element in the present algorithm, because the quality of the system's response is determined by the accuracy of this estimation. The main difficulty to obtain an accurate estimate of the step response around some operating point is to select an appropriate value for the amplitude and the sign of the step $du(t)$, because the "step response" of a nonlinear system is determined by the operating point, the size and the sign of the step signal. A good value of $du(t)$ should satisfy the following requirements:

- It is very important that the value $u(t-1)+du(t)$ do not saturate the "actuators" in the model. Even though the constraints are not taken into account for the optimization, the saturation constraints on the actuators must be taken into account. Failing to consider these saturation points can lead to an underestimated step response (with smaller gain) and this response will generate a response from the controller larger than needed, thereby possibly generating unstable behavior.
- Also important is the fact that $du(t)$ should be "close" to the predicted $\Delta u(t)$. Since this value is only known after the optimization, a good choice is to select $du(t) = \delta u(t|t-1)$ [$\delta u(t)$ obtained in the previous optimization step $t-1$], and it corresponds to the second element of the vector $\Delta \mathbf{U}(t-1)$. When the system reaches the steady state $\Delta \mathbf{U} \rightarrow 0$, it means that if we select $du(t)$ as stated before, Equation (6.22) will be badly conditioned. This imposes a minimum step size du_{min} in order to avoid this bad conditioning.

Summary:
The problem of nonlinear unconstrained predictive control can be solved without online optimization. Only the calculation of a simple matrix equation where the matrices are extracted from the model is sufficient to generate an optimal trajectory.

6.2.2 Example Predictive Control of a CSTR Using a Fuzzy Model

As example of the application of this control strategy, a plant representing a continuous-stirred tank reactor was chosen (see Figure 6.3). The model of this plant was presented in [77][78]. The model is described by the following differential equations:

$$\begin{aligned} \dot{C}_a(t) &= \tfrac{q}{v}(C_{a0} - C_a(t)) - k_0 C_a(t) e^{-\frac{E}{RT(t)}} \\ \dot{T}(t) &= \tfrac{q}{v}(T_0 - T(t)) + k_1 C_a(t) e^{-\frac{E}{RT(t)}} \\ &\quad + k_2 q_c(t)\left(1 - e^{-\frac{k_3}{q_c(t)}}\right)(T_{c0} - T(t)) \end{aligned} \tag{6.23}$$

The process describes the reaction where product A is converted into product B, the concentration $C_a(t)$ is the concentration of product A, $T(t)$ is the

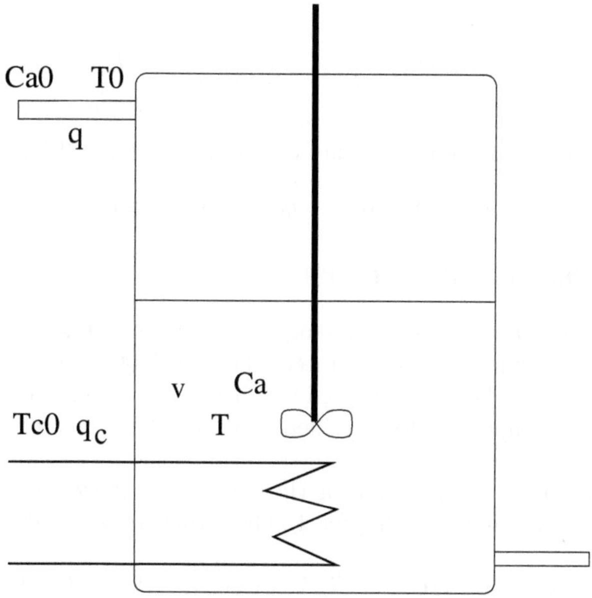

Figure 6.3. Continuous Stirred Tank Reactor

temperature of the mixture. The reaction is exothermic and it is controlled by a coolant flow whose rate is represented by $q_c(t)$. So, by changing the coolant flow, the temperature is controlled and by controlling the temperature, the concentration is controlled. C_{a0} is the inlet feed concentration, q is the process flow rate, T_0 and T_{c0} are the inlet feed and coolant temperatures; all these values are assumed constant at nominal values. In the same way, k_0, E/R, v, k_1, k_2 and k_3 are thermodynamic and chemical constants. The numerical values of these parameters are given in Table 6.1.

Table 6.1. CSTR Model Parameters

Parameter	Description	Nominal value
q	Process flow-rate	100 l/min
v	Reactor volume	100 l
k_0	Reaction rate constant	7.2×10^{10}min^{-1}
E/R	Activation energy	1×10^4K
T_0	Feed temperature	350 K
T_{c0}	Inlet coolant temp.	350 K
ΔH	Heat of reaction	2×10^5 cal/mol
C_p, C_{pc}	Specific heats	1 cal/g/K
ρ, ρ_c	Liquid densities	1×10^3 g/l
h_a	Heat transfer coeff.	7×10^5 cal/min/K
C_{a0}	Inlet feed concentration	1 mol/l

$$k_1 = \frac{\Delta H k_0}{\rho C_p} \quad k_2 = \frac{\rho_c C_{pc}}{\rho C_p v} \quad k_3 = \frac{h_a}{\rho_c C_{pc}}$$

The nominal conditions for a product concentration $C_a = 0.1$ mol/l are

$$T = 438.54\text{K} \qquad q_c = 103.41\text{l/min}$$

Fuzzy Identification of the CSTR

For the identification procedure a sampling time $Ts = 6$ sec was chosen. A sequence of 7500 samples was generated; the first 5000 samples were used as training set and the last 2500 as validation set. Figure 6.4 shows the input and the output signal used for the identification (This dataset can be found in [79].)

The regressors were selected using the regularity criterion and the heuristic search method explained in Chapter 4. The structure selected for the model is

$$\hat{C}_a(k+1) = f(\hat{C}_a(k), \hat{C}_a(k-1), \hat{C}_a(k-2), q_c(k-1)) \qquad (6.24)$$

The identified model is an NOE model; the fuzzy model has three triangular membership functions distributed regularly on the universes of discourse. The number of rules extracted was $3 \times 3 \times 3 \times 3 = 81$ rules. The membership functions can be observed in Figure 6.6 and the results of the validation in Figure 6.5. Figure 6.5 shows the validation error on simulation. It can be observed that the quality of the model is very good and in fact the two signals appear overlapped; the error is plotted at the bottom of Figure 6.5 in another scale to make it visible.

Implementation of the Controller and Performance Results

The predictive controller was implemented using the following parameters: prediction horizon $N_p = 10$, control horizon $Nc = 2$, cost matrices $Q = I^{N_p}$, $R = 10^{-4} I^{N_c}$, where I^n is an identity matrix such that $I^n \in \Re^{n \times n}$, $C(z^{-1}) = 1$ and $D(z^{-1}) = 1 - z^{-1}$, $du_{min} = 0.1(q_c^{max} - q_c^{min}) = 2$ l/min. Figure 6.8 shows the response of the system with a "stair" reference; each step has a duration of 10 min and an amplitude of 0.005 mol/l. The reference starts on 0.08 mol/l and it goes up to 0.12 mol/l. Perturbation steps of 0.005 mol/l are applied in the middle of each step. The reference and the perturbation signal are shown in Figure- 6.7. To compare the performance of the designed controller three other control strategies were implemented.

- A PID strategy optimized for the current reference. The parameters of the PID are

$$K_p = 91.2 \quad T_i = 0.178\,\text{sec} \quad T_d = 0.208\,\text{sec}$$

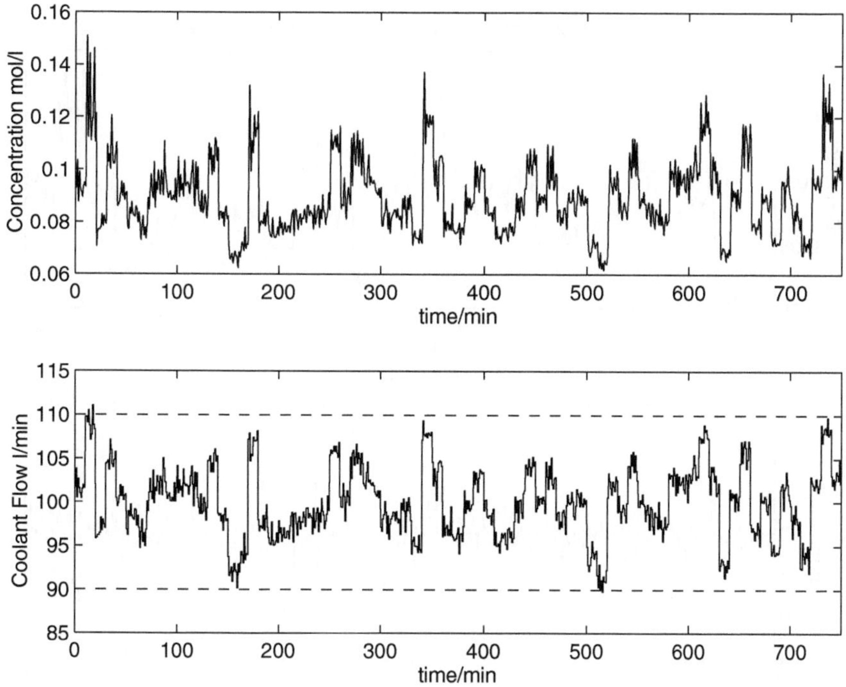

Figure 6.4. Identification experiments for the CSTR (Courtesy of Springer-Verlag [76])

- A Generalized Predictive Controller (GPC) designed with the same parameters as our predictive controller but using an Output Error linear model described as:

$$\hat{C}_a(k+1) = \frac{1.653 \times 10^{-4} z^{-1}}{1 - 2.43z^{-1} + 2.4z^{-2} - 1.189z^{-3} + 0.269z^{-4}} q_c(k) \qquad (6.25)$$

- An "optimal" nonlinear control strategy calculated using the Branch and Bound (B & B) optimization algorithm [80][81]. The Branch and Bound method is a discrete optimization algorithm. For this reason, the input range is discretized in 10 equidistant points. At each iteration the Branch and Bound algorithm will search in a space of 100 (10^{N_c}) possible solutions. The optimization is refined with a local Newton optimization. The model used for the controller is the same model used to simulate the plant, so that there is no mismatch between the model and the plant.

Figure 6.10 shows details of the performance comparison between the predictive controller based on the fuzzy model and the GPC and the PID. It is important to remark the degradation of the performance experimented by the systems with the linear GPC controller and the PID controller when the concentration is close to 0.12 mol/l. Observe that the PID controller is not

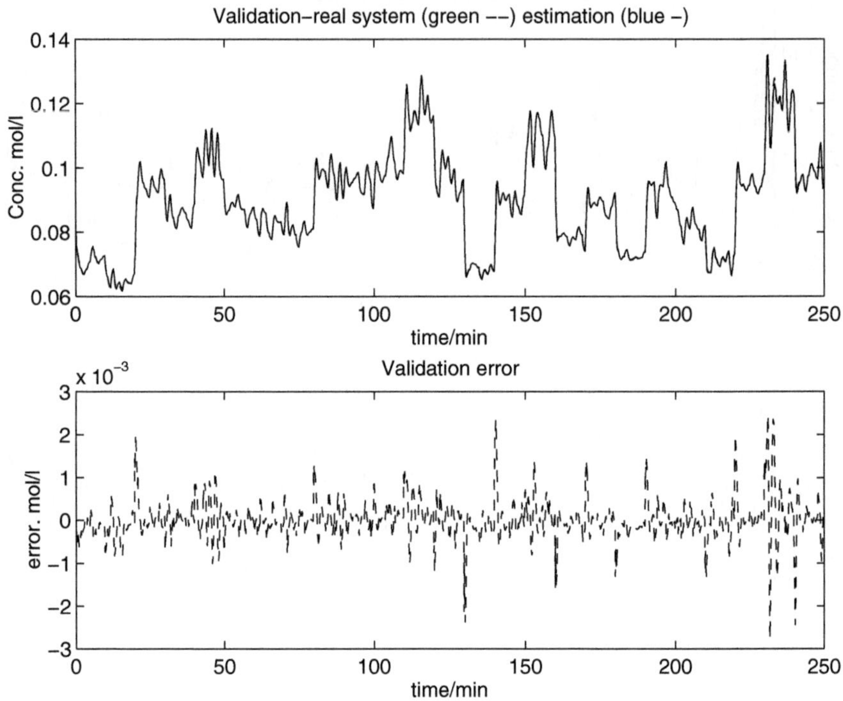

Figure 6.5. Validation results for the CSTR (Courtesy of Springer-Verlag [76])

able to stabilize the plant near this set point. In addition, the performance of the GPC is also degraded.

Figures- 6.9 and- 6.11 show the comparison between the fuzzy model predictive controller and the "optimal" control strategy. Observe that, when the change in the operating point is big (between $t = 0, ..., 2$), the two solutions are different, but when the changes are smaller the two solutions are almost the same, which indicates that the "suboptimal" solution is very close to the "optimal" solution. The execution time of the algorithm is a very important parameter in order to evaluate its applicability. The execution time was evaluated using a Pentium IV 2.8 GHz, and the algorithm was implemented as an s-function in Simulink. Observe that this execution time gives the possibility of sampling frequencies of the order of 100 Hz. It is important to observe that the execution time grows in a linear form with respect to the length of the prediction horizon and it changes very little with the increase of the control horizon. Other interesting comparisons with other fuzzy model based predictive control strategies using the present model are presented in [73].

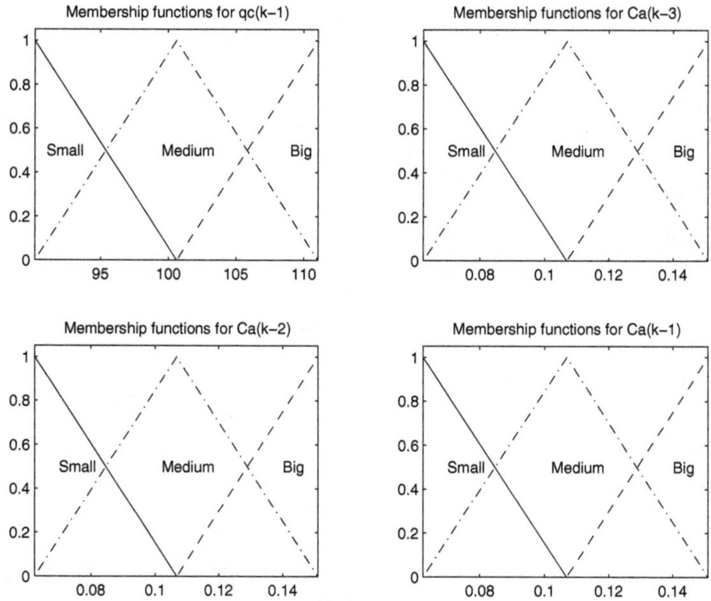

Figure 6.6. Membership function of the fuzzy model of the CSTR (Courtesy of Springer-Verlag [76])

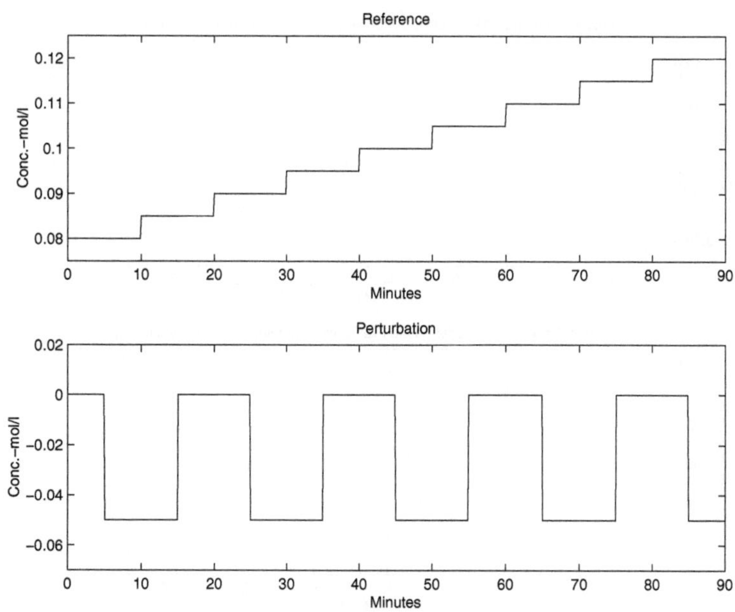

Figure 6.7. Reference and perturbation for the CSTR (Courtesy of Springer-Verlag [76])

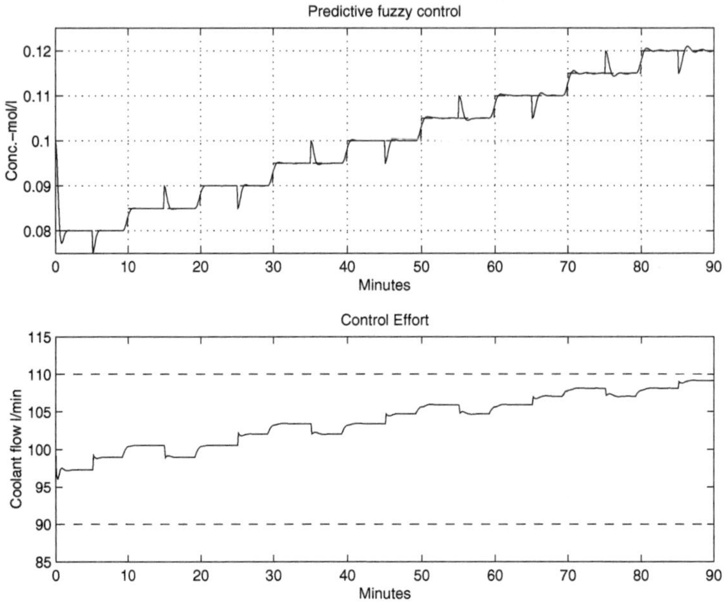

Figure 6.8. Response of the CSTR using an unconstrained fuzzy model-based predictive control (Courtesy of Springer-Verlag [76])

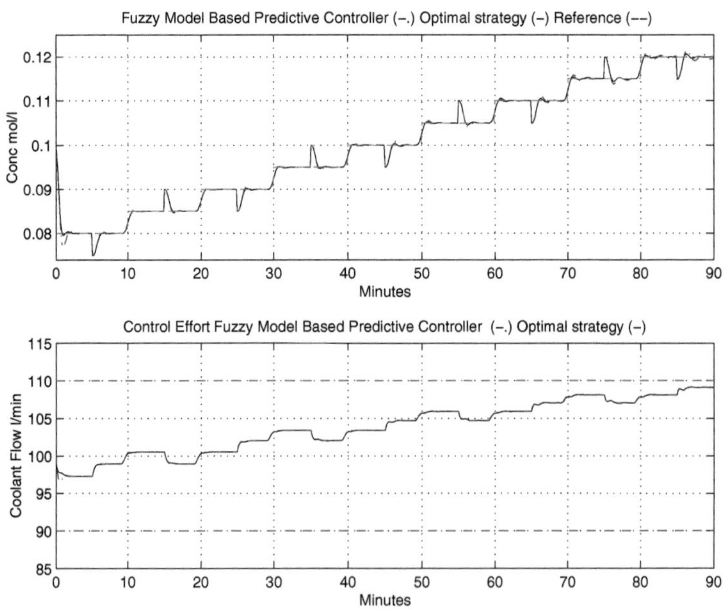

Figure 6.9. Response of the CSTR using an unconstrained fuzzy model-based predictive control. Comparison against the global "optimal" strategy

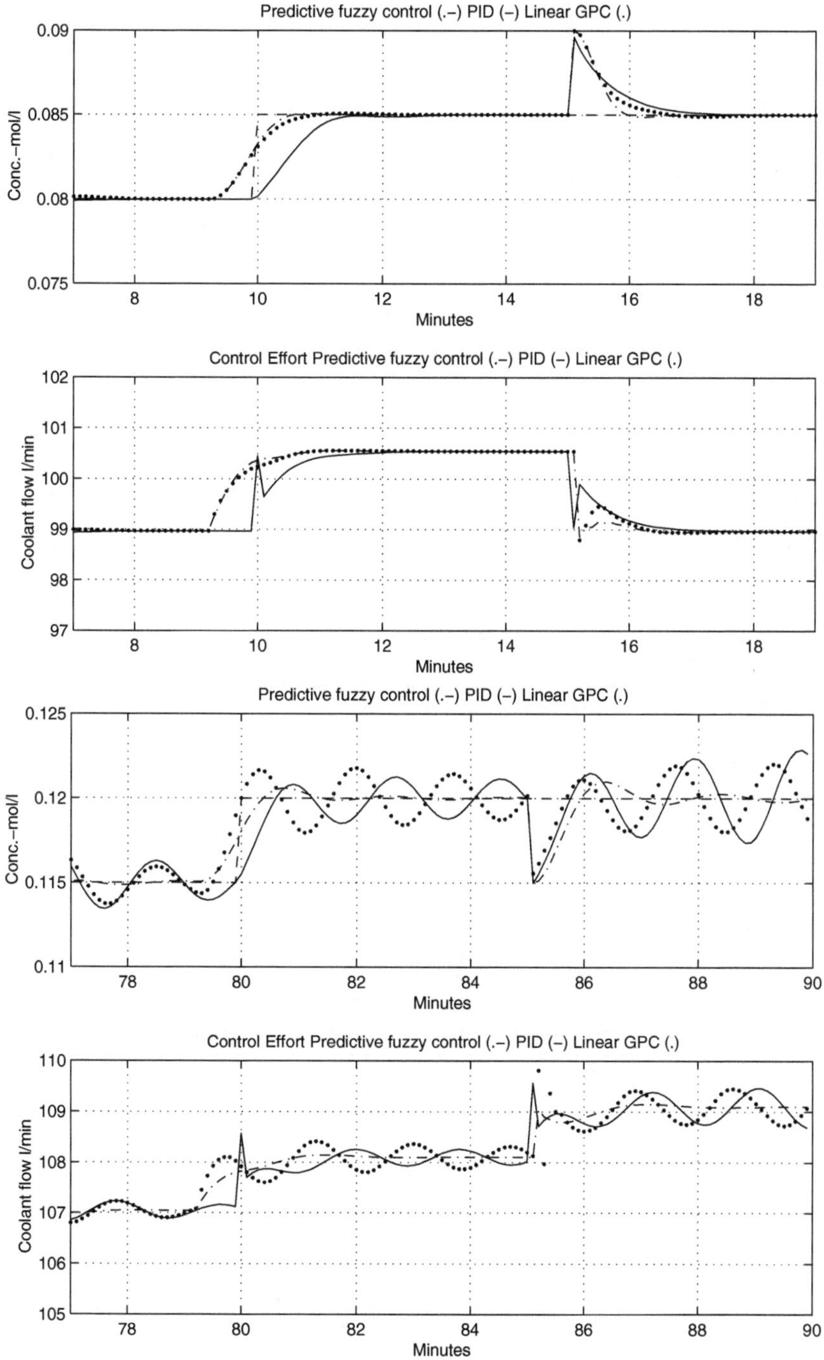

Figure 6.10. Response of the CSTR using an unconstrained fuzzy model-based predictive control. Detailed comparison for two different operating points (Courtesy of Springer-Verlag [76])

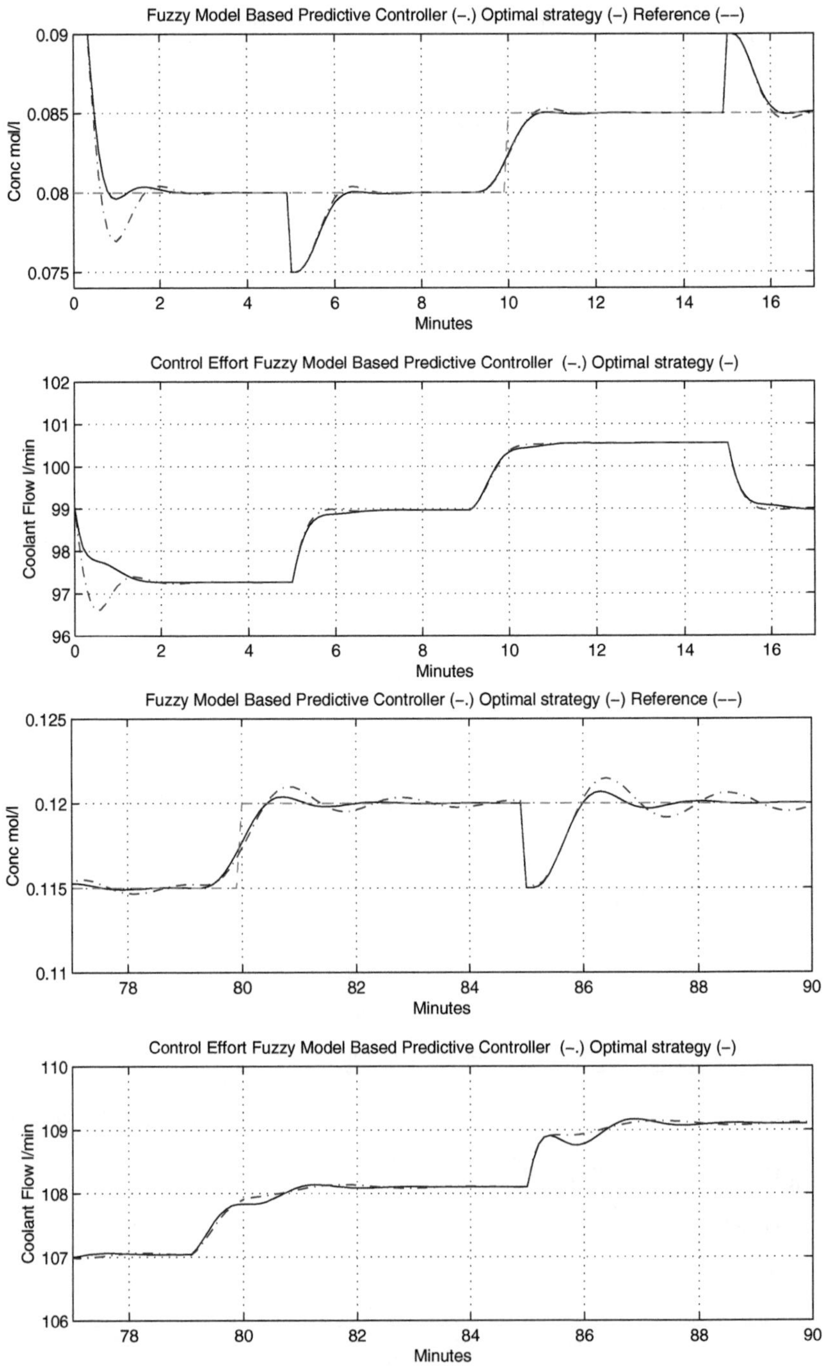

Figure 6.11. Response of the CSTR using an unconstrained fuzzy model-based predictive control. Detailed comparison against the "optimal" strategy for two different operating points

Table 6.2. Execution Time of the Controller for the CSTR Using an Unconstrained Fuzzy Model-Based Predictive Control. Execution Time (msec.) VS. Horizons

Nu/Np	10	11	12	13	14	15
2	5.60	6.25	6.58	7.07	7.43	7.83
4	5.90	6.30	6.68	7.12	7.50	7.88
6	5.95	6.41	6.74	7.17	7.54	7.95
8	5.95	6.39	6.81	7.17	7.54	7.99

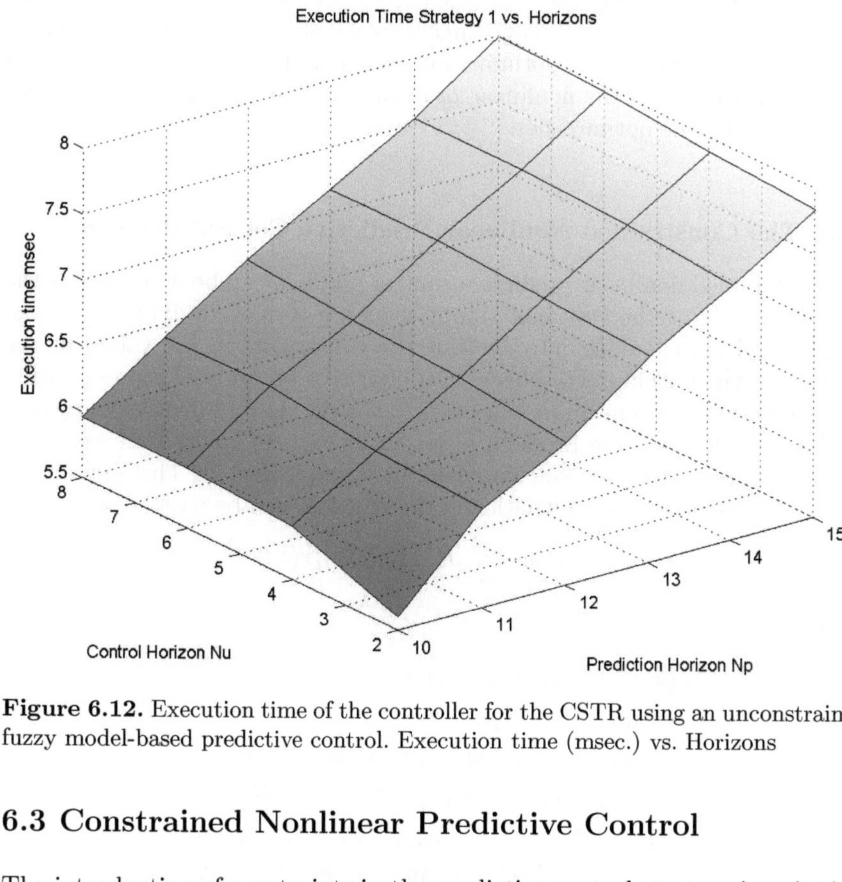

Figure 6.12. Execution time of the controller for the CSTR using an unconstrained fuzzy model-based predictive control. Execution time (msec.) vs. Horizons

6.3 Constrained Nonlinear Predictive Control

The introduction of constraints in the predictive control strategy is quite important. Constraints give more "realism" to the control actions by modeling saturation and slew rate constraints present in the actuators. It can also improve the formulation of a control system in terms of acceptable quality values (output constraints). Perhaps the most important feature from the economical point of view is that it guarantees maximal exploitation of the plant facility by driving the system closer to its maximum productivity without compromising the safety of the plant.

In the next lines, three algorithms for constrained nonlinear predictive control based on fuzzy models are presented. Initially, in Section 6.3.1 the constrained nonlinear predictive control problem is presented and a solution based on local linearization is derived. Sections 6.3.2, 6.3.3 and 6.3.4 present three algorithms to solve the constrained nonlinear predictive control problem. The strategy presented in Section 6.3.2 is an extension of the algorithm presented in a previous section for the unconstrained case; this control strategy can be applied on any type of fuzzy model. The strategies presented in Sections 6.3.3 and 6.3.4 exploit the structure of a Takagi–Sugeno fuzzy model to extract the information needed to solve the optimization problem. The strategy presented in Section 6.3.4 converts the nonlinear problem into a linear time-variant problem to simplify the optimization.

6.3.1 The Constrained Nonlinear Predictive Control Problem

The constrained nonlinear predictive control problem can be described as the problem of finding the "optimal" input sequence to move a dynamic system to a desired state, taking into account the constraints in the values of the inputs and the outputs. To achieve this objective an internal representation of the dynamical system (in this case a fuzzy model) is used in order to predict the future behavior. The problem is solved at each sampling time and only the first movement of the calculated input is implemented. This guarantees disturbance rejection. Mathematically the problem can be written as

$$\min_{u(k),\ldots,u(k+Nc)} J(u(k), x(k), r(k)) \tag{6.26}$$

subject to

$$\begin{aligned}
x(k+1) &= f(x(k), u(k)) \\
y(k) &= g(x(k), u(k)) \\
y_{\min} &\leq y(k) \leq y_{\max},\ \forall k = 1, \ldots, Np \\
u_{\min} &\leq u(k) \leq u_{\max},\ \forall k = 1, \ldots, Nc \\
\Delta y_{\min} &\leq \Delta y(k) \leq \Delta y_{\max},\ \forall k = 1, \ldots, Np \\
|\Delta u(k)| &\leq \Delta u_{\max},\ \forall k = 1, \ldots, Nc
\end{aligned}$$

where $J(.)$ is the "the cost function" and it is of the form

$$\begin{aligned}
J(u(k), x(k), r(k)) = &\sum_{t=k}^{k+Np} (r(t) - x(t))^T Q (r(t) - x(t)) \\
&+ \sum_{t=k}^{k+Nc} (u(t)^T R u(t) + \delta u(t)^t S \delta u(t))
\end{aligned} \tag{6.27}$$

or

$$J(u(k), y(k), r(k)) = \sum_{t=k}^{k+Np} (r(t) - y(t))^T Q(r(t) - y(t)) + \sum_{t=k}^{k+Nc} (u(t)^T Ru(t) + \delta u(t)^t S \delta u(t)) \quad (6.28)$$

where $\delta u(t) = u(t) - u(t-1)$, Q is positive definite matrix and R and S are positive semidefinite matrices.

The solution to this problem is computationally very expensive because it involves the solution of a constrained nonlinear quadratic program. The exact solution to this problem ("finding the global minima") is very complicated and it cannot be done in real time. The next section shows three approaches to solve this problem in real time generating a suboptimal solution with a performance quite close to the "optimal solution."

The design of a predictive controller demands the construction of a predictor. In this section, the formulation is presented in state-space form to generalize the problem in the multivariable form. The construction of the predictor in the state-space representation can be conducted in the following form: given the linearization of the system for the trajectory point $[x^*, u^*]$:

$$\begin{aligned} x(k+1) &= E + A(x(k) - x^*) + B(u(k) - u^*) \\ y(k) &= F + C(x(k) - x^*) + D(u(k) - u^*) \end{aligned} \quad (6.29)$$

where

$$\begin{aligned} E &= f(x^*, u^*) \\ F &= g(x^*, u^*) \\ A &= \frac{\partial f(x,u)}{\partial x}|_{x^*,u^*} \\ B &= \frac{\partial f(x,u)}{\partial u}|_{x^*,u^*} \\ C &= \frac{\partial g(x,u)}{\partial x}|_{x^*,u^*} \\ D &= \frac{\partial g(x,u)}{\partial u}|_{x^*,u^*} \end{aligned} \quad (6.30)$$

Using this simplification the prediction of the state and the output n steps ahead with a given input sequence U^n will be given by

$$x(k+n|k) = \sum_{i=0}^{n-1} A^i L + A^n x(k) + [A^{n-1}BA^{n-2}B \dots ABB] \begin{bmatrix} u(k|k) \\ u(k+1|k) \\ \vdots \\ u(k+n|k) \end{bmatrix}$$

$$y(k+n|k) = M + Cx(k+n|k) + Du(k+n|k) \tag{6.31}$$

where

$$L = E - Ax^* - Bu^*$$
$$M = F - Cx^* - Du^*$$

In matrix representation the prediction can be seen as

$$X^n = \underbrace{\Psi + \Phi x(k)}_{\text{Free response}} + \underbrace{H\overline{U}^n}_{\text{Forced response}} \tag{6.32}$$

$$Y^n = M^n + C^n X^n + D^n \underline{U}^n \tag{6.33}$$

$$Y^n = \underbrace{M^n + C^n[\Psi + \Phi x(k)]}_{\text{Free response}} + \underbrace{C^n H\overline{U}^n + D^n \underline{U}^n}_{\text{Forced response}}$$

$$Y^n = \underbrace{M^n + C^n[\Psi + \Phi x(k)]}_{\Gamma} + \underbrace{[C^n H\, 0] + [0\, D^n]}_{\Lambda} U^n \tag{6.34}$$

where

$$X = \begin{bmatrix} x(k+1|k) \\ x(k+2|k) \\ \vdots \\ x(k+n|k) \end{bmatrix} \quad \Psi = \begin{bmatrix} L \\ L + AL \\ \vdots \\ L + AL + \ldots + A^{n-1}L \end{bmatrix} \quad \Phi = \begin{bmatrix} A \\ A^2 \\ \vdots \\ A^n \end{bmatrix}$$

$$H = \begin{bmatrix} B & 0 & \ldots & 0 \\ AB & B & \ldots & 0 \\ \vdots & \vdots & \ddots & \vdots \\ A^{n-1}B & A^{n-2}B & \ldots & B \end{bmatrix} \quad M^n = \begin{bmatrix} M \\ M \\ \vdots \\ M \end{bmatrix}$$

$$U^n = \begin{bmatrix} u(k|k) \\ u(k+1|k) \\ \vdots \\ u(k+n|k) \end{bmatrix} \quad \overline{U}^n = \begin{bmatrix} u(k|k) \\ u(k+1|k) \\ \vdots \\ u(k+n-1|k) \end{bmatrix} \quad \underline{U}^n = \begin{bmatrix} u(k+1|k) \\ u(k+2|k) \\ \vdots \\ u(k+n|k) \end{bmatrix}$$

$$C^n = \begin{bmatrix} C & 0 & \ldots & 0 \\ 0 & C & \ldots & 0 \\ \vdots & \vdots & \ddots & \vdots \\ 0 & 0 & \ldots & C \end{bmatrix} \quad D^n = \begin{bmatrix} D & 0 & \ldots & 0 \\ 0 & D & \ldots & 0 \\ \vdots & \vdots & \ddots & \vdots \\ 0 & 0 & \ldots & D \end{bmatrix}$$

When the predictor is constructed with an input sequence N_c and prediction horizon N_p the matrices of the predictor are converted to

$$X = \begin{bmatrix} x(k+1|k) \\ x(k+2|k) \\ \vdots \\ x(k+N_p|k) \end{bmatrix} \quad \Psi = \begin{bmatrix} L \\ L + AL \\ \vdots \\ L + AL + \ldots + A^{N_p-1}L \end{bmatrix}$$

$$\Phi = \begin{bmatrix} A \\ A^2 \\ \vdots \\ A^{N_p} \end{bmatrix} \quad M^n = \begin{bmatrix} M \\ M \\ \vdots \\ M \end{bmatrix}$$

$$H = \begin{bmatrix} B & 0 & \dots & 0 \\ AB & B & \dots & 0 \\ \vdots & \vdots & \ddots & \vdots \\ A^{N_p-2}B & A^{N_p-3}B & \dots & \sum_{i=0}^{N_p-N_c-1} A^i B \\ A^{N_p-1}B & A^{N_p-2}B & \dots & \sum_{i=0}^{N_p-N_c} A^i B \end{bmatrix}$$

$$U^{N_u} = \begin{bmatrix} u(k|k) \\ u(k+1|k) \\ \vdots \\ u(k+N_c|k) \end{bmatrix} \quad \overline{U}^n = \begin{bmatrix} u(k|k) \\ u(k+1|k) \\ \vdots \\ u(k+N_c-1|k) \end{bmatrix} \quad \underline{U}^n = \begin{bmatrix} u(k+1|k) \\ u(k+2|k) \\ \vdots \\ u(k+N_c|k) \end{bmatrix}$$

$$C^{N_p} = \begin{bmatrix} C & 0 & \dots & 0 \\ 0 & C & \dots & 0 \\ \vdots & \vdots & \ddots & \vdots \\ 0 & 0 & \dots & C \end{bmatrix}, \in R^{N_p n_o \times N_p n_s}$$

$$D^n = \begin{bmatrix} D & 0 & \dots & 0 \\ 0 & D & \dots & 0 \\ \vdots & \vdots & \ddots & \vdots \\ 0 & 0 & \dots & D \\ 0 & 0 & \dots & D \end{bmatrix}, \in R^{N_p n_o \times N_c n_i}$$

where n_s is the number of states, n_o the number of outputs and n_i the number of inputs.

Once this local linear representation has been obtained, the optimization problem can be written as a quadratic program (QP). In fact, now the problem is a classical linear constrained predictive control problem.

Using the vector notation, the cost function will be written as

$$J(U^n, Y^n) = \\ (Y^n_{\text{ref}} - Y^n)^T Q (Y^n_{\text{ref}} - Y^n) + U^{nT} R U^n + (\Delta U^n - \bar{U}_{k-1})^T S (\Delta U^n - \bar{U}_{k-1}) \tag{6.35}$$

where

$$Y^n_{\text{ref}} = [y_{\text{ref}}(k+1), \dots, y_{\text{ref}}(k+n)]^T$$

is the vector of the reference. Introducing (6.34) in (6.35):

$$J(U^n) = \\ J_{\min} + 2[(\Gamma - Y^n_{\text{ref}})^T Q \Lambda - \bar{U}^T_{k-1} S \Delta] U^n + U^{nT} [\Lambda^T Q \Lambda + R + \Delta^T S \Delta] U^n \tag{6.36}$$

where

$$J_{\min} = Y_{\text{ref}}^{nT} Q Y_{\text{ref}}^{n} + \Gamma^T Q \Gamma - 2 Y_{\text{ref}}^{nT} Q \Gamma + \bar{U}_{k-1}^T S \bar{U}_{k-1}$$

is the minimum cost due to the reference and the free response and it cannot be modified by any control input. The constraints can be written as follows:

$$\begin{bmatrix} I^{(n_i N_c)} \\ -I^{(n_i N_c)} \\ I^{(n_i)} 0^{(n_i \times n_i N_c)} \\ -I^{(n_i)} 0^{(n_i \times n_i N_c)} \\ \Delta \\ -\Delta \\ \Lambda \\ -\Lambda \end{bmatrix} U^{N_c} \leq \begin{bmatrix} U_{\max}^{N_c} \\ -U_{\min}^{N_c} \\ \Delta U_{\max}^{N_c} + \bar{U}_{k-1} \\ \Delta U_{\max}^{N_c} - \bar{U}_{k-1} \\ \Delta U_{\max}^{N_c} \\ \Delta U_{\max}^{N_c} \\ Y_{\max}^{N_p} - \Gamma \\ -Y_{\min}^{N_p} + \Gamma \end{bmatrix} \tag{6.37}$$

where

$$I = \begin{bmatrix} 1\,0 \ldots 0 \\ 0\,1 \ldots 0 \\ \vdots\,\vdots \ddots \vdots \\ 0\,0 \ldots 1 \end{bmatrix} \qquad \Delta = \begin{bmatrix} I & 0 & 0 & \ldots & 0 & 0 \\ -I & I & 0 & \ldots & 0 & 0 \\ 0 & -I & I & \ldots & 0 & 0 \\ \vdots & \vdots & \vdots & \ddots & \vdots & \vdots \\ 0 & 0 & 0 & \ldots & -I & I \end{bmatrix}$$

$$\bar{U}_{k-1}^{N_c} = \begin{bmatrix} u(k-1) \\ 0 \\ \vdots \\ 0 \end{bmatrix} \quad U_{\max}^{N_c} = \begin{bmatrix} u_{\max}(k) \\ u_{\max}(k+1) \\ \vdots \\ u_{\max}(k+N_c) \end{bmatrix} \quad U_{\min}^{n} = \begin{bmatrix} u_{\min}(k) \\ u_{\min}(k+1) \\ \vdots \\ u_{\min}(k+n) \end{bmatrix},$$

$$\Delta U_{\max}^{N_c} = \begin{bmatrix} \Delta u_{\max}(k) \\ \Delta u_{\max}(k+1) \\ \vdots \\ \Delta u_{\max}(k+N_c) \end{bmatrix} \quad Y_{\max}^{N_p} = \begin{bmatrix} Y_{\max}(k) \\ Y_{\max}(k+1) \\ \vdots \\ Y_{\max}(k+N_p) \end{bmatrix}$$

$$Y_{\min}^{N_p} = \begin{bmatrix} Y_{\min}(k) \\ Y_{\min}(k+1) \\ \vdots \\ Y_{\min}(k+N_p) \end{bmatrix}$$

where $I^{(n)} = I \in \Re^{n \times n}$ and $0^{(m \times n)} \in \Re^{m \times n}$.

At each sampling time, this QP is solved with new parameters. The three algorithms presented in the following sections are based on this concept, but they differ in the way the parameters of the QP are obtained from the nonlinear model. It is important to remark that these methods generate a suboptimal solution to the original nonlinear quadratic program (NLQP), because at each step a pseudo-linear approximation of the problem is used. This suboptimal solution will be very close to the exact solution of the NLQP if the nonlinearities of the plant are smooth such that along the prediction horizon the real trajectory is not very different from the trajectory generated by the proposed predictor. The algorithms also show some methods to improve the quality of the approximation.

Summary:
The problem of nonlinear constrained predictive control is formulated as a nonlinear quadratic optimization problem. By means of local linearization a relaxation can be obtained and the problem can be solved using quadratic programming. This is the solution of the linear constrained predictive control problem.

6.3.2 Approach Using Estimated Step Response

This approach can be applied to any type of nonlinear model, whether it be a fuzzy model (Takagi–Sugeno or Mamdani), a neural network model, a Volterra series model, wavelet model, gain scheduling, *etc.* The main idea is to solve also a QP by using an estimation of the step response as a linearized model. The formulation presented in this section is a generalization of the strategy presented for unconstrained predictive control. The advantage of this approach is that the linearization takes into account not only the current operating point (x^*, u^*) but also the direction of the control action in order to reach the next operating point $(x(k+1|k), u(k|k))$, by means of a similar reasoning as the one used in Equations (6.34) and (6.35). This comes from the fact that this method is based on the step response and the step response of a nonlinear system depends on the operating point, the amplitude of the step and the direction. In this method, the estimation of the step response uses the "prior knowledge" gained from the previously calculated control actions.

The prediction of the state and the output n steps ahead can also be represented as

$$
\begin{aligned}
x(k+n|k) &= \sum_{i=0}^{n-1} A^i L + A^n x(k) + \sum_{i=0}^{n-1} A^i B u_{k-1} + \\
&\quad + [\sum_{i=0}^{n-1} A^i B, \sum_{i=0}^{n-2} A^i B, \ldots, AB + B,\ B] \begin{bmatrix} \delta u(k|k) \\ \delta u(k+1|k) \\ \vdots \\ \delta u(k+n-1|k) \end{bmatrix} \\
y(k+n|k) &= M + Cx(k+n|k) + Du(k+n|k)
\end{aligned}
\tag{6.38}
$$

where $\delta u(k|k) = u(k|k) - u(k-1)$, in matrix representation

$$
X^n = \underbrace{\Psi + \Phi x(k) + G\bar{U}_{k-1}}_{\text{Free response}} + \underbrace{G\overline{\Delta U}^{n-1}}_{\text{Forced response}} \tag{6.39}
$$

$$Y^n = M^n + C^n X^n + D^n \underline{U}^n \tag{6.40}$$

$$Y^n = \underbrace{M^n + C^n\Psi + C^n\Phi x(k) + ([C^n G\, 0] + D^n_\Delta)\bar{U}_{k-1}}_{\text{Free response}} + \underbrace{([C^n G\, 0] + D^n_\Delta)\overline{\Delta U^n}}_{\text{Forced response}}$$

$$Y^n = \underbrace{M^n + C^n\Psi + C^n\Phi x(k)}_{\Gamma} + \underbrace{([C^n G\, 0] + D^n_\Delta)}_{\Xi}\bar{U}_{k-1} + \underbrace{([C^n G\, 0] + D^n_\Delta)}_{\Xi}\overline{\Delta U^n}$$

$$Y^n = \underbrace{\Gamma + \Xi\bar{U}_{k-1}}_{\Sigma} + \Xi(\Delta U^n - \bar{U}_{k-1}) = \Sigma + \Xi(\Delta U^n - \bar{U}_{k-1}) \tag{6.41}$$

$$Y^n = \Gamma + \Xi\Delta U^n \tag{6.42}$$

where $M^n, X^n, \Phi, \Psi, C^n$ are equivalent to the ones previously described and the *free response* Γ is modified by the addition of the term $\Xi\bar{U}_{k-1}$,

$$G = \begin{bmatrix} B & 0 & \dots 0 \\ AB + B & B & \dots 0 \\ \vdots & \vdots & \ddots \vdots \\ \sum_{i=0}^{n-1} A^i B & \sum_{i=0}^{n-2} A^i B & \dots B \end{bmatrix}$$

$$\overline{\Delta U^n} = \Delta U^n - \bar{U}_{k-1} = \begin{bmatrix} \delta u(k|k) \\ \delta u(k+1|k) \\ \vdots \\ \delta u(k+n|k) \end{bmatrix} \qquad D^n_\Delta = \begin{bmatrix} D & D & 0 & \dots & 0 \\ D & D & D & \dots & 0 \\ \vdots & \vdots & \vdots & \ddots & \vdots \\ D & D & D & \dots & D \end{bmatrix}$$

$$\Xi = \begin{bmatrix} CB + D & D & \dots & 0 & 0 \\ CAB + CB + D & CB + D & \dots & 0 & 0 \\ \vdots & \vdots & \ddots & \vdots & \vdots \\ \sum_{i=0}^{n-1} CA^i B + D & \sum_{i=0}^{n-2} CA^i B + D & \dots & CB + D & D \end{bmatrix}$$

Observe that the first column of the matrix Ξ corresponds to the *step response* of the system. So the response of the system can be described as the sum of the *free response* with the product of the Toeplitz matrix constructed with the "local" *step response* and the incremental input ΔU^n. The cost function (6.37) written with the response described by (6.41) is

$$\begin{aligned} J(U^n) = J_{\min} + 2[(\Sigma - \Xi\bar{U}_{k-1} - Y^n_{\text{ref}})^T Q\Xi\Delta - \bar{U}^T_{k-1} S\Delta]U^n + \dots \\ + U^{nT}[\Delta^T\Xi^T Q\Xi\Delta + R + \Delta^T S\Delta]U^n \end{aligned} \tag{6.43}$$

where

$$\begin{aligned} J_{\min} = Y^{nT}_{\text{ref}} Q Y^n_{\text{ref}} + (\Sigma - \Xi\bar{U}_{k-1})^T Q(\Sigma - \Xi\bar{U}_{k-1}) - \\ - 2Y^{nT}_{\text{ref}} Q(\Sigma - \Xi\bar{U}_{k-1}) + \bar{U}^T_{k-1} S\bar{U}_{k-1} \end{aligned}$$

is the minimum cost due to the reference and the free response. The QP program will be written using now the cost function (6.43) and the constraint definition (6.37).

In this procedure, the parameters of the QP are updated at each sampling time. The parameters that are updated at each sampling time are the matrices Σ and Ξ. The Σ matrix is the *free response* constructed by evolving the system form $k+1$ to $k+N_p$ as

$$\begin{aligned} \bar{x}(k+1) &= f(\bar{x}(k), u^*) \\ \bar{y}(k) &= g(\bar{x}(k), u^*) \end{aligned} \tag{6.44}$$

where $f(.,.)$ and $g(.,.)$ are the description of the plant given by the fuzzy model, $u^* = u(k)$ is the current input of the system, $x(\bar{k})$ and $\bar{y}(k)$ are, respectively, the "estimated" state and *free response*. Observe that this *free response* is different from the one calculated in the previous section. The difference is that the *free response* of the previous section is calculated by evolving the fuzzy system with $u^* = 0$. The Σ vector will look like

$$\Sigma = \begin{bmatrix} \bar{y}(k+1|k) \\ \bar{y}(k+2|k) \\ \vdots \\ \bar{y}(k+N_p|k) \end{bmatrix} \tag{6.45}$$

The Ξ matrix is obtained by making step experiments. A step input is equivalent to $\overline{\Delta U}^n_{\text{step}} = [du, 0, \ldots, 0]^T$, where du is a vector in $\Re^{n_i}$ where only one entry (i) is different from zero. The $||du||$ is the size of the applied step. The *free response* of the system will be

$$Y^n_{\text{free}} = \Gamma + \Xi \bar{U}_{k-1}$$

and the response with the step:

$$Y^n_{\text{step}} = \Gamma + \Xi \bar{U}_{k-1} + \Xi \overline{\Delta U}^n_{\text{step}}$$

operating:

$$Y^n_{\text{step}} - Y^n_{\text{free}} = \Xi \overline{\Delta U}^n_{\text{step}} = \begin{bmatrix} (CB+D)du \\ (CAB+CB+D)du \\ \vdots \\ (\sum_{j=0}^{n-1} CA^j B + D)du \end{bmatrix}$$

because only the i entry of du is different from zero, then the i column of Ξ will be obtained as

$$\begin{bmatrix} (CB+D)_i \\ (CAB+CB+D)_i \\ \vdots \\ (\sum_{j=0}^{n-1} CA^j B + D)_i \end{bmatrix} = \frac{1}{du_i} \begin{bmatrix} (CB+D)_i du_i \\ (CAB+CB+D)_i du_i \\ \vdots \\ (\sum_{j=0}^{n-1} CA^j B + D)_i du_i \end{bmatrix} \tag{6.46}$$

Observe that only the first n_i columns (n_i is the number of inputs) are needed because Ξ is a block Toeplitz matrix.
An important remark is that the amplitude and the direction of the du_i element will determine the obtained step response. This is due to the fact that the step response of a nonlinear system is a function of the operating point, the amplitude and the direction of the step. Our estimation had already dealt with obtaining the step response at the operating point, but the amplitude of the step and its direction have not been determined. A good method to determine the size and direction of du_i is to apply the most likely input that will be applied by the control action. This input will be obtained from the solution obtained in the previous optimization. So the most suitable value for du_i at time k will be

$$du_i^k = \delta u(k|k-1)_i \tag{6.47}$$

Observe that, when a steady state is reached, then $u(k|k-1)_i \approx u(k-1)_i$, making the value of $du_i \approx 0$ and the estimation of Ξ a badly conditioned operation. For this reason, a du_{min} vector should be defined with the minimum value of du, so that the estimation of Ξ is reliable. To sum up, the control algorithm will be

Algorithm

At each sampling time:

1. Read the current output of the system and update $y(k)$.
2. With the input $u(k-1)$ calculate the*free response* Σ using Equation (6.44).
3. Calculate the entries of the du vector for the n_i step experiments using Equation (6.47).
4. If $du_i < du^i_{\text{min}}$ make $du_i = du^i_{\text{min}}$.
5. Make the step experiments and calculate the Ξ matrix.
6. With Ξ and Σ construct the matrices for the cost [Equation (6.43)] and the constrains [Equation (6.37)] of the QP problem.
7. Solve the QP problem.
8. Apply only the first control action $u(k|k)$.

The advantage of this method is that no special structure is needed for the model to perform prediction task and to extract the step response. The price paid for this is that every time a simulated step response must be calculated for each input. On the other hand, it is also an advantage since the "linearization" provided by the step response will have a validity region matching the next most likely movement of the plant.

Summary:
This control algorithm solves the nonlinear constrained predictive control problem solving a quadratic programming problem using parameters such as the *free response* obtained by simulation over the nonlinear fuzzy model and the extraction of a local step response also derived from the same model by simulation.

6.3.3 Approach Using Takagi–Sugeno Fuzzy Models

This approach exploits the properties of Takagi–Sugeno fuzzy models with rules of the form

$$\text{If } x(k) \text{ is } \mathcal{A}_i \text{ AND } u(k) \text{ is } \mathcal{U}_i \text{ THEN } x(k+1) = E_i + A_i x(k) + B_i u(k)$$
$$y(k) = F_i + C_i x(k) + D_i u(k)$$

The evaluation of this model will generate at each step not only the output $y(k)$, but also a linear models of the form

$$\begin{aligned} x(k+1) &= E(k) + A(k)x(k) + B(k)u(k) \\ y(k) &= F(k) + C(k)x(k) + D(k)u(k) \end{aligned} \tag{6.48}$$

where

$$E(k) = \sum_{i=0}^{L} w_i E_i \quad A(k) = \sum_{i=0}^{L} w_i A_i \quad B(k) = \sum_{i=0}^{L} w_i B_i$$
$$F(k) = \sum_{i=0}^{L} w_i F_i \quad C(k) = \sum_{i=0}^{L} w_i C_i \quad D(k) = \sum_{i=0}^{L} w_i D_i$$

such that $\sum_{i=0}^{L} w_i = 1$.
L is the number of rules, the term w_i is the result of the inference process in the rule i and $E_i, A_i, B_i, F_i, C_i, D_i$ are the matrices of the dynamic system which are the consequences of the rule i.

If the membership functions are normal and the inference uses the product as the AND operation, the result of the inference is a convex combination of the linear models of the "active" rules. Under this assumption, the result of each inference will be considered a local linearization of the nonlinear model. The representation given by Equation (6.48) is the representation of a linear time-variant system.

With this representation at each moment, we can directly obtain from the model a local linearization that can be used to construct the QP. The parameters $A(k), B(k), C(k), D(k)$ are used to construct the matrices C^n, H, D^n at each sampling time and, with this, to construct the matrix Λ. The vector Γ corresponding to the *free response* is constructed by evolving the system (6.44) from $k+1$ to $k+N_p$ with input $u^* = 0$.

The control algorithm can be summarized in the following actions:

Algorithm

At each sampling time:

1. Read the current output of the system and update $y(k)$.
2. Make $u^* = 0$ and calculate the *free response* Γ [Equation (6.44)].
3. With the current states and inputs obtain from the fuzzy system the current $A(k), B(k), C(k), D(k)$ [Equation (6.49)].
4. With $A(k), B(k), C(k), D(k)$ compose the Λ matrix [Equation (6.34)].
5. With Λ and Γ construct the matrices for the cost [Equation (6.36)] and the constraints [Equation (6.37)] of the QP problem.
6. Solve the QP problem.
7. Apply only the first control action $u(k|k)$.

The contribution of this approach is the capacity to exploit the information given directly by the Takagi–Sugeno fuzzy model. This approach is very attractive for systems of high order because no simulation is needed to obtain the parameters to solve the optimization; the matrices can be generated directly from the inference of the fuzzy system. The use of this approach is very attractive to the industry for practical reasons related with the capacity of this model structure to combine local models identified in experiments around the different operating points.

Summary:
This control algorithm solves the nonlinear constrained predictive control problem solving a quadratic programming problem using parameters such as the *free response* obtained by simulation using the nonlinear fuzzy model and the extraction of a "local" linear model obtained from the inference process of a Takagi–Sugeno fuzzy model.

6.3.4 Approach Using Takagi–Sugeno Fuzzy Models and Multiple Models in the Predictor

This approach increases the nonlinear information included on the matrices operating in the QP. This increase of information is achieved by formulating the predictor using a time-variant model over the prediction horizon. In the previous formulation, the model was assumed to change only at each iteration of the controller. The model is fixed along the prediction horizon in the predictor. The current approach assumes a change in the model of the predictor not only at each iteration of the controller, but also at each time step in the prediction horizon.

The formulation of the predictor will be similar to the formulation presented in the previous section, but observe that the matrices used to build the predictor include the time-variant information:

$$X^n = \underbrace{\Psi + \Phi x(k)}_{\text{Free response}} + \underbrace{H\overline{U}^n}_{\text{Forced response}} \tag{6.49}$$

$$Y^n = M^n + C^n X^n + D^n \underline{U}^n \tag{6.50}$$

$$Y^n = \underbrace{M^n + C^n[\Psi + \Phi x(k)]}_{\text{Free response}} + \underbrace{C^n H\overline{U}^n + D^n\underline{U}^n}_{\text{Forced response}}$$

$$Y^n = \underbrace{M^n + C^n[\Psi + \Phi x(k)]}_{\Gamma} + \underbrace{[C^n H\, 0] + [0\, D^n]}_{\Lambda} U^n \tag{6.51}$$

where

$$X = \begin{bmatrix} x(k+1|k) \\ x(k+2|k) \\ \vdots \\ x(k+N_p|k) \end{bmatrix} \quad \Phi = \begin{bmatrix} A(k) \\ A(k+1)A(k) \\ \vdots \\ \prod_{j=0}^{N_p-1} A(k+N_p-1-j) \end{bmatrix}$$

$$M^{N_p} = \begin{bmatrix} M(k+1) \\ M(k+2) \\ \vdots \\ M(k+N_p) \end{bmatrix}$$

$$\Psi = \begin{bmatrix} L(k) \\ L(k+1) + A(k+1)L(k) \\ L(k+2) + A(k+2)L(k+1) + A(k+2)A(k+1)L(k) \\ \vdots \\ L(k+N_p-1) + A(k+N_p-1)L(k+N_p-2) + \dots \\ \dots + \prod_{j=0}^{N_p-2} A(k+N_p-1-j)L(k) \end{bmatrix}$$

$$H = \left[\begin{matrix} B(k) & 0 \\ A(k+1)B(k) & B(k+1) \\ A(k+2)A(k+1)B(k) & A(k+2)B(k+1) \\ \vdots & \vdots \\ \prod_{j=0}^{N_p-1} A(k+N_p-1-j)B(k) & \prod_{j=0}^{N_p-2} A(k+N_p-1-j)B(k+1) \end{matrix}\right.$$

$$\left.\begin{matrix} \dots & 0 \\ \dots & 0 \\ \dots & 0 \\ \ddots & \vdots \\ \dots & B(k+N_p-1) + \sum_{i=0}^{N_p-N_c-1} \prod_{j=0}^{i} A(k+N_p-1-j)B(k+N_p-i-2) \end{matrix}\right]$$

$$U^{N_u} = \begin{bmatrix} u(k|k) \\ u(k+1|k) \\ \vdots \\ u(k+N_c|k) \end{bmatrix} \quad \overline{U}^n = \begin{bmatrix} u(k|k) \\ u(k+1|k) \\ \vdots \\ u(k+N_c-1|k) \end{bmatrix} \quad \underline{U}^n = \begin{bmatrix} u(k+1|k) \\ u(k+2|k) \\ \vdots \\ u(k+N_c|k) \end{bmatrix}$$

$$C^n = \begin{bmatrix} C(k+1) & 0 & \dots & 0 \\ 0 & C(k+2) & \dots & 0 \\ \vdots & \vdots & \ddots & \vdots \\ 0 & 0 & \dots & C(k+N_p) \end{bmatrix}, \in R^{N_p n_o \times N_p n_s}$$

$$D^n = \begin{bmatrix} D(k+1) & 0 & \dots & 0 \\ 0 & D(k+1) & \dots & 0 \\ \vdots & \vdots & \ddots & \vdots \\ 0 & 0 & \dots & D(k+N_p-1) \\ 0 & 0 & \dots & D(k+N_p) \end{bmatrix}, \in R^{N_p n_o \times N_c n_i}$$

where n_s is the number of states, n_o the number of outputs and n_i the number of inputs.
These matrices are then replaced in the cost function (6.36) and in the constraints (6.37) and the QP is solved.
The algorithm work as follows:

Algorithm

1. Read the current output of the system and update $y(k)$.
2. Make $u^* = 0$ and calculate the *free response* Γ [Equation (6.44)].
3. With the current states and inputs obtain from the fuzzy system the current $A(k), B(k), C(k), D(k)$ [Equation (6.51)].
4. With $A(k), B(k), C(k), D(k)$ compose the Λ matrix [Equation (6.34)].
5. With Λ and Γ construct the matrices for the cost [Equation (6.36)] and the constraints [Equation (6.37)] of the QP problem.
6. Solve the QP problem.
7. Apply the generated input sequence to the fuzzy model and generate the set of $E(k), F(k), A(k), B(k), C(k), D(k)$ matrices over the prediction horizon.
8. Construct the Γ and the Λ matrices and using the cost [Equation (6.36)] and the constrains [Equation (6.37)] solve the QP problem.
9. If the time for calculation has not expired and the difference between two consecutive solutions is larger than a given value ϵ, repeat the iteration from step 7.
10. Apply only the first control action $u(k|k)$.

For the first iteration the constructed predictor assumes that the plant is linear time-invariant in the term corresponding to the forced response. From this iteration, a suboptimal solution is obtained. This solution is equivalent to the solution presented in Section 6.3.3.

With the solution from the first iteration, it is possible to construct a predictor by using time-variant linear models included in the term of the forced response. A new QP problem will be solved and a new input sequence will be generated. The process will be repeated until the sampling time is over or the difference among two consecutive input sequences is smaller than a given value ϵ.

This method can deliver an additional improvement on the quality of the solution since extra information over the nonlinearity is introduced in the optimization. The price paid for this improvement is computational time since each entry of the columns of the Λ matrix must be calculated since it is no longer a Toeplitz matrix.

Summary:
This control algorithm improves the nonlinear calculation by introducing extra information on the simplification such that a solution of the approximated QP method resembles more the actual solution of the nonlinear optimization problem.

6.3.5 Example Predictive Control of a Steam Generator Using a Fuzzy Model

This application is based on the simulation model presented by Pelegrinetti *et al.* [82]. There they obtain a model of the boiler at Abbott Power Plant in Champaign, Illinois. The model is a multivariable model, which includes four inputs (fuel, air, water flow, and steam demand) and four outputs (pressure, oxygen, and level in the drum, steam flow). The obtained model includes perturbation and measurement noises. This model is used as the real plant for the simulations. The model of the steam generator (see Figure 6.13) is described by the following system of equations:

$$\dot{x}_1(t) = c_{11}x_4(t)x_1^{9/8}(t) + c_{12}u_1(t-\tau_1) - c_{13}u_3(t-\tau_3) + c_{14} \tag{6.52}$$

$$\dot{x}_2(t) = -c_{21}x_2 + \frac{c_{22}u_2(t-\tau_2) - c_{23}u_1(t-\tau_1 - c_{24}u_1(t-\tau_1 x_2(t)}{c_{25}u_2(t-\tau_2 + c_{26}u_1(t-\tau_1} \tag{6.53}$$

$$\dot{x}_3(t) = c_{31}x_1(t) - c_{32}x_4(t)x_1(t) + c_{33}u_3(t-\tau_3) \tag{6.54}$$

$$\dot{x}_4(t) = -c_{41}x_4(t) + c_{42}u_1(t-\tau_1) + c_{43} + u_4(t) + n_5 \tag{6.55}$$

$$y_1(t) = c_{51}x_1(t-\tau_4) + n_1(t) \tag{6.56}$$

$$y_2(t) = c_{61}x_2(t-\tau_5) + n_2(t) \tag{6.57}$$

$$\begin{aligned} y_3(t) &= c_{70}x_1(t-\tau_6) + c_{71}x_3(t-\tau_6) + c_{72}x_4(t-\tau_6) + c_{73}u_3(t-\tau_3-\tau_6) \\ &\quad + c_{74}u_1(t-\tau_1-\tau_6) + \frac{[c_{75}x_1(t-\tau_6) + c_{76}][1 - c_{77}x_3(t-\tau_6)]}{x_3(t-\tau_6)[x_1(t-\tau_6) + c_{78}]} \\ &\quad + c_{79} + n_3(t) \end{aligned} \tag{6.58}$$

$$y_4(t) = [c_{81}x_4(t-\tau_7) + c_{82}]x_1(t-\tau_7) + n_4(t) \tag{6.59}$$

where x_1 is the drum pressure state (kgf/cm^2); y_1 is the measured drum pressure (PSI); y_2 and x_2 are the measured excess oxygen level and its state,

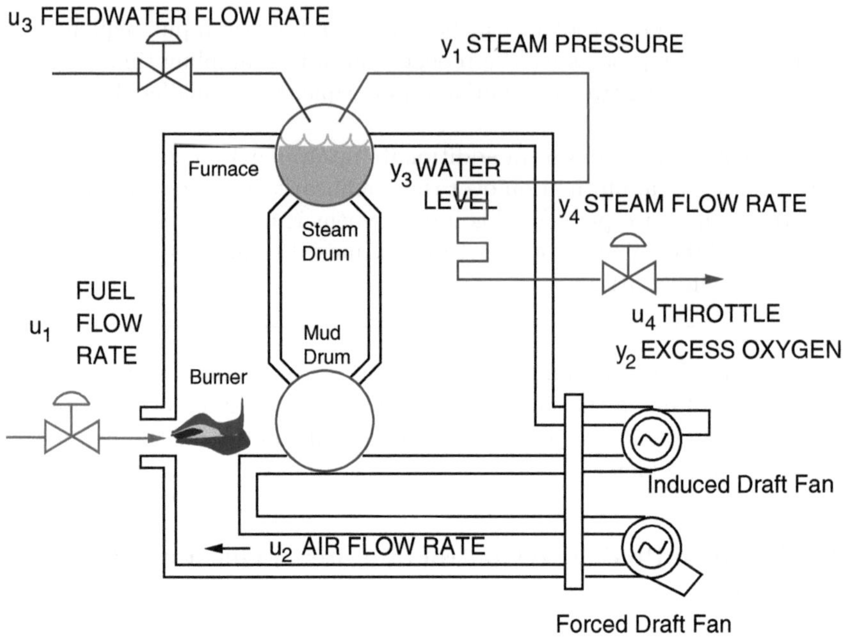

Figure 6.13. Steam generator

respectively (percent); x_3 is the system fluid's density (kg/m^3); y_3 is the drum water level (in.); y_4 is the steam flow rate (kg/s); u_1, u_2, u_3 are, respectively, the fuel, air, and feed water flow rate inputs, which take values between 0 and 1; x_4 is the exogenous variable related to the steam demanded. The constants are shown in Table 6.3. The variables n_i are colored noise sequences generated by first-order models driven by zero mean, unit variance white noise.

$$n_1 = \frac{0.75s + 0.1}{s + 0.001} w_1 \quad n_2 = \frac{0.019s + 0.001}{s + 0.024} w_2 \quad n_2 = \frac{0.105s + 0.038}{s + 0.010} w_3$$

$$n_4 = \frac{0.01s + 0.0001}{s + 0.001} w_4 \quad n_5 = \frac{0.003s + 0.003}{s + 0.0075} w_5$$

where n_i, $i = 1, \ldots, 5$, are colored noise and w_i is the unit variance white noise. A linear model was obtained by linearizing around the operating point:

$$\begin{aligned} x^0 &= [22.5\,2.5\,621.17\,0.6941]^T \\ y^0 &= [320\,2.5\,0.0\,9.3053]^T \\ u^0 &= [0.32270\,0.39503\,0.37404\,0]^T \end{aligned}$$

The linearized model is described by

$$\begin{aligned} \dot{x} &= A_p x + B_p u \\ y &= C_p x + D_p u \end{aligned}$$

Table 6.3. Coefficients of the Nonlinear Equation of the Power Plant Model

$c_{11} = -0.00478$	$c_{31} = 0.00533176$	$c_{70} = -0.1048569$
$c_{12} = 0.280$	$c_{32} = 0.0251950$	$c_{71} = 0.15479$
$c_{13} = 0.01348$	$c_{33} = 0.7317058$	$c_{72} = 0.4954961$
$c_{14} = 0.02493$	$c_{41} = 0.04$	$c_{73} = -0.20797$
$c_{21} = 0.1540357$	$c_{42} = 0.0299886$	$c_{74} = 1.2720$
$c_{22} = 103.5462$	$c_{43} = 0.018088$	$c_{75} = -324212.7805$
$c_{23} = 107.4835$	$c_{51} = 14.214$	$c_{76} = -99556.24778$
$c_{24} = 1.95150$	$c_{61} = 1.00$	$c_{77} = 0.0011850$
$c_{25} = 29.04$	$c_{81} = 0.85663$	$c_{78} = -1704.50476$
$c_{26} = 1.824$	$c_{82} = -0.18128$	$c_{79} = -103.7351$
$\tau_1 = 2, \tau_2 = 2, \tau_3 = 3, \tau_4 = 3, \tau_5 = 4, \tau_6 = 10, \tau_7 = 2.$		

where

$$A_p = \begin{bmatrix} -0.005509 & 0 & 0 & -0.1588 \\ 0 & -0.2062 & 0 & 0 \\ -0.01216 & 0 & 0 & -0.5672 \\ 0 & 0 & 0 & -0.040 \end{bmatrix}$$

$$B_p = \begin{bmatrix} 0.2800 & 0 & -0.01348 & 0 \\ -9.375 & 7.658 & 0 & 0 \\ 0 & 0 & 0.7317 & 0 \\ 0.02999 & 0 & 0 & 0.040 \end{bmatrix}$$

$$C_p = \begin{bmatrix} 14.21 & 0 & 0 & 0 \\ 0 & 1.0 & 0 & 0 \\ 0.3221 & 0 & 0.1434 & 11.16 \\ 0.4133 & 0 & 0 & 19.28 \end{bmatrix}$$

$$D_p = \begin{bmatrix} 0 & 0 & 0 & 0 \\ 0 & 0 & 0 & 0 \\ 1.272 & 0 & -0.2080 & 0 \\ 0 & 0 & 0 & 0 \end{bmatrix}$$

Fuzzy Identification of the Steam Generator

To perform the identification of the model, the level control is first stabilized by means of a feed-forward control plus a PID control. The feed-forward control signal is proportional to the steam flow to compensate the mass of water in the drum; a PID controller performs further control. The representation of the system with this stabilizing control actions can be observed in the Figure 6.14. With this modification the system will have the following inputs: fuel rate (u_1), air rate (u_2), reference level (r_{Level}), steam demand

(u_4). A sampling time of $T_s = 3$ sec was chosen and several experiments were performed. A total of five multiple-input–single-output fuzzy models was extracted. The details of the extracted models are presented in Table 6.4. The data points used for this identification are available at [83]. The inputs for

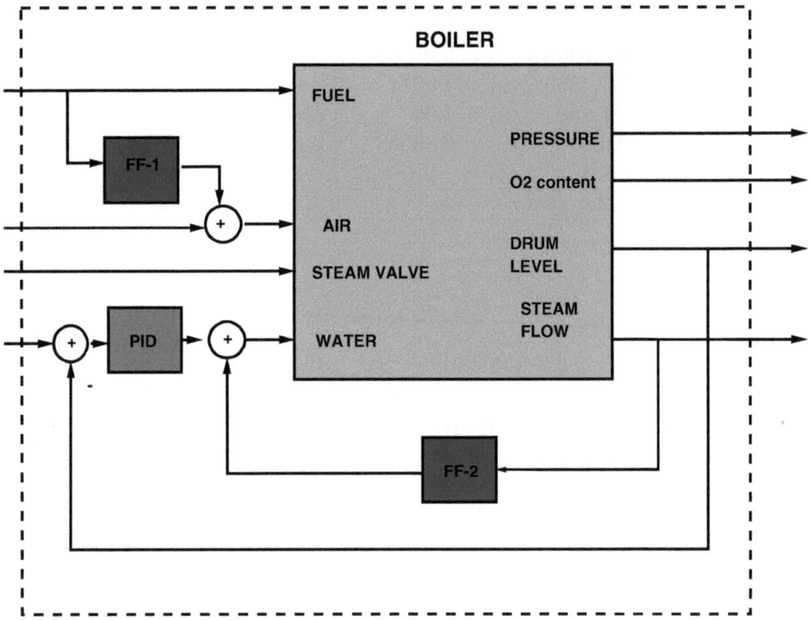

Figure 6.14. Stabilization scheme for the steam generator

Table 6.4. *Steam Generator. Characteristics of the Identified Fuzzy Model*

Model	No. rules
$\hat{y_1}(k) = f_1(\hat{y_1}(k-2), \hat{x_4}, u_1(k-2), u_1(k-3))$	81
$\hat{y_2}(k) = f_2(\hat{y_2}(k-2), u_1(k-2), u_2(k-2))$	27
$\hat{y_3}(k) = f_3(r_{Level}(k-8), \hat{y_4}(k-5), \hat{y_4}(k-6), u_1(k-4))$	81
$\hat{y_4}(k) = f_4(\hat{y_1}(k-1), \hat{x_4}(k-1))$	9
$\hat{x_4}(k) = f_5(\hat{x_4}(k-1), u_1(k-1), u_4(k-1))$	27
Total rules	225

each model were selected by using the RC index. The five models obtained can be described as

$$\hat{y_1}(k) = f_1(\hat{y_1}(k-2), \hat{x_4}, u_1(k-2), u_1(k-3)) \tag{6.60}$$

$$\hat{y_2}(k) = f_2(\hat{y_2}(k-2), u_1(k-2), u_2(k-2)) \tag{6.61}$$

$$\hat{y_3}(k) = f_3(r_{Level}(k-8), \hat{y_4}(k-5), \hat{y_4}(k-6), u_1(k-4)) \quad (6.62)$$

$$\hat{y_4}(k) = f_4(\hat{y_1}(k-1), \hat{x_4}(k-1)) \quad (6.63)$$

$$\hat{x_4}(k) = f_4(\hat{x_4}(k-1), u_1(k-1), u_4(k-1)) \quad (6.64)$$

where y_i represent the outputs (Pressure, Oxygen level, Drum level and Steam Flow, respectively) and x_4 is an internal state variable. Figure 6.15 shows the validation results of the model. Another model (a Takagi–Sugeno fuzzy

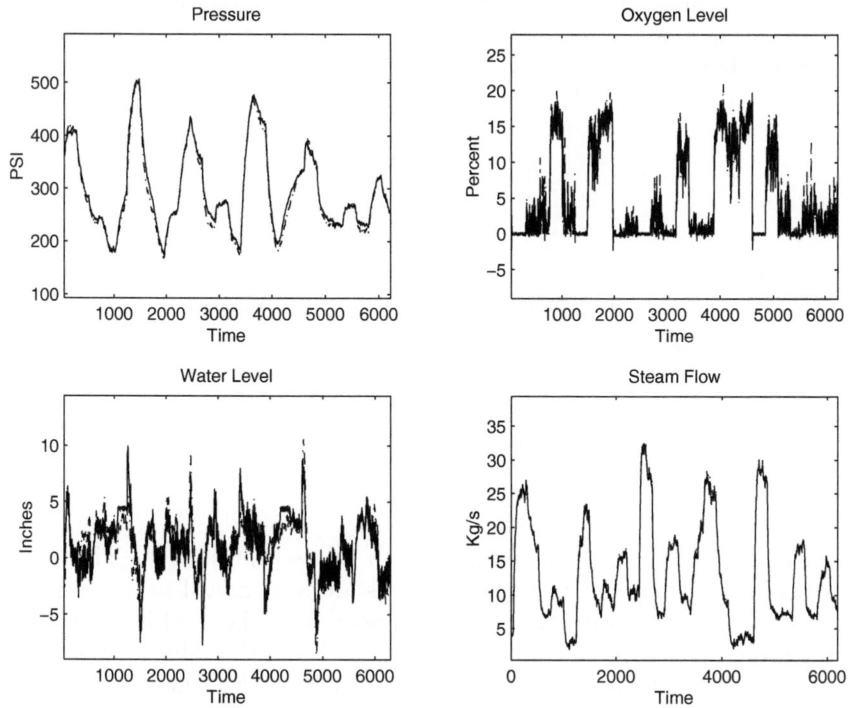

Figure 6.15. Identification of the steam generator. Validation experiment: (-) Plant data, (–) Fuzzy model data

model) was generated with only five rules, equally distributed over the domain of the Steam flow. This second model only describes the dynamics around the nominal operating points of the Pressure, Level and Excess oxygen and uses five local models for different values of steam flow. The rules of this system look as follows:

$$\text{IF } \textbf{Steam Flow } (y_4(k)) \text{ is } \mathfrak{F}_i \text{ THEN } \begin{array}{l} X(k+1) = E_i + A_i X(k) + B_i U(k) \\ Y(k+1) = F_i + C_i X(k) + D_i U(k) \end{array}$$

Implementation of the Controller and Performance Results

Four predictive controllers were implemented. The first one corresponds to a linear MPC constructed using the linearized model presented in the previous section (*MPC with linear model*); the second MPC corresponds to an MPC controller using the estimated step response method (*MPC with Fuzzy Model*; see Section 6.3.2); the third one corresponds to the strategy using Takagi–Sugeno fuzzy models without time-variant information (*MPC with TS Fuzzy Model* see Section 6.3.3) and the fourth controller corresponds to the strategy using Takagi–Sugeno models with time-variant information (see Section 6.3.4).

The following parameters were used in the construction of the controllers: prediction horizon $N = 5$, control horizon $Nc = 2$, $C_j(z^{-1}) = 1$ and $D_j(z^{-1}) = 1 - z^{-1}\ \forall j = 1, \ldots, 4$, $\alpha = 0.9$ the Q matrix

$$Q = \begin{bmatrix} q_1 & 0 & 0 & 0 \\ 0 & q_2 & 0 & 0 \\ 0 & 0 & q_3 & 0 \\ 0 & 0 & 0 & q_4 \end{bmatrix}$$

where q_i are $\Re^{N \times N}$ diagonal matrices with values [0.0014, 0.0001, 0.0002, 0.0503] and the R matrix is

$$R = \begin{bmatrix} r_1 & 0 & 0 & 0 \\ 0 & r_2 & 0 & 0 \\ 0 & 0 & r_3 & 0 \\ 0 & 0 & 0 & r_4 \end{bmatrix}$$

where r_i are $\Re^{Nc \times Nc}$ diagonal matrices with values [50 0.1 1 10^4]. The inputs corresponding to the fuel and the air valves are constrained to be in the interval [0 1] and with a $du_{\max} = 0.1$; the reference of the level is constrained to be in the interval [-4 4] and with a $du_{\max} = 1$ and the valve for the steam is constrained in the interval [-0.15 0.03] (the reason that this valve was not constrained in the interval [0 1] is that the valve was originally modeled as a disturbance variable) and with a $du_{\max} = 1$. The solution with the linear MPC is not subject to constraints.

The performance of the controllers is evaluated by using an experiment where the energy generated by the plant is first reduced from the 39% of the maximum power to the 27%; then it is raised to the 100% and finally reduced to the 27%. This change should be done while *Pressure*, *Oxygen level* and *Water level* are kept constant with values 320 PSI, 2.5% and 0 inches, respectively, and the *Steam flow* is changed from 9.3 Kg/s to 6.6 Kg/s, then it is raised to 22.096 Kg/s and finally reduced again to 6.6 Kg/s. Figure 6.16 and Table 6.5 show the comparative results of the experiments.

The comparison shows some interesting results. It is clear that the proposed control strategies outperforms the classical linear MPC solution. Their capacity to reject disturbances can be clearly observed when the system is

Table 6.5. Comparative Results- Performance Calculated for the Steam Flow

	MPC with fuzzy model Strategy 1	MPC with TS fuzzy Model Strategy 2	MPC with TS fuzzy Model Strategy 3	MPC with linear model
Variance on 27% of the capacity	0.0011	0.0055	0.0055	0.0119
Variance on 100% of the capacity	4.42×10^{-4}	0.00430	0.00427	0.0096
Overshoot on transition from 27%–100%	1.06%	1.83%	1.83%	1.32%
Overshoot on transition from 100%–27%	0.25%	1%	1%	2.58%
Settling time on transition from 27%–100%	103 sec	150 sec	150 sec	423 sec
Settling time on transition from 100%–27%	120 sec	190 sec	190 sec	558 sec

operated at 100% of the capacity. The linear MPC is not capable of keeping the level at the desired value. In addition, it is important to observe that the settling time has been reduced to one quarter of the settling time for the system with the linear MPC without any increase in overshoot. The reduction of the variance once more compared with the linear MPC strategy is around one order of magnitude for the MPC strategy with estimated step response and half order of magnitude for the strategy based on the Takagi–Sugeno model. The MPC based on the Takagi–Sugeno model has the best disturbance rejection capacity. It does make a better task controlling the drum pressure. The settling time of the MPC strategy with estimated step response is the shortest.

6.3.6 Example: Nonlinear Predictive Control of a Gas-Phase High-Density Polyethylene (HDPE) Reactor

This example[1] shows the control of a gas-phase reactor used in the production of polyethylene. The example is based on a model built in gPROMS by the company IPCOS N.V. in Belgium, and the model is based on the models proposed by Choi and Ray [85] and the Ph.D. thesis of McAuley [86]. The process is shown in Figure 6.17. In this process the monomer ethylene reacts with the

[1] Part of these results were obtained in the framework of the a master thesis in collaboration with IPCOS N.V. and the Katholieke Universiteit Leuven, Belgium [84].

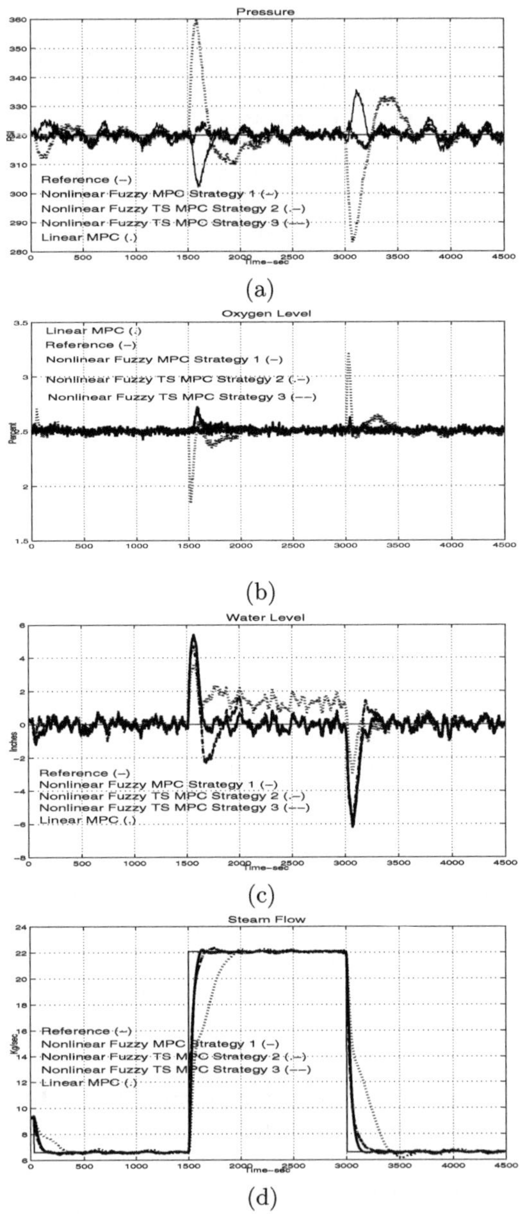

Figure 6.16. Comparative results of the control of the steam-generating unit: -Strategy 1 (-) MPC with fuzzy model using step response, -Strategy 2 (.-) MPC using Takagi–Sugeno fuzzy model, -Strategy 3 (- -) MPC using Takagi–Sugeno fuzzy model, (.)MPC with linear model-(a) Pressure (b) Oxygen level (c) Water level (d) Steam flow

co-monomer butylene to produce the HDPE. The ratios butylene/ethylene (CH_4/CH_2) (**MV1**) and hydrogen/ethylene (H_2/CH_2)(**MV2**) are very important to guarantee the quality of the HDPE. The quality of the HDPE is defined by two parameters, melt index (**CV1**) and density (**CV2**). The unreacted ethylene is recycled. Nitrogen (**MV3**) is used as transportation medium and cooling agent. It does not take part in the reaction. The system has three

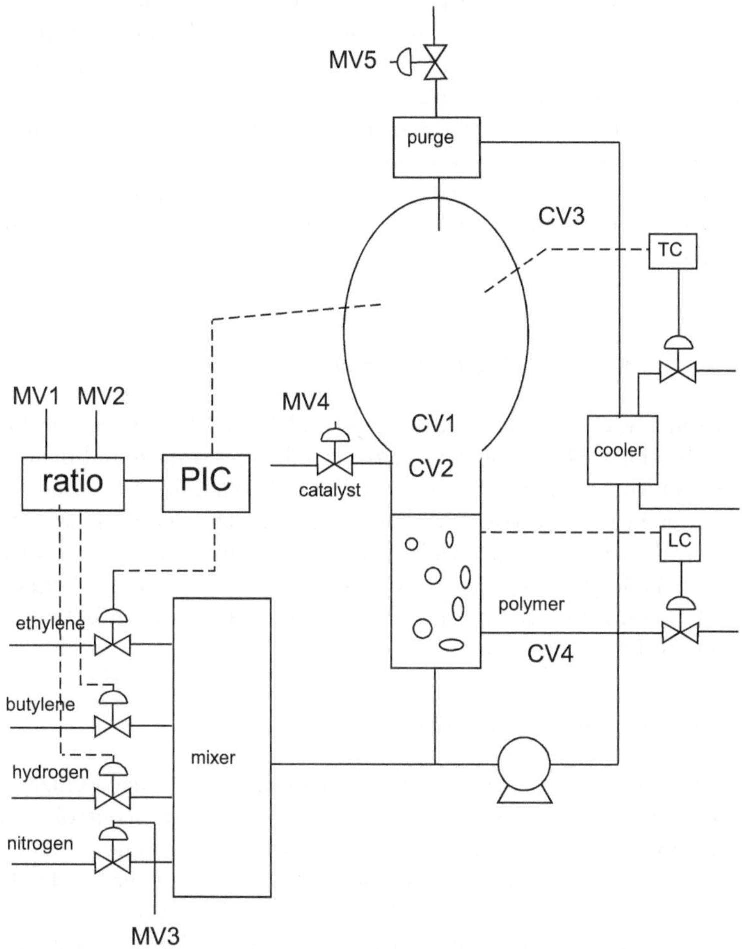

Figure 6.17. High-density polyethylene gas-phase reactor

PID controllers to guarantee the safety of the process: temperature control for the reactor, level control for the liquid phase inside the reactor and a pressure control. Additionally, two ratio controls for CH_4/CH_2 and H_2/CH_2 are used in the operation.

The predictive controller will manipulate four variables: butylene/ethylene ratio (CH_4/CH_2) (**MV1**) and hydrogen/ethylene ratio (H_2/CH_2)(**MV2**), nitrogen (**MV3**) and catalyst (**MV3**) in order to obtain some desired properties in the HDPE melt index (**CV1**) and density (**CV2**) while maintaining the temperature (**CV3**) and the production (**CV4**) at acceptable levels.

This system should operate over a wide range since a flexible production is expected. By flexible we mean multiple density and melt index are expected. Nowadays the automatic transitions between different qualities are considered a challenge since serious savings can be achieved by reducing the time used to perform a transition. Traditionally these transitions are performed manually or automatically but using very conservative and far from optimal operations. The use of a nonlinear controller is motivated in this case by the multiple dynamics encountered along the so-called production slate, which defines the set of qualities that can be produced. In this case the set of products mentioned in the slate is 104, represented by 8 density and 13 melt indexes and their respective combinations.

Fuzzy Model for the HDPE Reactor

The controller was implemented using a model with 104 rules, one for each grade on the slate. The model was constructed using trapezoidal membership functions, which are shown in Figure 6.18, and dynamic state-space models of 45th order in the consequences. However large this model is a significant reduction compared with the original rigorous model represented (with more than 800 variables with several hundreds of algebraic differential equations). Figure 6.19 shows the validation of the model for a transition between grade K2 and grade L4.

Controller Design and Results

A predictive controller was constructed using the Takagi–Sugeno fuzzy model described in the previous section. The parameters of the controller were a prediction horizon $N_p = 5$ and a control horizon $N_u = 2$. The controller was built with a weight equal for all inputs and 1000 times smaller for the density. All the inputs were constrained by their respective maximum amounts.

Figure 6.20 shows the results given by the controller compared with an offline precalculated optimal trajectory obtained over the rigorous model. The objective of the precalculated optimal trajectory was to maximize the added value [87]. The objective of the fuzzy predictive controller was to achieve the change as fast as possible. Observe that the results generated by the predictive controller are quite close to the optimal trajectory.

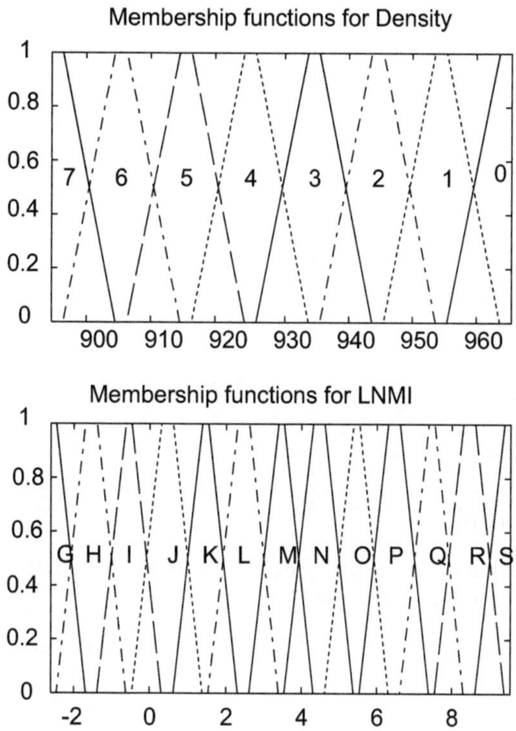

Figure 6.18. Membership functions for the Takagi–Sugeno fuzzy model of the HDPE reactor

6.4 Conclusions

This chapter has presented a novel approach to fuzzy model-based control. The combination of fuzzy modeling and predictive control is the result of the maturation of the modeling and identification techniques using fuzzy structures. The goal was to design a simple control strategy. –Simple for the designer and simple for the end user of the control systems.

The simplicity for the designer of the control system is reflected on the fact that the only task during the tuning process is the definition of the goal of the control systems in terms of a quadratic cost function and the constraints of the elements involved in the system. Other tuning parameters such as control and prediction horizons are closely related with the settling times and the model order. These parameters exhibit a monotonic tendency making the tuning of the controller a simple task. Of course, a price must be paid to achieve this simplicity and this cost can be divided in two: (1) the effort that must be put to build the model; and (2) the computational cost of calcu-

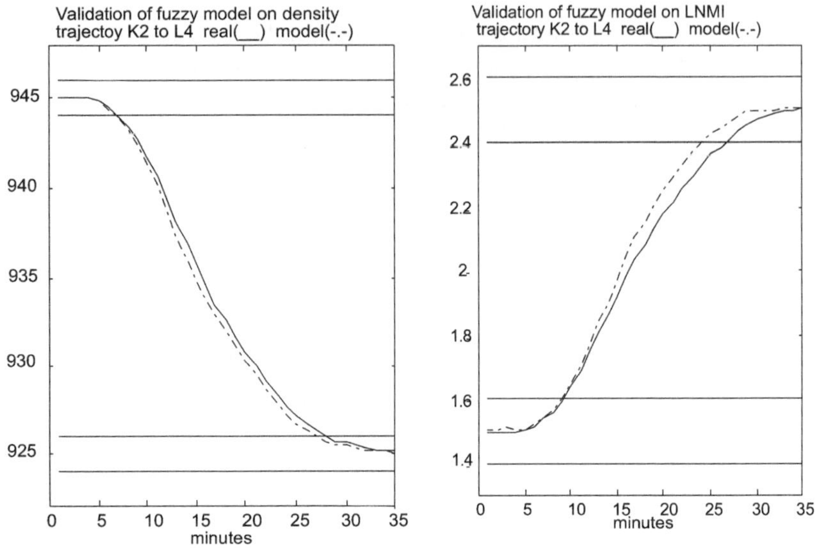

Figure 6.19. Validation of the Takagi–Sugeno model for the HDPE reactor on the transition between grade K2 and grade L4

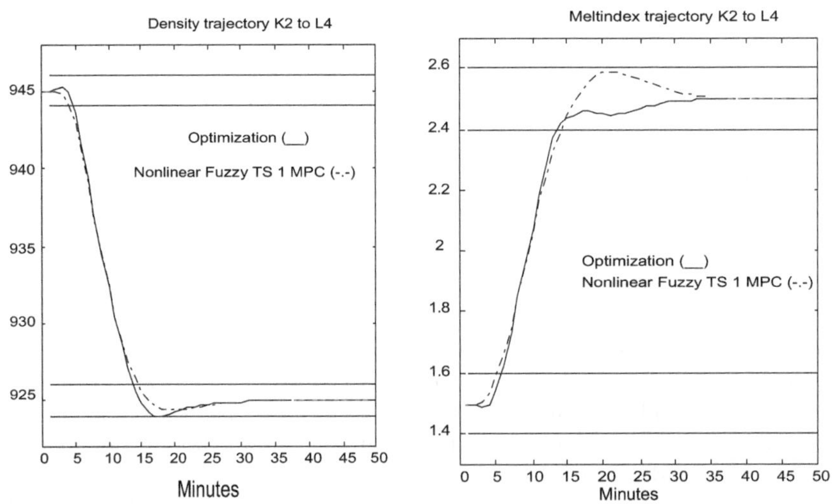

Figure 6.20. Result of the Takagi–Sugeno model-based predictive control for the HDPE reactor on the transition between grade K2 and grade L4

lating the control actions at each sampling time. The fact that the described algorithms are computationally intensive limits the sampling time with the current computational technology (Pentium IV 2.8 GHz) to 7×10^{-3} seconds for unconstrained system and around 0.2 seconds for constrained multivariable systems of low order. This will limit the applications where these strategies can be applied today to applications to process control problems where sampling time is about 1 sec. The unconstrained strategy can be applied even to some mechanic systems with already fast dynamics (sampling times of the order of 150 Hz).

This chapter includes four control algorithms with their respective derivations. Two of the proposed algorithms are particularly interesting for their applicability to other types of nonlinear models and not only to fuzzy systems. These algorithms are the strategies based on the estimated step response.

It is also important to note the advantages of the other two contributed strategies. The two strategies based on the use of Takagi–Sugeno fuzzy models exploit the structural and the simulation information provided by the model. Control engineers are particularly attracted by the fact that the model "looks alike" gain scheduling models.

The key point of the strategies presented in the current chapter is the reduction of the original complex nonlinear program (with no possibility for online implementation) to a simple quadratic program (online implementable) by exploiting the structure of the problem and the structure of the models and by relaxing the constraints imposed by the problem. The relaxation introduced by the methods presented in this chapter is a relaxation of the equality constraints, which are in fact the plant description. In some cases the solutions can be quite close to the "global optimal solution," especially for systems with monotonic nonlinearities. The assumptions used to reduce the problem demand some smoothness in the nonlinearities, which are, of course, related with the prediction capability of the model when "strong" nonlinearities are involved.

The three examples presented are realistic simulations of industrial processes. The improvement in the performance is clear from the simulations. The systems used in the examples are characterized by strong changes in gain and dynamics at the different operating points.

7

Robust Nonlinear Predictive Control Using Fuzzy Models

7.1 Introduction

The increasing popularity of the fuzzy models for nonlinear system identification and modeling can be explained by the possibility to extract "local" information about the dynamics of the system. Different algorithms for identification of Takagi–Sugeno fuzzy models have been proposed in recent years [71] [57]. The representation of locally linearized models in the form of a Takagi–Sugeno fuzzy model has shown its advantages in simplifying the design of nonlinear controllers.

The previous chapter showed the advantages of the use of model-based predictive control techniques in the process control industry as a technique to design multivariable controllers with direct performance specifications.

The complexity of the solution of predictive controllers for nonlinear systems was discussed in the previous chapter. Relaxations to the problem and its reduction to a quadratic program were the strategies applied to obtain a solution. This solution is very close to the real "optimal" solution (see results of the example in Section 6.2.2), but it demands a very accurate description of the nonlinear plant to guarantee such a performance.

An accurate description of the plant is not always an achievable goal. The limitations are (1) the impossibility of performing extensive experiments on the plant and (2) the fact that the aging process of the plant generates a mismatch with respect to the model.

For stability and performance reasons, it is very important to take into account this mismatch during the optimization process. The solution obtained will guarantee a minimum performance that will be reflected in the quality of the product.

This chapter presents an algorithm for nonlinear predictive control using the description of the nonlinear plant provided by Takagi–Sugeno fuzzy models. The method solves the problem by formulating a robust quadratic program. The solution of the robust quadratic program is obtained by transforming the program into a second–order cone program [88]. It is important to remark that the solution presented in [88] only takes into account uncertainties in the Hessian matrix. In this chapter, the uncertainties are extended to the linear term of the cost function and the problem is once more reformulated as a second-order cone program. Perhaps the most important benefit of the method is the fact that constraints are always respected by the solution despite the uncertainties of the model. In this way robust performance can be guaranteed. The chapter is organized as follows: Section 7.2 presents the formulation of a robust quadratic program as a second-order cone program including the contribution where uncertainties in the linear term are included; Section 7.3 describes the problem analytically; Section 7.4 shows the nominal solution when the problem has no uncertainties (it is the same solution presented in the previous chapter but included here for completeness); Section 7.5 presents an approximated solution to the problem of robust predictive control using robust quadratic programming; Section 7.7 shows some possible ways to describe the uncertainty on the fuzzy model and Section 7.8 concludes the chapter.

Summary:
This chapter introduces a method to construct robust nonlinear predictive controllers based on a robust quadratic programming method.

7.2 Robust Quadratic Programming

Robust programming has been a subject of study in recent years [89] [90] [91]. Boyd *et al.* [92] have shown the practical applications of these techniques in control and filter design. Other applications include robust antenna array design and truss topologies. Robust programming is a class of optimization problems where the parameters such as coefficients of cost functions and/or constraints are uncertain and prescribed into a defined set.

Efficient interior point methods have been created for some of these problems. A very interesting property of these methods is that their computational complexity is polynomial with respect to the number of constraints. For instance, the computational complexity for the second-order cone program solution is a value proportional to $\sqrt{l}$, where l is the number of constraints. This fact makes these algorithms very promising for large-scale optimization. This property opens wide possibilities in the application of these methods not only for robust optimization but also for plant-wide optimizers. Some authors have already explored the applications of interior point methods to solve the nominal problem of linear predictive control [93].

This section will focus its attention into the problem of robust quadratic programming (RQP). The problem is formulated as follows:

$$\min_{x} \max_{P \in \mathcal{E}, q \in \mathcal{F}} x^T P x + 2 q^T x + r \tag{7.1}$$

The objective is to minimize the cost function (7.1), where $x \in \Re^N$ is the optimization variable, $P \in \Re^{N\times N}$ is a symmetric positive definite matrix also known as the Hessian matrix, the vector $q \in \Re^N$ is also known as the linear term of the cost function and the scalar value r. Observe that in this case the Hessian matrix P and the linear term q are considered uncertain and their uncertainties are described by the sets $\mathcal{E}$ and $\mathcal{F}$. This set $\mathcal{E}$ describes the uncertainty in the following form for the P matrix:

$$\mathcal{E} = \left\{ P_0 + \sum_{i=1}^{m} P_i u_i \mid ||u|| \leq 1 \right\}$$

where P_0 and P_i are matrices in $\Re^{N\times N}$, u is a vector in $\Re^m$ and $||.||$ is the Euclidean norm. The set $\mathcal{F}$ describes the uncertainty of the linear term q as follows:

$$\mathcal{F} = \left\{ q_0 + \sum_{i=1}^{n} q_i v_i \mid ||v|| \leq 1 \right\}$$

where q_0 and q_i are vectors in $\Re^N$, v is a vector in $\Re^n$ and $||.||$ is the Euclidean norm.

Lobo *et al.* [88] show that a robust quadratic program can be written as a second-order cone program (SOCP). However, in that formulation, the uncertainty was restricted to the Hessian matrix. This text presents an extension to the formulation presented in [88] by including the uncertainty in the linear term q, described by the set $\mathcal{F}$. The next lines will describe the basics of a second-order cone program.

A second-order cone program is defined as follows:

$$\begin{aligned} &\min_{x} \ f^T x \\ &\text{subject to} \ ||A_i x + b_i|| \leq c_i^T x + d_i, \ \ i = 1, \ldots, N \end{aligned} \tag{7.2}$$

where $x \in \Re^n$ is the optimization variable, $||.||$ is the Euclidean norm and the problem parameters are $f \in \Re^n, A_i \in \Re^{(n_i-1)\times n}, b_i \in \Re^{n_i-1}, c_i \in \Re^n$ and $d_i \in \Re$.

As mentioned previously, there are very efficient methods to solve this optimization problem with complexity proportional to $\sqrt{l}$, where l is the number of constraints.

To transform the robust quadratic program into a second-order cone program, the description of the uncertainty sets $\mathcal{E}$ and $\mathcal{F}$ is introduced in the cost function (7.1),

$$\min \max_{\substack{||u||\leq 1 \\ ||v||\leq 1}} x^T P_0 x + \sum_{i=1}^{m} x^T P_i x u_i + 2q_0^T x + 2\sum_{j=1}^{n} q_j^T x v_j + r \tag{7.3}$$

applying the triangle inequality:

$$\min[x^T P_0 x + 2q_0^T x + \max_{||u||\leq 1} ||x^T P_i x|| ||u|| + 2\max_{||v||\leq 1} ||q_j^T x|| ||v|| + r] \\ \leq \min[x^T P_0 x + 2q_0^T x + ||x^T P_i x|| + 2||q_j^T x|| + r] \tag{7.4}$$

Using this representation, the problem can be converted into a problem with linear cost with quadratic constraints,

$$\min f + 2q_0^T x + t + 2d + r \tag{7.5}$$

subject to

$$\begin{aligned} x^T P_0 x &\leq f \\ x^T P_i x &\leq w_i \\ ||q_j^T x|| &\leq d \\ ||w|| &\leq t \end{aligned}$$

This optimization can be formulated as an SOCP as follows:

$$\min f + 2q_0^T x + t + 2d + r \tag{7.6}$$

subject to

$$\begin{aligned} \left\| \begin{bmatrix} 2P_0^{\frac{1}{2}} x \\ f - 1 \end{bmatrix} \right\| &\leq f + 1 \\ \left\| \begin{bmatrix} 2P_i^{\frac{1}{2}} x \\ w_i - 1 \end{bmatrix} \right\| &\leq w_i + 1 \\ ||q_j^T x|| &\leq d \\ 0 &\leq f \\ 0 &\leq w_i \\ ||w|| &\leq t \end{aligned} \tag{7.7}$$

Observe that with this new formulation the search space is extended from the space of x to the spaces of f, w_i, t and d.

Summary:
The problem of robust quadratic programming can be reduced to a second-order cone program, which can be solved very efficiently using interior point optimization algorithms.

7.3 Problem Description

The formulation of the robust nonlinear predictive control problem is expressed in the following lines as

$$\min_{u(k),\ldots,u(k+Nc)} \max_{\delta\in\Delta} J(Y_{\text{ref}}, y, u, \delta) \tag{7.8}$$

subject to

$$\begin{aligned} x(k+1) &= f(x(k), u(k), \delta) \\ y(k) &= g(x(k), u(k), \delta) \\ &u_{\min} \le u \le u_{\max} \\ &|u(k) - u(k-1)| \le \Delta u_{\max} \end{aligned}$$

where $J(...)$ is a cost function (typically quadratic) which penalizes the deviation of the output of the plant y with respect to the reference signal Y_{ref}, in a prescribed period (N_p samples-prediction horizon). The minimization searches for a sequence of inputs $(u(k+1), u(k+2), \ldots, u(k+N_c))$ subject to constraints such as the plant dynamics described by $f(...)$ and $g(...)$ and input constraints related with saturation and slew rate. The parameter $||\delta|| \le \rho$ represents a bounded uncertainty in the plant dynamics. For the present case the dynamics of the plant $[f(...)$ and $g(...)]$ will be represented using dynamic fuzzy models, in a state-space form or in an input–output form.

This optimization problem is a very complex problem of robust nonlinear programming (RNLP), and with the actual computational resources it is impossible to guarantee that a solution is found in a prescribed number of steps.

Summary:
The problem of robust nonlinear predictive control demands the solution on real time of a problem of robust nonlinear programming (RNLP). Such a task is not feasible with the current state-of-the-art optimization techniques.

7.4 Nominal Solution

The nominal problem (without uncertainty, $||\delta|| = 0$) is simpler than the robust one, but still it is a nonlinear program (NLP). In the previous chapter, some approximated solutions were found by reducing the problem to a quadratic program (QP). . Other approaches for general classes of nonlinear systems have also been proposed in the literature [65] The approaches presented in the previous chapter explore the use of three different pseudo-linearization methods to convert the NLP into a QP.

The nominal problem assumes no uncertainty:

$$\min_{u(k),\ldots,u(k+Nc)} J(Y_{\text{ref}}, y, u) \tag{7.9}$$

subject to

$$\begin{aligned} x(k+1) &= f(x(k), u(k)) \\ y(k) &= g(x(k), u(k)) \\ &\quad u_{\min} \le u(k) \le u_{\max} \\ &\quad |u(k) - u(k-1)| \le \Delta u_{\max} \end{aligned}$$

using a pseudo-linearization[1] given by

$$Y = \underbrace{\Gamma}_{\text{Free response}} + \underbrace{\Lambda u}_{\text{Forced response}} \tag{7.10}$$

and a quadratic cost function given by

$$\begin{aligned} J(Y, U^n) = (Y_{\text{ref}} - Y)^T Q (Y_{\text{ref}} - Y) + \\ + U^{nT} R U^n + (\Delta U^n - \bar{U}_{k-1})^T S (\Delta U^n - \bar{U}_{k-1}) \end{aligned} \tag{7.11}$$

The problem can be written as the following QP:

$$\begin{aligned} J(U^n) = J_{\min} + \\ + 2[(\Gamma - Y_{\text{ref}})^T Q \Lambda - \bar{U}_{k-1}^T S \Delta] U^n \\ + U^{nT} [\Lambda^T Q \Lambda + R + \Delta^T S \Delta] U^n \end{aligned} \tag{7.12}$$

where

$$J_{\min} = Y_{\text{ref}}^t Q Y_{\text{ref}} + \Gamma^T Q \Gamma - 2 Y_{\text{ref}}^T Q \Gamma + \bar{U}_{k-1}^T S \bar{U}_{k-1}$$

is the minimum cost that cannot be modified by any control input. The constraints will be written as follows:

$$\begin{bmatrix} I^{(n_i N_c)} \\ -I^{(n_i N_c)} \\ I^{(n_i)} 0^{(n_i \times n_i N_c)} \\ -I^{(n_i)} 0^{(n_i \times n_i N_c)} \\ \Delta \\ -\Delta \end{bmatrix} U^{N_c} \le \begin{bmatrix} U_{\max}^{N_c} \\ -U_{\min}^{N_c} \\ \Delta U_{\max}^{N_c} + \bar{U}_{k-1} \\ \Delta U_{\max}^{N_c} - \bar{U}_{k-1} \\ \Delta U_{\max}^{N_c} \\ \Delta U_{\max}^{N_c} \end{bmatrix} \tag{7.13}$$

where I is an identity matrix.

$$U_{\max}^{N_c} = \begin{bmatrix} u_{\max}(k) \\ u_{\max}(k+1) \\ \vdots \\ u_{\max}(k+N_c) \end{bmatrix}$$

[1] The term "pseudo-linearization" comes from the fact that Γ is obtained by simulation of the *free response* in the nonlinear model and Λ is the result of a linearization (time invariant or variant).

$$U_{\min}^{N_c} = \begin{bmatrix} u_{\min}(k) \\ u_{\min}(k+1) \\ \vdots \\ u_{\min}(k+N_c) \end{bmatrix} \qquad \Delta U_{\max} = \begin{bmatrix} \Delta u_{\max}(k) \\ \Delta u_{\max}(k+1) \\ \vdots \\ \Delta u_{\max}(k+N_c) \end{bmatrix}$$

where $I^{(n)} = I \in \Re^{n \times n}$ and $0^{(m \times n)} \in \Re^{m \times n}$.

This method works well when the nonlinearities are smooth, the model is a good representation of the plant and the control actions do not move the system far away from the region where the "pseudo-linearization" is valid.

7.5 Formulation of the Predictive Control Problem as a robust quadratic program

This section expands the nominal solution presented in Section 7.4 by introducing uncertainty in the parameters of the cost function of the QP [see Equation (7.12)].

For the current formulation the uncertainty will be restricted to the forced response term $(\Lambda + \delta\Lambda)U$. The problem described in Equation (7.12) will be converted to

$$\begin{aligned} J(U^n) &= J_{\min} + \\ &+ 2[(\Gamma - Y_{\text{ref}})^T Q\Lambda - \bar{U}_{k-1}^T S\Delta + (\Gamma - Y_{\text{ref}})^T Q\delta\Lambda] U^n \\ &+ U^{nT}[\Lambda^T Q\Lambda + R + \Delta^T S\Delta + \delta\Lambda^T Q\delta\Lambda + \Lambda^T Q\delta\Lambda + \delta\Lambda^T Q\Lambda +] U^n \end{aligned} \tag{7.14}$$

where $J_{\min}$ has the same description presented in Equation (7.12).

Using this description, we can give the P matrix for the RQP by

$$P = P_0 + \sum_{i=1}^{m} P_i u_i$$

$$P_0 = \Lambda^T Q\Lambda + R + \Delta^T S\Delta \tag{7.15}$$

$$\sum_{i=1}^{m} P_i u_i = \delta\Lambda^T Q\delta\Lambda + \Lambda^T Q\delta\Lambda + \delta\Lambda^T Q\Lambda \tag{7.16}$$

where $\delta\Lambda$ is an n-dimensional ellipsoidal uncertainty in the matrix used to build the forced response. Observe that the n-dimensional ellipsoidal uncertainty in $\delta\Lambda$ is reflected as an m-dimensional ellipsoidal uncertainty in P with $m = n + \sum_{i=1}^{n} i = n^2 + \frac{3}{2}n$.

The q vector will be described as:

$$q = q_0 + \sum_{i=1}^{n} q_i v_i$$

$$q_0^T = (\Gamma - Y_{\text{ref}})^T Q\Lambda - \bar{u}_{k-1}^T S\Delta \tag{7.17}$$

$$\sum_{i=1}^{n} q_i^T v_i = (\Gamma - Y_{\text{ref}})^T Q \delta \Lambda \tag{7.18}$$

Observe that no uncertainty is considered in the *free response* Γ. The main reason for this simplification is the complexity to obtain an analytical expression that can be introduced into the RQP. The addition of uncertain terms in the *free response* makes the solution computationally expensive and very conservative, destroying the advantages of the "pseudo-linearization." On the other hand, the uncertainty introduced by this term is only due to the mismatch between the plant and the model and there is no uncertainty due to the linearization process.

Summary:
Using the concept of "pseudo-linearization" the problem of robust nonlinear predictive control can be reduced to a problem of robust quadratic programming.

7.6 The Control Algorithm

The control algorithm can be described (see Figure 7.1) at each sampling time by the following steps:

Algorithm

1. Read the current output of the system and update $y(k)$.
2. With the input $u(k-1)$ calculate the*free response* Γ using Equation (6.44).
3. Construct the Λ matrix.
4. With the Λ and the Γ matrices construct the matrix P_0 and the vector q_0.
5. Using the uncertainty description build the set of matrices P_i and the vectors q_i.
6. Solve the SOCP described in Equation (7.6).
7. Apply the first control action, which is the first entry of the solution of the SOCP.

7.7 Uncertainty Description in Fuzzy Models

The following lines include some ideas about the way to represent the uncertainty by using the structure of the Takagi–Sugeno fuzzy models.

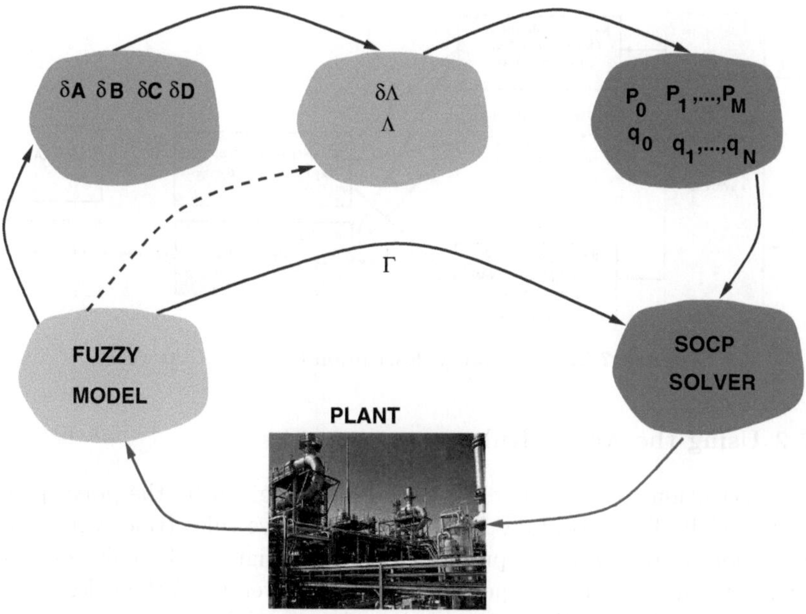

Figure 7.1. Diagram with the stages of the control algorithm

7.7.1 Local Uncertainty Described on Each Rule

This is perhaps the simplest method to describe the uncertainty. The idea is to use a Takagi–Sugeno description of the plant where the consequences of the rules are linear dynamic systems with their respective uncertainties (see Figure 7.2). In this case, the uncertainty and the linearization of the plant will be assumed to be the convex combination of the local uncertainties.

$$\begin{gathered} \dots \\ \text{IF } x(k) \text{ is } \mathcal{A}_i \\ \text{THEN } x(k+1) = L_i + A_i x(k) + B_i u(k) \,|\{\delta A_i, \delta B_i\} \\ \dots \end{gathered}$$

Among the advantages of this method are the simplicity of the description and the possibility to describe locally the uncertainty based on the real knowledge about the quality of each of the local models (for instance, the covariance of the parameters during identification). It is important to remark that this description will be valid only if the fuzzy sets are complementary in the antecedents such that the local description will be equivalent to the local linearization. It is probably the less conservative description, but also it demands more effort during the design phase.

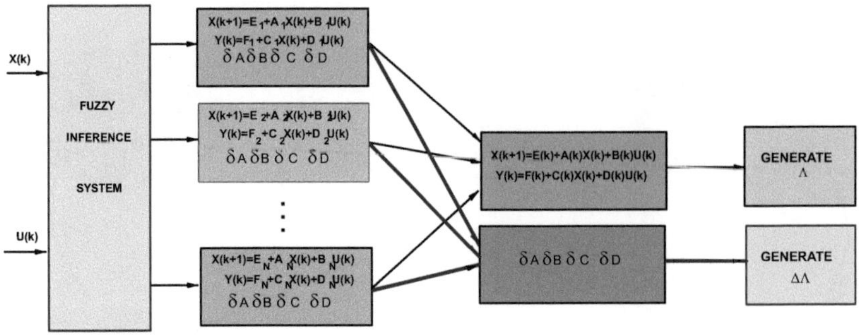

Figure 7.2. Uncertainty description on each of the rules

7.7.2 Using the Active Rules

This description assumes that the uncertainty is given by the polytope constructed with all the local descriptions of the active rules (rules with a firing value different from 0) at the present instant (see Figure 7.3). This description is efficient but inaccurate, especially when the coverage of the rules is small. As a result of the small coverage of the rules, it might happen that the system reaches rules that are beyond the set of active rules within the prediction horizon. In this case, the uncertainty will be underestimated. For rules with wide coverage, it is a very interesting solution because the system will tend to remain within the set of active rules.

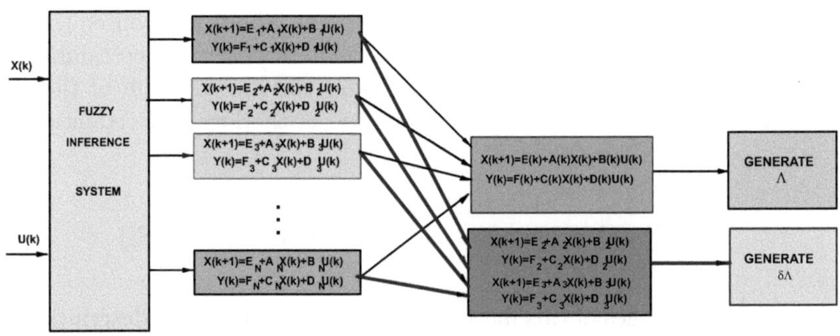

Figure 7.3. Uncertainty description using the "polytope" of the active rules

7.7.3 Using All the Rules

The polytope constructed with all the models present in the rules is a quite conservative approach, but it can also be constructed quickly because it can be precomputed (see Figure 7.4). The conservativeness can lead to infeasibility in the solution. For rules with wide coverage, it is a very interesting but very conservative solution and it is recommended only for systems with gentle nonlinearities.

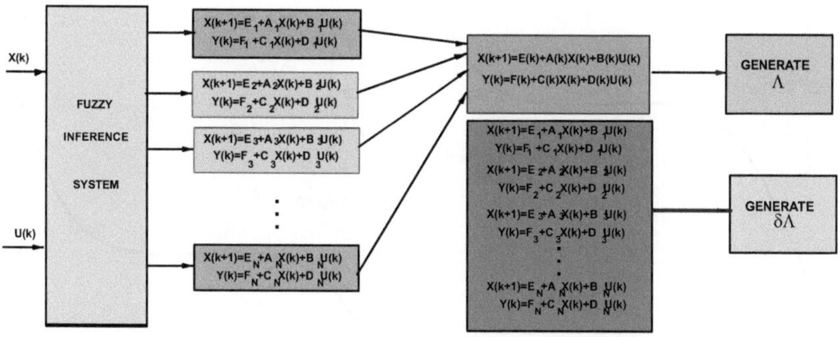

Figure 7.4. Uncertainty description using the "polytope" of the complete set of rules

7.7.4 Using the Reachable Set

In this method, the uncertainty is described also as a polytope, generated with all rules that can be reached in the control horizon (see Figure 7.5). The regions of the state-space that can be reached in N_c steps (control horizon) from the current state $x(k)$ are limited, because the input is constrained on its maximum value and increment. These regions compose the reachable set at time k. This description can be prepared in advance during the design phase and can be converted in the first description proposed to improve the performance. This method is a good compromise between the two previously described. Observe that the reachable set can be generated dynamically according not only to the maximum and minimum value of the inputs, but also using the information about the predictions of the input sequence. This input sequence will tell which rules will be activated in the prediction horizon, and this information will be used to build an uncertainty description, *i.e.*, the polytope of all active rules in the prediction horizon.

Summary:
The uncertainty in the model can be described in many different ways. Takagi–Sugeno fuzzy models offer a natural mechanism to obtain a description of the uncertainty of the model.

7.8 Conclusions and Perspectives

This chapter has presented a formulation for robust nonlinear predictive control based on Takagi–Sugeno fuzzy models. The structure of the Takagi–

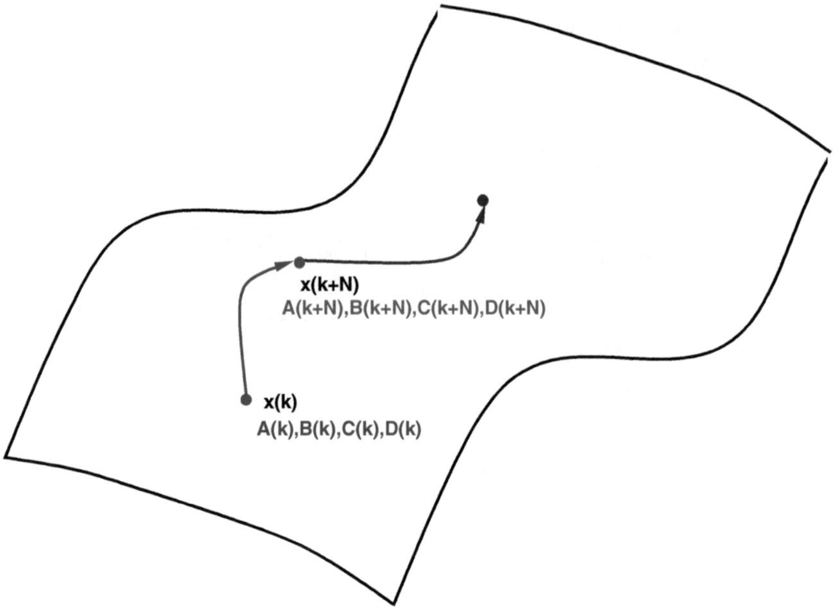

Figure 7.5. Graphic representation of the reachable set

Sugeno fuzzy models has been exploited to generate the formulation of the problem. The formulation converts the Nonlinear min-max program into a robust quadratic program (RQP). The chapter shows the way to convert the (RQP) into a second-order cone program. This type of problem can be solved in a very efficient way by using interior point methods.

The chapter shows that by using Takagi–Sugeno fuzzy models it is possible to design a robust nonlinear predictive controllers with a certain degree of transparency given by the local linear representations of the system and its uncertainties.

Future work must be devoted to refine the uncertainty description such that it will be more compact. The stability of these control schemes can be guaranteed by using end constraints; however, there is no formal proof to this statement.

The main contributions included in this chapter are the formulation of the problem of robust nonlinear predictive control in terms of a robust quadratic program and the extension of the formulation of the robust quadratic program as a second-order program by extending the uncertainties to the linear term of the cost function. Another contribution is the formulation of possible ways to obtain and represent the uncertainty.

8

Conclusions and Future Perspectives

8.1 Conclusions and Summary

This book has presented several contributions to a multitude of different topics of fuzzy systems. However, to cover all the issues related to fuzzy control in detail will demand many more pages. Chapter 1 presents the capabilities of fuzzy logic systems to approach nonlinear functions. The chapter includes a detailed mathematical description of the constructing units and graphical examples used to expand the comprehension of the subject.

Chapter 2 shows the main techniques to approximate nonlinear functions by using fuzzy models, trained with input–output data. Expressions of the gradients applied during the optimization process to adjust the parameters of the models have been derived. The problem of generalization is addressed and a technique to improve the generalization capabilities of the fuzzy models as well as to overcome the lack of excitation is included. This technique guarantees a lower bound in the quality of the model (the fuzzy model will be at least as good as the best multilinear approximation).

The fact that one of the comparative advantages of fuzzy systems in comparison with other "universal approximators" is its linguistic interpretability explains the formulation of the AFRELI algorithm included in Chapter 3. The AFRELI algorithm in combination with the FuZion algorithm has been designed to guarantee a good trade-off between numerical accuracy and interpretability. The method exploits some successful elements proposed in other methods to reduce the complexity of the constructed model.

The algorithm generates automatically the fuzzy sets from the data and the interactive labeling process (with intervention of the designer) guarantees an agreement between the fuzzy set and its semantic meaning.

The algorithm generates a rule base covering all the possible cases; this guarantees the completeness of the rule base, but the associated drawback is the exponential growth of the rule base as the number of inputs increases. However, this is only a storage problem because the description of the fuzzy sets guarantees that only 2^N rules (N number of inputs) are activated on each inference. This fact makes the inference process fast because only a limited number of rules are evaluated.

The numerical performance of the model can be improved by making a "fine" tuning of the parameters of the antecedents by means of constrained gradient descent techniques.

Chapter 4 presents the formulation of the problem of system identification in the framework of fuzzy systems. The chapter discusses the main issues in this area such as regressors selection, experiment design, structure, parameter adjusting and validation.

The regressor's selection is a very important task needed in order to counteract the "curse" of dimensionality that arises as the number of inputs of the model increases. Some efficient methods are presented in this chapter to trade off the complexity of the calculation with the accuracy of the solution.

The optimization of the parameters of the models is a task that involves the use of gradient descent techniques. The recursive structure of the dynamic models demands the use of gradients generated dynamically. The derivation of the dynamic systems that generate the gradients for the most common membership functions is also an important element contributed in this book.

Chapter 5 opens the second part of the book, which has been devoted to the subject of control using fuzzy logic. This chapter presents several control synthesis techniques. The chapter begins with the description of the controller designed using pure expert knowledge. The chapter also explains the "paradoxical" success of this type of controller in process control applications. An interesting contribution is a method to tune fuzzy controllers using the parameters of a previously designed PID controller. This contribution is very important because it is possible to guarantee that the fuzzy controller will be able to achieve at least the same performance as obtained with the PID controller. Other linear controllers can be converted into an equivalent fuzzy controller, making smoother the migration from linear control toward nonlinear control. A proof of this statement is a contribution, which can be found in the Appendix D. Other techniques are also shown in the chapter. Among those, it is important to mention techniques such as inverse modeling and model referenced adaptive control. Techniques such as feedback linearization are also explained. This technique is limited to a certain class of nonlinear systems (affine nonlinear systems) and has been criticized for its lack of robustness. However, a more elaborate control technique based on the same principle has been presented. This technique is the sliding mode control. This technique deserves special attention; its robustness makes it a very good candidate for applications where the dynamics of plant are not very well known. Only rough bounds on the gain are needed to achieve an acceptable performance. Finally, the fuzzy gain scheduling technique is presented. This design technique is quite "elegant" from the analytical point of view. The use of advanced algorithms for semidefinite programming facilitates the solution of the LMIs generated by the synthesis problem. More advanced synthesis techniques include H^{∞} criterion for disturbance rejection; robust synthesis and observer design can be developed by the same type of methods.

In Chapter 6 the book presents a set of fuzzy model-based predictive control strategies. These strategies combine two ideas that have been accepted by the industry for their simplicity and good performance. The ideas combined in this strategy are predictive control and fuzzy modeling. The development of these strategies has been preceded by a maturation of the modeling and identification techniques using fuzzy structures. This is simple control strategy, simple from the designer's point of view and simple to operate and maintain.

The simplicity during the design phase comes from the fact that the design task is focused in the formulation of the goal of the control systems in terms of a quadratic cost function and the physical or the safety constraints imposed to the elements involved in the system. Other tuning parameters such as control and prediction horizons are closely related with the settling time and model order and typically exhibit a monotonic tendency, making the tuning of the controller a simple task. Of course, the simplicity does not come for free. The strategy concentrates the effort in two tasks. One task is the effort that must be put to build the model and the other task is the operation of the control system, which is computationally intensive, compared with more traditional controllers. Today the limit of the sampling frequencies lies around the 100 Hz, which are already good not only for process control but also for some mechanical systems.

The algorithms presented in this chapter are applicable for unconstrained and constrained systems. Also, these algorithms presented can be extended in some cases to other model structures such as neural networks, Volterra series, splines, support vector machines, physical models, *etc.* The central aspect of the strategies presented is the reduction of the original complex nonlinear program (with no possibility for online implementation), to a simple quadratic program (online implementable) by exploiting the structure of the problem and the structure of the models. The relaxation introduced by the methods presented in this chapter can be explained as a relaxation of the equality constraints imposed by the plant description. In some cases, these approximated solutions can be quite close to the "global optimal solution" of the original nonlinear program. The assumptions used to reduce the problem demand some smoothness in the nonlinearities, which are directly related with the prediction capability of the model.

Finally, the book closes with a chapter dedicated to the solution of the problem of robust nonlinear predictive control based on Takagi–Sugeno fuzzy models. The structure of the Takagi–Sugeno fuzzy models has been exploited to generate a formulation of the problem as a robust optimization problem. The robust solution guarantees that under the uncertainty described the constraints will not be violated. This fact is very important to guarantee stability, especially when it is enforced by means of end constraints. The formulation generates a big set of constraints; however, the performance of the optimization procedure is only slightly degraded because the solution of the optimization problem is found by means of interior point optimization algorithms. The chapter also presents some possible ways to formulate the uncertainty in the problem.

8.2 Perspectives and Future Work

This section is included as a guideline for the new researchers and attempts to point out some research topics that are important to be pursued in the coming years.

In the area of modeling, the curse of dimensionality is an open problem that has limited the application of fuzzy systems to systems with a small number of inputs. The solution to this problem must keep in mind the importance of the linguistic relevance of the solution. A promising strategy is the formulation of hierarchies; however, there is no systematic method to define the priority of one variable with respect to the others. Heuristic methods are the only available tools. Some type of measurements should be defined to guide the selection and the position in the hierarchy of the variables to guarantee the maximum generalization with the minimal description.

Another important problem that deserves attention is the design of clustering techniques with variable shapes such that the number of rules obtained after projection is optimal.

The identification problem presents a series of open problems. Important problems are model order estimation and more constructive regressor selection methods. New validation methods and a better link between the results given by the validation method and the way to improve the quality of the model are also very important.

When the identification problem includes Takagi–Sugeno models, the selection of the variables governing the scheduling of the local models is also an open problem. So far, only heuristic methods have been proposed.

Even if the models obtained via identification are consistent from the linguistic point of view, the use of delayed variables ($x(k-n)$) is not very intuitive. A more intuitive description is in terms of tendencies of the variables ($\Delta x(k)$). A method to convert a model described in terms of delayed variables into a model described with tendencies of the variables, and vice versa, will be very useful to extract more linguistic information from the dynamic model.

Experiment design is a very important issue that must be investigated. This issue has special relevance when the system to be identified is very complex. Performing experiments in some processes can be very expensive or take such a long time that their number will be very limited. Identification along trajectories is a very interesting issue in the chemical process industry. Here good experiment design is paramount due to the high costs of production.

Even though one of the earliest developments on the application of fuzzy systems has been the design of controllers using expert knowledge, still there is no test for stability for this kind of controllers. Probably the main difficulty comes from the fact that mathematical knowledge about the plant is very limited, and a formal test of stability demands some knowledge about the plant. The proposed technique to design fuzzy controllers from PID controllers has no formal proof for stability. However, if the plant is a linear system of second order, it is possible to design a synthesis method that guarantees the stability of the system. The author is more skeptical as to whether such a method can be constructed for a linear plant with any order.

The sliding mode control strategy using fuzzy systems is a very promising strategy; so far, the limitation is that such a technique is applicable only to affine systems. Further research must be oriented to extend this method to more general nonlinear plants. Finally, the design of fuzzy controllers using LMIs has been considered conservative for the difficulty to find a common P matrix. Johansson *et al.* [59] have proposed new alternatives. The relaxation of the problem is obtained by means of formulating a piecewise quadratic Lyapunov function; the potential of this idea seems very wide.

The idea of predictive control using fuzzy models has been studied extensively in this book. However, there is enough space for improvement and future research. One very important issue is to study the performance of the algorithms and their possible modification such that other cost functions involving direct economic cost can be taken into account. Such a cost function can be in some cases quite nonlinear and even discontinuous. New research must be oriented to find ways to introduce other types of inequality constraints. So far, only linear constraints have been treated. Linear inequality constraints cover a big spectrum of constraints such as saturation and slew rate constraints. The use of nonlinear constraints can make possible the introduction of energy-related aspects.

The authors have explored the use of active set methods and interior point methods to solve the quadratic optimization programs generated by the presented algorithms. Interior point methods are very promising for their capacity to handle a large number of constraints. However, some problems must be solved before these techniques can be applied exploiting their full potential. The matrices involved in the optimization exhibit a regular structure that can be exploited to improve the performance of the optimization algorithms.

The robust predictive control strategy demands more research about the way to define the uncertainty and how to obtain such information from the plant and/or the model. Finally, the stability of this control schemes can be guaranteed by using end constraints; however, there is no formal proof to this statement.

resting from the fact that mathematical knowledge of the [illegible] and [illegible] of stability demands some knowledge about the plant. [illegible] required technique to design fuzzy controllers [illegible] no formal proof for stability. However [illegible] [illegible] of the system. The [illegible] [illegible] not intended for the [illegible].

[illegible]

[illegible] With [illegible] and [illegible] good [illegible] stability [illegible] of the [illegible] [illegible] [illegible] and [illegible] however, [illegible] [illegible] techniques.

Part III

Appendices

A

Fuzzy Set Theory

A.1 Introduction

The purpose of this appendix is to present an introduction to the main concepts of fuzzy sets. The concept of fuzzy sets arises as an answer to the problems of paradoxes, uncertainties and absence of precision found in crisp sets. Fuzzy sets are more inherent to nature than crisp sets. Crisp sets only consider elements with very well-defined characteristics, such that a clear set boundary can be established. Many authors try to relate fuzziness to probabilities. In some cases fuzziness and probabilities can be treated with similar rules, but it is very important to recall that fuzziness presents the degree of belonging of one element to a certain set, while probabilities describe the behavior of many elements that belong to a certain set.

A.2 Fuzzy Sets

Let X denote the universal set. A conventional (crisp) set A is defined by a characteristic (membership) function $\mu(x)\,(x \in X)$ that assigns the values 1 or 0 to each element $x \in X$, respectively, if x belongs or does not belong to A. $\mu_A : X \rightarrow \{0, 1\}$

A fuzzy set A is defined by a *membership function* $\mu_A : X \rightarrow [0, 1]$ that describes the membership degree of the elements of A. Values of $\mu_A(x)$ closer to 1 denote a higher degree of set membership.

A.2.1 Some Examples of Membership Functions

Let $X = \Re$. For the statement "x is around M" we can define the following membership functions:

- Delta membership:

$$\mu(x) = \begin{cases} 1 & x = M \\ 0 & \text{otherwise} \end{cases}$$

- Step membership:

$$\mu(x) = \begin{cases} 1 & M - M_1 \leq M \leq M + M_1 \\ 0 & \text{otherwise} \end{cases}$$

- Ramp or triangular membership:

$$\mu(x) = \begin{cases} 0 & x < M - M_1 \text{ or } x > M + M_1 \\ 1 + \frac{x-M}{M_1} & M - M_1 \leq x \leq M \\ 1 - \frac{x-M}{M_1} & M \leq x \leq M + M_1 \end{cases}$$

- Exponential membership:

$$\mu(x) = exp(-t|x - M|)$$

- Gaussian membership:

$$\mu(x) = exp(-t(x - M)^2)$$

A.3 Basic Definitions of Fuzzy Sets

A.3.1 Support

The support of a fuzzy set A over the universe X is defined as the crisp subset where the membership function $\mu_A(x)$ is larger than zero.

$$\text{Supp}(A) = \{x|\mu(x) > 0\}$$

A.3.2 Core

The core of a fuzzy set A over the universe X is defined as the crisp subset where the membership function $\mu_A(x)$ is equal to 1.

$$\text{Core}(A) = \{x|\mu(x) = 1\}$$

A.3.3 Height

The height of a fuzzy set A is the supremum of its membership function:

$$\text{hgt}(A) = \sup_{x \in X} \mu_A(x)$$

A.3.4 Normal Fuzzy Set

A fuzzy set A is "normal" if there exists at least one value of $x \in X$ such that $\mu_A(x) = 1$.

A.3.5 α-Cut

The α-cut of a fuzzy set A is defined as the crisp subset of X where $\mu_A(x) \leq \alpha$.

$$A_\alpha = \{x | \mu_A(x) \leq \alpha\}$$

A.3.6 Strict α-Cut

The strict α-cut of a fuzzy set A is defined as the crisp subset of X where $\mu_A(x) < \alpha$.

A.3.7 Convexity

A fuzzy set A defined in $\Re^n$ is convex if each of its α-cuts is convex.

A.4 Operations on Fuzzy Sets

A.4.1 *A* Is Contained in *B*

A set A is contained in the set B or A is a subset of B, denoted by $A \subseteq B$, if

$$\mu_A(x) \leq \mu_B(x) \quad \forall x \in X$$

A subset is proper $(A \subset B)$ if $(\mu_A(x) < \mu_B(x))$.

A.4.2 Complement, Negation

The membership function $\mu_{\bar{A}}(x)$ of the complement of A (denoted by $\bar{A}$) is defined by

$$\mu_{\bar{A}}(x) = 1 - \mu_A(x), \quad \forall x \in X$$

The relative complement of the set A with respect to a set B is defined by:

$$\mu_{\bar{A}_B}(x) = \mu_B(x) - \mu_A(x), \; x \in X \;\text{ if } \mu_B(x) > \mu_A(x)$$

A.4.3 Intersection

The intersection of a set A with a set B is defined by

$$A \cap B = \{x | x \in A \wedge x \in B\}$$

The most important operators for the intersection are

- Extreme operator

$$\mu_{A\cap_1 B}(x) = \mu_A(x) \wedge \mu_B(x) = min\{\mu_A(x), \mu_B(x)\} \qquad \forall x \in X$$

- Product operator

$$\mu_{A\cap_2 B}(x) = \mu_A(x)\mu_B(x) \qquad \forall x \in X$$

For crisp sets:

$$\mu_{A\cap_1 B}(x) = \mu_{A\cap_2 B}(x)$$

For fuzzy sets:

$$\mu_{A\cap_1 B}(x) \geq \mu_{A\cap_2 B}(x)$$

Characteristics of the intersection operators:

- $\cap_1$ and $\cap_2$ are commutative.
- $\cap_1$ and $\cap_2$ are associative.
- Identity: $\mu_A \cap_i 1 = \mu_A$.
- Absorption: $\mu_A \cap_i 0 = 0$.
- $\cap_1$ is an idempotent operator. $\mu \cap_1 \mu = \mu$.
- $\cap_2$ is not an idempotent operator. $\mu \cap_2 \mu \neq \mu$, $\mu \cap_2 \mu \subset \mu$.
- For fuzzy sets the law of non-contradiction ($\mu \cap \bar{\mu} = 0$) does not hold. Example: Let $\mu = 0.5$, $\mu \cap_1 \bar{\mu} = 0.5$ or $\mu \cap_2 \bar{\mu} = 0.25$.
- Product intersection is a subset of minimum intersection: $(\mu_A \cap_1 \mu_B) \subset (\mu_A \cap_2 \mu_B)$.
- For $\mu_A \subset \mu_B$, $\mu_A \cap_1 \mu_B = \mu_A$ and $\mu_A \cap_2 \mu_B \subset \mu_A$.

A.4.4 Union

The union of a set A with a set B is defined by

$$A \cup B = \{x | x \in A \vee x \in B\} \qquad \forall x \in X$$

The most important operators for the union operation are

- Extreme operator

$$\mu_{A\cup_1 B}(x) = \mu_A(x) \vee \mu_B(x) = max\{\mu_A(x), \mu_B(x)\} \qquad \forall x \in X$$

- Sum operator

$$\mu_{A\cup_2 B}(x) = \mu_A(x) + \mu_B(x) - \mu_A(x)\mu_B(x) \qquad \forall x \in X$$

For crisp sets:

$$\mu_{A\cup_1 B}(x) = \mu_{A\cup_2 B}(x)$$

For fuzzy sets:

$$\mu_{A\cup_1 B}(x) \leq \mu_{A\cup_2 B}(x)$$

Characteristics of the union operators:

- $\cup_1$ and $\cup_2$ are commutative.
- $\cup_1$ and $\cup_2$ are associative.
- Identity: $\mu_A \cup_i 0 = \mu_A$.
- Absorption: $\mu_A \cup_i 1 = 1$.
- $\cup_1$ is an idempotent operator. $\mu \cup_1 \mu = \mu$.
- $\cup_2$ is not an idempotent operator. $\mu \cup_2 \mu \neq \mu$, $\mu \cup_2 \mu \subset \mu$.
- For fuzzy sets the law of excluded middle ($\mu \cup \bar{\mu} = X$) does not hold. Example: Let $\mu = 0.5$, $\mu \cup_1 \bar{\mu} = 0.5$ or $\mu \cup_2 \bar{\mu} = 0.75$.
- $(\mu_A \cup_1 \mu_B) \subset (\mu_A \cup_2 \mu_B)$.
- For $\mu_A \subset \mu_B$, $\mu_A \cup_1 \mu_B = \mu_B$ and $\mu_A \cup_2 \mu_B \supset \mu_B$.

A.5 Fuzzy relations

A relation represents the presence or absence of association, interaction or interconnection between the elements of two or more sets. A binary relation is any relation between two sets. For example:

$$R(x, y) = (X \text{ AND } Y) \qquad x \in X, y \in Y$$

A fuzzy relation $R(x, y)$ is a fuzzy subset of $X \times Y$.

For membership function $\mu(x, y)$

$$R = \{\mu(x, y) : X \times Y \rightarrow [0, 1]\}$$

or

$$R = \{(x, y), \mu_R(x, y)\} = \bigcup (x, y)\mu_R(x, y)$$

A fuzzy relation on sets $X_1, X_2, ..., X_n$ is a fuzzy subset of $X_1 \times X_2 \times ... \times X_n$

$$R_n = \{\mu(x_1, x_2, ..., x_n) : X_1 \times X_2 \times ... \times X_n \to [0,1]\}$$

or

$$R_n = \bigcup\{(x_1, x_2, ..., x_n), \mu_R(x_1, x_2, ..., x_n)\} : X_1 \times X_2 \times ... \times X_n \to [0,1]$$

A fuzzy relation can be represented by

- Membership function
- Matrix (table) if the number of elements is finite
- Picture with gray shades corresponding to μ value

A.5.1 Projection of Fuzzy Relations

In a binary fuzzy relation one can define two projections:

- First projection: $R^{(1)} = \{x, \max_y \mu(x,y)\}\ x, y \in X \times Y$
- Second projection: $R^{(2)} = \{y, \max_x \mu(x,y)\}\ x, y \in X \times Y$

The global projection (also known as the height of the relation) is given by

$$R^g = \max_x \max_y \mu(x,y)$$

A fuzzy relation is called normal if the global projection is 1. The operations between fuzzy sets can be extended to relations, for instance, intersection, union, *etc.*

A.5.2 Composition of Relations

The combination of fuzzy sets and fuzzy relations is called the *composition.* Given:

$$\begin{array}{rll} R(x,y) & (x,y) \in X \times Y & R : X \times Y \to [0,1] \\ S(y,z) & (y,z) \in Y \times Z & S : Y \times Z \to [0,1] \end{array}$$

Composition $C(x,z)$

- Max-min composition:
  $\mu_c(x,z) = \max\{\min(\mu_R(x,y), \mu_S(y,z))\} \quad x \in X, y \in Y, z \in Z$
- Max product composition:
  $\mu_c(x,z) = \max\{\mu_R(x,y)\mu_S(y,z)\} \quad x \in X, y \in Y, z \in Z$

A.6 Approximate Reasoning

A.6.1 Introduction

Approximate reasoning is a well-known form of fuzzy logic and uses a group of inference rules. These inference rules use fuzzy propositions as premises. The output of the inference rules is a set of conclusions. In the case of crisp sets, these are either true or false, but when the inference rules use fuzzy premises the conclusions have a certain degree of truth.

A.6.2 Linguistic Variables

Lotfi Zadeh defines a linguistic variable as follows: "By a linguistic variable we mean a variable whose values are words or sentences in a natural or artificial language. For example, *Age* is a linguistic variable if its values are linguistic rather than numerical, *i.e.young, not young, very young, quite young, old, not very old and not very young, etc.*, rather than 20, 21, 22,..." [5]

Driankov[6] represents a linguistic variable and the framework as

$$\langle X, \mathcal{L}X, \mathcal{X}, M_x \rangle$$

where X denotes the name of the symbolic variable, *e.g., age, height, speed, pressure, error, change of error, etc.* $\mathcal{L}X$ is the set of linguistic values that X can take. In the case of the variable *age* A we have

$$\mathcal{L}A = \{\textit{very young, young, adult, mature, old, very old}\}$$

In control systems, the linguistic variables are usually the *error* and the *change of error* and the set is usually represented by the set $\{NB, NM, NS, ZO, PS, PM, PB\}$ where NB means "Negative Big," NM "Negative Medium," ZO means "Zero," *etc.* $\mathcal{L}X$ is also called the term set of X or the reference set of X. $\mathcal{X}$ is the physical domain of the variables; for instance, for the variable *age*, the domain could be $[0, 95]$. $\mathcal{X}$ is also called U or the universe of discourse and can be continuous or discrete. M_x is a semantic function that gives an interpretation of the linguistic value in terms of the quantitative elements of $\mathcal{X}$. It is the set of the membership function in the universe of discourse $\mathcal{X}$, *i.e.*,

$$M_X : LX \rightarrow \widetilde{LX}$$

A.7 General Structure of a Fuzzy Inference System

The purpose of this section is to give an overview about the structure, characteristics and functioning of the fuzzy inference systems. The fuzzy inference system is an inference system based on linguistic rules sometimes generated by empirical knowledge. First the crisp values coming from quantitative measurements are converted to linguistic values ($\mathcal{L}X$). This process is called fuzzification. The fuzzification process uses the membership functions to make such a conversion. Figure A.1 shows the membership functions for the fuzzification of a crisp temperature.

A.7.1 Control Rules as a Knowledge Representation

After the fuzzification, the control rules can be applied. These rules are presented in an *IF .. AND.. THEN ..OR ELSE* form and represent the set of

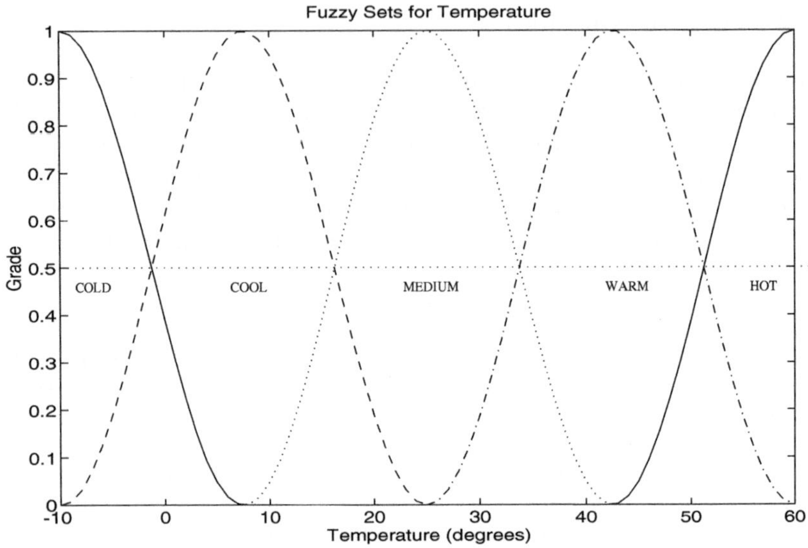

Figure A.1. Fuzzy sets for temperature

decisions when the input variables belong to certain fuzzy sets. The consequence of the rule is calculated as a max-min composition.

if $\mathfrak{L}A_1$ and $\mathfrak{L}B_1$ and .. and $\mathfrak{L}N_1$ then $\mathfrak{L}U_1$
or else
if $\mathfrak{L}A_1$ and $\mathfrak{L}B_1$ and .. and $\mathfrak{L}N_2$ then $\mathfrak{L}U_2$
or else
$\vdots$
or else
if $\mathfrak{L}A_N$ and $\mathfrak{L}B_N$ and .. and $\mathfrak{L}N_N$ then $\mathfrak{L}U_N$

Represented in the form of a fuzzy relation:

$$R : R_1 \cup R_2 \cup \ldots \cup R_N = \bigcup_{i=1}^{N} (\mathfrak{L}A_i \times \mathfrak{L}B_i \times \ldots \times \mathfrak{L}N_i \times \mathfrak{L}U_i)$$

Because the *AND* operation can be represented as a *min* operation and the *OR* can be represented as the *max* operation, a rule can be represented by

$$R(\mu_A(l), \mu_B(j), \ldots, \mu_U(k)) = \max_{1 \le i \le N} \{\mathfrak{L}A_l(\mu_A(l)) \wedge \mathfrak{L}B_j(\mu_B(j)) \wedge \ldots \wedge \mathfrak{L}U_i(\mu_U(k))\}$$

A.7.2 Defuzzification

After the inference process, the result of the process has to be converted to a crisp value. This procedure is called *defuzzification*. Some defuzzification methods are

- Center-of-area/ gravity
- Center-of-sums
- Height

The influence of the mentioned defuzzification method on the controller performance is negligible. The next lines present the different defuzzification methods.

Center-of-Area/Gravity

For discrete systems:

$$x^* = \frac{\sum_{i=1}^{n} x_i \mu_U(x_i)}{\sum_{i=1}^{n} \mu_U(x_i)}$$

For continuous membership function:

$$x^* = \frac{\int_U x \mu_U(x) dx}{\int_U \mu_U(x) dx}$$

Center-of-Sums

For discrete systems:

$$x^* = \frac{\sum_{i=1}^{n} x_i \sum_{k=1}^{l} \mu_k(x_i)}{\sum_{i=1}^{n} \sum_{k=1}^{l} \mu_k(x_i)}$$

For continuous membership function:

$$x^* = \frac{\int_U x \sum_{k=1}^{l} \mu_k(x) dx}{\int_U \sum_{k=1}^{l} \mu_k(x) dx}$$

Height

This method takes the peak value of each consequence and makes a weighted sum of these peak values, where the weights are the degree of membership of the fired rule. The method is equal to the center of gravity when the consequence membership functions are singletons. Using a singleton as the membership functions of the consequences, the membership function of the ith consequence is

$$\mu_i(x) = 1 \quad \text{if } x = x_i$$
$$\mu_i(x) = 0 \quad \text{if } x \neq x_i$$

The defuzzification using this method will generate the following expression:

$$x^* = \frac{\sum_{i=1}^{L} x_i \mu_i(x)}{\sum_{i=1}^{L} \mu_i(x)}$$

This method is computationally fast and generates continuous values, making it very useful for function approximation.

B

Clustering Methods

Clustering methods are a set of techniques to reduce groups of information X represented as p-dimensional vectors into characteristic sets A_i characterized by feature vectors $v_i \in \Re^p$ and membership functions μ_A. The applications of these techniques include pattern recognition, classification and the applications presented in this book for fuzzy modeling and identification. This appendix presents only the methods that are used in this book, and it is by no means a complete survey of these techniques.

B.1 Fuzzy C-Means [2]

The fuzzy C-means clustering algorithm is based on the minimization of the cost function:

$$\min_{(U,V)} \left\{ J_m(U,V;X) = \sum_{k=1}^{n} \sum_{i=1}^{c} (\mu_{ik})^m ||x_k - v_i||_A^2 \right\} \tag{B.1}$$

where X is the set of vectors, $x_k \in \Re^p$ with the information, $V = [v_1, \ldots, v_c]$ is the set of feature vectors, $||.||_A$ is the norm of the vector defined as $x^T Ax$, where A is assumed to be the identity matrix, and $U \in M_{fc}$ is the fuzzy partition matrix, defined as an element of the set:

$$M_{fc} = \left\{ U \in \Re^{c \times N} | \mu_{ik} \in [0,1], \forall i,k; \sum_{i=1}^{c} \mu_{ik} = 1, \forall k; 0 < \sum_{k=1}^{N} \mu_{ik} < N, \forall i \right\} \tag{B.2}$$

The ith row of the fuzzy partition matrix contains the membership values of the vectors x to the A_i fuzzy set. The elements of U are calculated as

$$\mu_{ik} = \left[\sum_{j=1}^{c} \left(\frac{||x_k - v_i||_A}{||x_k - v_j||_A} \right)^{\frac{2}{m-1}} \right]^{-1} \quad \forall i,k \tag{B.3}$$

The prototypes v_i are calculated as

$$v_i = \frac{\sum_{k=1}^{n}(\mu_{ik})^m x_k}{\sum_{k=1}^{n}(\mu_{ik})^m} \; \forall i \tag{B.4}$$

The fuzzy C-means algorithm works as follows:

Algorithm Fuzzy C-Means

Given the data set X with N vectors, select the number of clusters $1 < c < N$, the exponent m, the termination tolerance $\epsilon > 0$ and the matrix A to calculate the induced norm, and initialize the matrix U randomly such that $U^{(0)} \in M_{fc}$.

- **Repeat for** $j = 1, 2, \ldots$.
- *Step 1:* **Calculate the prototypes:**

$$v_i^{(j)} = \frac{\sum_{k=1}^{n}(\mu_{ik}^{(j-1)})^m x_k}{\sum_{k=1}^{n}(\mu_{ik})^m} \quad 1 \le i \le c \tag{B.5}$$

- *Step 2:* **Calculate fuzzy partition matrix:**

$$\mu_{ik}^{(j)} = \left[\sum_{l=1}^{c}\left(\frac{||x_k - v_i^{(j)}||_A}{||x_k - v_l^{(j)}||_A}\right)^{\frac{2}{m-1}}\right]^{-1} \quad 1 \le i \le c,\, 1 \le k \le N \tag{B.6}$$

- **Until** $||U^{(j)} - U^{(j-1)}|| < \epsilon$.

It is important to remark that when the vector x_k is equal to one of the prototypes v_i the expression (B.6) becomes singular. For this case the membership value μ_{ik} for this vector is equal to one and zero for all the other entries in the kth row of U.

The parameter m is a very important parameter. As $m \to \infty$, the means of the clusters tend to the mean of the set X.

B.2 Using Fuzzy Covariance Matrix: Gustafson and Kessel Algorithm [3]

The fuzzy Gustafson and Kessel clustering algorithm is based on the minimization of the cost function:

$$\min_{(U,V,\mathbf{A})}\left\{J_m(U,V,\mathbf{A};X) = \sum_{k=1}^{n}\sum_{i=1}^{c}(\mu_{ik})^m (x_k - v_i)^T A_i (x_k - v_i)\right\} \tag{B.7}$$

where X is the set of vectors, $x_k \in \Re^p$ with the information, $V = [v_1, \ldots, v_c]$ is the set of feature vectors, $\mathbf{A} = [A_1, \ldots, A_c]$ is a set of c norm-inducing matrices and $U \in M_{fc}$ is the fuzzy partition matrix, defined as an element of the set:

$$M_{fc} = \left\{ U \in \Re^{c \times N} | \mu_{ik} \in [0,1], \forall i,k; \sum_{i=1}^{c} \mu_{ik} = 1, \forall k; 0 < \sum_{k=1}^{N} \mu_{ik} < N, \forall i \right\} \tag{B.8}$$

The ith row of the fuzzy partition matrix contains the membership values of the vectors x to the A_i fuzzy set.

Observe that the cost function can be arbitrarily small by reducing the norm of each A_i. For this reason a constraint is introduced to preserve the norm of A_i:

$$|A_i| = \rho_i \qquad \rho_i > 0$$

Applying the *Lagrange multipliers* to the above mentioned optimization problem generates the following expression for A_i:

$$A_i = [\rho_i \det(P_i)]^{1/n} P_i^{-1} \tag{B.9}$$

where P_i is the fuzzy covariance matrix:

$$P_i = \frac{\sum_{k=1}^{n} (\mu_{ik})^m (x_k - v_i)(x_k - v_i)^T}{\sum_{k=1}^{n} (\mu_{ik})^m} \tag{B.10}$$

The elements of U are calculated as

$$\mu_{ik} = \left[\sum_{j=1}^{c} \left(\frac{(x_k - v_i)^T A_i (x_k - v_i)}{(x_k - v_j)^T A_i (x_k - v_j)} \right)^{\frac{2}{m-1}} \right]^{-1} \quad \forall i,k \tag{B.11}$$

The prototypes v_i are calculated as

$$v_i = \frac{\sum_{k=1}^{n} (\mu_{ik})^m x_k}{\sum_{k=1}^{n} (\mu_{ik})^m} \quad \forall i \tag{B.12}$$

The Gustafson and Kessel algorithm works as follows:

Algorithm Fuzzy Covariance Matrix – Gustafson and Kessel

Given the data set X with N vectors, select the number of clusters $1 < c < N$, the exponent m, the termination tolerance $\epsilon > 0$ and the volumes ρ_i of the matrices A_i to calculate the induced norms, and initialize the matrix U randomly such that $U^{(0)} \in M_{fc}$.

- **Repeat for** $j = 1, 2, \ldots$.
- *Step 1:* **Calculate the prototypes:**

$$v_i^{(j)} = \frac{\sum_{k=1}^{n} (\mu_{ik}^{(j-1)})^m x_k}{\sum_{k=1}^{n} (\mu_{ik}^{(j-1)})^m} \quad 1 \le i \le c \tag{B.13}$$

- *Step 2:* **Calculate the fuzzy covariance matrices:**

$$P_i^{(j)} = \frac{\sum_{k=1}^{n} (\mu_{ik}^{(j-1)})^m (x_k - v_i^{(j)})(x_k - v_i^{(j)})^T}{\sum_{k=1}^{n} (\mu_{ik}^{(j-1)})^m} \tag{B.14}$$

- *Step 3:* **Calculate the induced-norm matrices:**

$$A_i = [\rho_i \det(P_i)]^{1/n} P_i^{-1} \quad 1 \le i \le c \tag{B.15}$$

- *Step 4:* **Calculate the fuzzy partition matrix:**

$$\mu_{ik}^{(j)} = \left[\sum_{l=1}^{c} \left(\frac{(x_k - v_i^{(j)})^T A_i^{(j)} (x_k - v_i^{(j)})}{(x_k - v_l^{(j)})^T A_i^{(j)} (x_k - v_l^{(j)})} \right)^{\frac{2}{m-1}} \right]^{-1} \quad 1 \le i \le c, 1 \le k \le N \tag{B.16}$$

- **Until** $||U^{(j)} - U^{(j-1)}|| < \epsilon$.

It is important to remark that when the vector x_k is equal to one of the prototypes v_i the expression (B.16) becomes singular. For this case the membership value μ_{ik} for this vector is equal to one and zero for all the other entries in the kth row of U.

The parameter m is a very important parameter. As $m \to \infty$, the means of the clusters tend to the mean of the set X.

B.3 Mountain Clustering [4]

In this algorithm a super set of the feature vectors V is proposed in advance, then some vectors are selected according to the value of the mountain function calculated for the given vector v_i. The mountain function is defined as

$$M(v_i) = \sum_{k=1}^{N} e^{-(\alpha d(v_i, x_k))} \tag{B.17}$$

where α is a positive constant and $d(v_i, x_k)$ is a distance measure from v_i to x_k and it is typically but not necessarily:

$$d(v_i, x_k) = ||v_i - x_k||_2 \tag{B.18}$$

Two of the most popular elections of the feature vector candidates are

- Take the set of feature candidates equal to the set of data points $V = X$. This is not very efficient for large data sets.
- Take a grid (arbitrary) defined in the interval where the points of X are defined. This is not very efficient for vectors defined on a large-dimensional space.

Compared with other clustering methods the mountain clustering method has as its main advantage the fact that the number of clusters does not need to be defined in advance. The algorithm works as follows:

Algorithm Mountain Clustering

Given the data set X with N vectors, select the parameters α and β, the termination tolerance ϵ, and the set of feature vector candidates V with c elements.

- *Step 1:* **Calculate the initial mountain function**

$$M^{(0)}(v_i) = \sum_{k=1}^{N} e^{-(\alpha d(v_i, x_k))} \quad 1 \leq i \leq c \tag{B.19}$$

- **Repeat for** $j = 1, 2, \ldots$.
- *Step 2:* **Find the largest mountain value:**

$$M^{(j-1)*} = \max_{v_i} M^{(j-1)}(v_i) \tag{B.20}$$

- *Step 3:* **Define the location of the maximum of the mountain value as the center of the** $j - 1$ **cluster:**

$$v^{(j-1)*} = \arg\max_{v_i} M^{(j-1)}(v_i) \tag{B.21}$$

- *Step 4:* **Calculate the revised mountain function** $M^(j)(v_i)$**:**

$$M^{(j)}(v_i) = M^{(j-1)}(v_i) - M^{(j-1)*} \sum_{k=1}^{N} e^{-(\beta d(v^{(j-1)*}, x_k))} \quad 1 \leq i \leq c \tag{B.22}$$

- **Until** $M^{(j-1)*} < \epsilon$.

C

Gradient Expressions Used in Identification with Fuzzy Models

The gradients derived on this appendix are useful in the optimization of dynamic models. The cost function to be minimized is the quadratic cost function, which is defined as

$$V_N(\theta) = \frac{1}{2N} \sum_{t=1}^{N} |y(t) - \hat{y}(t|\theta)|^2 \tag{C.1}$$

where $y(t)$ is the output of the "real" system at time t,

$$\hat{y}(t|\theta) = f(\varphi(t), \theta) \tag{C.2}$$

$\hat{y}(t|\theta)$ is the output of the constructed model parameterized by the vector θ. The vector θ describes the membership functions and the position of the singletons in the consequences and

$$\varphi(t) = [y(t-1), \ldots, y(t-m), \hat{y}(t-1), \ldots, \hat{y}(t-n), \ldots \\ u(t), \ldots, u(t-k), \varepsilon(t-1), \ldots, \varepsilon(t-l)]$$

is the set of regressors of the model. The derivations shown in this appendix can be applied to the structures NFIR, NARX, NOE, NARMAX and NBJ.

This appendix initially shows the derivation of the gradient for the consequences of the rules; in the second part it shows the derivation for the parameters of different types of membership functions.

C.1 Gradient for the Singleton Consequences

The expression for the gradient is given by

$$\frac{\partial V_N(\theta)}{\partial \bar{y}^l} = \frac{1}{N} \sum_{t=1}^{N} (y(t) - \hat{y}(t|\theta))(-\frac{\partial \hat{y}(t|\theta)}{\partial \bar{y}^l}) \tag{C.3}$$

with

$$\frac{\partial \hat{y}(t|\theta)}{\partial \bar{y}^l} = \bar{y}^l \frac{\partial w_l(\varphi(t))}{\partial \bar{y}^l} + w_l(\varphi(t)) \tag{C.4}$$

Observe that if the model is NARX or NFIR the term $\partial w_l(\varphi(t))/\partial \bar{y}^l = 0$ and the expression for the gradient will be the same used for static function approximation. The term $\partial w_l(\varphi(t))/\partial \bar{y}^l$ is dependent from previous gradient values and must be generated dynamically.

$$\frac{\partial w_l(\varphi(t))}{\partial \bar{y}^l} = \frac{\frac{\partial \mu_l(\varphi(t))}{\partial \bar{y}^l} \sum_{i=1}^{L} \mu_i(\varphi(t)) - \mu_l(\varphi(t)) \frac{\partial \sum_{i=1}^{L} \mu_i(\varphi(t))}{\partial \bar{y}^l}}{\sum_{i=1}^{L} \mu_i(\varphi(t))} \tag{C.5}$$

where

$$\begin{aligned} \frac{\partial \mu_l(\varphi(t))}{\partial \bar{y}^l} &= \sum_{i \in \mathcal{Y}} \frac{\mu_l(\varphi(t))}{\mu_l^i(\hat{y}(t-k(i)))} \frac{\partial \mu_l^i(\hat{y}(t-k(i)))}{\partial \bar{y}^l} \\ &+ \sum_{j \in \mathcal{E}} \frac{\mu_l(\varphi(t))}{\mu_l^j(\varepsilon(t-m(j)))} \frac{\partial \mu_l^j(\varepsilon(t-m(j)))}{\partial \bar{y}^l} \end{aligned} \tag{C.6}$$

where $\mathcal{Y}$ represents the set of inputs related with the regressors $\hat{y}(.)$ with delay $k(i)$ and $\mathcal{E}$ represents the set of inputs related with the regressors $\varepsilon(.)$ with delay $m(j)$, and

$$\begin{aligned} \frac{\sum_{i=1}^{L} \mu_i(\varphi(t))}{\partial \bar{y}^l} &= \sum_{l=1}^{L} \left\{ \sum_{i \in \mathcal{Y}} \frac{\mu_l(\varphi(t))}{\mu_l^i(\hat{y}(t-k(i)))} \frac{\partial \mu_l^i(\hat{y}(t-k(i)))}{\partial \bar{y}^l} \right. \\ &\left. + \sum_{j \in \mathcal{E}} \frac{\mu_l(\varphi(t))}{\mu_l^j(\varepsilon(t-m(j)))} \frac{\partial \mu_l^j(\varepsilon(t-m(j)))}{\partial \bar{y}^l} \right\} \end{aligned} \tag{C.7}$$

Finally, according to the type of membership functions used in the terms $\partial \mu_l^j(\varepsilon(t-m(j)))/\partial \bar{y}^l$ and $\partial \mu_l^i(\hat{y}(t-k(i)))/\partial \bar{y}^l$ will have the following expressions:

C.1.1 With Trapezoidal Membership Functions

For *trapezoidal membership functions* using the parameterization given in Equation (2.17):

$$\frac{\partial \mu_l^i(\hat{y}(t-k(i)))}{\partial \bar{y}^l} = \begin{cases} 0 & \text{if } \hat{y}(t-k(i)) < a_l^i \\ \frac{1}{b_l^i - a_l^i} \frac{\partial \hat{y}(t-k(i))}{\partial \bar{y}^l} & \text{if } a_l^i < \hat{y}(t-k(i)) < b_l^i \\ 0 & \text{if } b_l^i < \hat{y}(t-k(i)) < c_l^i \\ \frac{-1}{d_l^i - c_l^i} \frac{\partial \hat{y}(t-k(i))}{\partial \bar{y}^l} & \text{if } c_l^i < \hat{y}(t-k(i)) < d_l^i \\ 0 & \text{if } \hat{y}(t-k(i)) > d_l^i \end{cases} \tag{C.8}$$

$$\frac{\partial \mu_l^j(\varepsilon(t-m(j)))}{\partial \bar{y}^l} = \begin{cases} 0 & \text{if } \varepsilon(t-m(j)) < a_l^j \\ \frac{-1}{b_l^j - a_l^j}\frac{\partial \hat{y}(t-m(j))}{\partial \bar{y}^l} & \text{if } a_l^j < \varepsilon(t-m(j)) < b_l^j \\ 0 & \text{if } b_l^j < \varepsilon(t-m(j)) < c_l^j \\ \frac{1}{d_l^j - c_l^j}\frac{\partial \hat{y}(t-m(j))}{\partial \bar{y}^l} & \text{if } c_l^j < \varepsilon(t-m(j)) < d_l^j \\ 0 & \text{if } \varepsilon(t-m(j)) > d_l^j \end{cases} \tag{C.9}$$

C.1.2 With Polynomial Membership Functions

For *polynomial membership functions* using the parameterization given in Equation (2.31)

$$\frac{\partial \mu_l^i(\hat{y}(t-k(i)))}{\partial \bar{y}^l} =$$

$$\begin{cases} 0 & \text{if } \hat{y}(t-k(i)) < a_l^i \\ \frac{6[\hat{y}(t-k(i))^2 - (a_l^i + b_l^i)\hat{y}(t-k(i))] + a_l^{i\,3} - 3ba_l^{i\,2}}{(a_l^i - b_l^i)^3}\frac{\partial \hat{y}(t-k(i))}{\partial \bar{y}^l} & \text{if } a_l^i < \hat{y}(t-k(i)) < b_l^i \\ 0 & \text{if } b_l^i < \hat{y}(t-k(i)) < c_l^i \\ \frac{6[\hat{y}(t-k(i))^2 - (d_l^i + c_l^i)\hat{y}(t-k(i))] + d_l^{i\,3} - 3cd_l^{i\,2}}{(d_l^i - c_l^i)^3}\frac{\partial \hat{y}(t-k(i))}{\partial \bar{y}^l} & \text{if } c_l^i < \hat{y}(t-k(i)) < d_l^i \\ 0 & \text{if } \hat{y}(t-k(i)) > d_l^i \end{cases} \tag{C.10}$$

$$\frac{\partial \mu_l^j(\varepsilon(t-m(j)))}{\partial \bar{y}^l} =$$

$$\begin{cases} 0 & \text{if } \varepsilon(t-m(j)) < a_l^j \\ \frac{6[(a_l^j + b_l^j)\varepsilon(t-m(j)) - \varepsilon(t-m(j))^2] - a_l^{j\,3} + 3ba_l^{j\,2}}{(a_l^j - b_l^j)^3}\frac{\partial \hat{y}(t-m(j))}{\partial \bar{y}^l} & \text{if } a_l^j < \varepsilon(t-m(j)) < b_l^j \\ 0 & \text{if } b_l^j < \varepsilon(t-m(j)) < c_l^j \\ \frac{6[(d_l^j + c_l^j)\varepsilon(t-m(j)) - \varepsilon(t-m(j))^2] - d_l^{j\,3} + 3cd_l^{j\,2}}{(d_l^j - c_l^j)^3}\frac{\partial \hat{y}(t-m(j))}{\partial \bar{y}^l} & \text{if } c_l^j < \varepsilon(t-m(j)) < d_l^j \\ 0 & \text{if } \varepsilon(t-m(j)) > d_l^j \end{cases} \tag{C.11}$$

C.1.3 With Gaussian Membership Functions

For *Gaussian membership functions* using the parameterization presented in (2.45):

$$\frac{\partial \mu_l^i(\hat{y}(t-k(i)))}{\partial \bar{y}^l} = -2\frac{(\hat{y}(t-k(i)) - \bar{x}_l^i)}{\sigma_l^{i\,2}}\mu_l^i(\hat{y}(t-k(i)))\frac{\partial \hat{y}(t-k(i))}{\partial \bar{y}^l} \tag{C.12}$$

$$\frac{\partial \mu_l^j(\varepsilon(t-m(j)))}{\partial \bar{y}^l} = 2\frac{(\varepsilon(t-m(j)) - \bar{x}_l^j)}{\sigma_l^{j\,2}}\mu_l^j(\varepsilon(t-m(j)))\frac{\partial \hat{y}(t-m(j))}{\partial \bar{y}^l} \tag{C.13}$$

C.2 Gradient for the Parameters of the Membership Functions

The expression for the gradient is given by

$$\frac{\partial V_N(\theta)}{\partial \bar{y}^l} = \frac{1}{N}\sum_{t=1}^{N}(y(t) - \hat{y}(t|\theta))(-\frac{\partial \hat{y}(t|\theta)}{\partial \alpha}) \tag{C.14}$$

where $\alpha \subset \theta$ represents all the adjustable parameter in the model excluding the consequences, and

$$\frac{\partial \hat{y}(t|\theta)}{\partial \alpha_n} = \frac{1}{\sum_{l=1}^{L}\mu_l(\varphi(t))}\sum_{l\in\mathcal{U}}(\bar{y}^l - \hat{y}(t|\theta))\frac{\mu_l(\varphi(t))}{\mu_j^i(\hat{y}(t-k(l)))}\frac{\partial \mu_j^i(\hat{y}(t-k(l)))}{\partial \alpha_n} \tag{C.15}$$

where α_n is a parameter of the membership function $\mu_j^i(.)$ and $\mathcal{U}$ is the set of rules that include in the antecedents the membership function $\mu_j^i(.)$. According to the type of membership functions and the regressors the term $\partial\mu_j^i(\hat{y}(t-k(l)))/\partial\alpha_n$ will have expressions, which are presented in the following lines.

C.2.1 With Trapezoidal Membership Functions

For *trapezoidal membership functions* α_n could be $a_i^j, b_i^j, c_i^j, d_i^j$ and the gradients will be

$$\frac{\partial \mu_j^i(\hat{y}(t-k(l)))}{\partial a_i^j} = \begin{cases} 0 & \hat{y}(t-k(l)) < a_i^j \\ \dfrac{\frac{\partial \hat{y}(t-k(l))}{\partial a_i^j}(b_i^j - a_i^j) + (\hat{y}(t-k(l)) - b_i^j)}{(b_i^j - a_i^j)^2} & a_i^j < \hat{y}(t-k(l)) < b_i^j \\ 0 & \hat{y}(t-k(l)) > b_i^j \end{cases} \tag{C.16}$$

$$\frac{\partial \mu_j^i(\varepsilon(t-m(l)))}{\partial a_i^j} = \begin{cases} 0 & \varepsilon(t-m(l)) < a_i^j \\ \dfrac{-\frac{\partial \hat{y}(t-m(l))}{\partial a_i^j}(b_i^j - a_i^j) + (\varepsilon(t-m(l)) - b_i^j)}{(b_i^j - a_i^j)^2} & a_i^j < \varepsilon(t-m(l)) < b_i^j \\ 0 & \varepsilon(t-m(l)) > b_i^j \end{cases} \tag{C.17}$$

$$\frac{\partial \mu_j^i(\hat{y}(t-k(l)))}{\partial b_i^j} = \begin{cases} 0 & \hat{y}(t-k(l)) < a_i^j \\ \dfrac{\frac{\partial \hat{y}(t-k(l))}{\partial b_i^j}(b_i^j - a_i^j) - (\hat{y}(t-k(l)) - a_i^j)}{(b_i^j - a_i^j)^2} & a_i^j < \hat{y}(t-k(l)) < b_i^j \\ 0 & \hat{y}(t-k(l)) > b_i^j \end{cases} \tag{C.18}$$

$$\frac{\partial \mu_j^i(\varepsilon(t-m(l)))}{\partial b_i^j} = \begin{cases} 0 & \varepsilon(t-m(l)) < a_i^j \\ \dfrac{-\frac{\partial \hat{y}(t-m(l))}{\partial b_i^j}(b_i^j - a_i^j) - (\varepsilon(t-m(l)) - a_i^j)}{(b_i^j - a_i^j)^2} & a_i^j < \varepsilon(t-m(l)) < b_i^j \\ 0 & \varepsilon(t-m(l)) > b_i^j \end{cases} \tag{C.19}$$

$$\frac{\partial \mu_j^i(\hat{y}(t-k(l)))}{\partial c_i^j} = \begin{cases} 0 & \hat{y}(t-k(l)) < c_i^j \\ \frac{-\frac{\partial \hat{y}(t-k(l))}{\partial c_i^j}(d_i^j - c_i^j) + (d_i^j - \hat{y}(t-k(l)))}{(d_i^j - c_i^j)^2} & c_i^j < \hat{y}(t-k(l)) < d_i^j \\ 0 & \hat{y}(t-k(l)) > d_i^j \end{cases} \tag{C.20}$$

$$\frac{\partial \mu_j^i(\varepsilon(t-m(l)))}{\partial c_i^j} = \begin{cases} 0 & \varepsilon(t-m(l)) < c_i^j \\ \frac{\frac{\partial \hat{y}(t-m(l))}{\partial c_i^j}(d_i^j - c_i^j) + (d_i^j - \varepsilon(t-m(l)))}{(d_i^j - c_i^j)^2} & c_i^j < \varepsilon(t-m(l)) < d_i^j \\ 0 & \varepsilon(t-m(l)) > d_i^j \end{cases} \tag{C.21}$$

$$\frac{\partial \mu_j^i(\hat{y}(t-k(l)))}{\partial d_i^j} = \begin{cases} 0 & \hat{y}(t-k(l)) < c_i^j \\ \frac{\frac{\partial \hat{y}(t-k(l))}{\partial d_i^j}(d_i^j - c_i^j) - (d_i^j - \hat{y}(t-k(l)))}{(d_i^j - c_i^j)^2} & c_i^j < \hat{y}(t-k(l)) < d_i^j \\ 0 & \hat{y}(t-k(l)) > d_i^j \end{cases} \tag{C.22}$$

$$\frac{\partial \mu_j^i(\varepsilon(t-m(l)))}{\partial d_i^j} = \begin{cases} 0 & \varepsilon(t-m(l)) < c_i^j \\ -\frac{\frac{\partial \hat{y}(t-m(l))}{\partial d_i^j}(d_i^j - c_i^j) - (d_i^j - \varepsilon(t-m(l)))}{(d_i^j - c_i^j)^2} & c_i^j < \varepsilon(t-m(l)) < d_i^j \\ 0 & \varepsilon(t-m(l)) > d_i^j \end{cases} \tag{C.23}$$

C.2.2 With Polynomial Membership Functions

For *polynomial membership functions* α_n could be $a_i^j, b_i^j, c_i^j, d_i^j$ and the gradients will be

$$\frac{\partial \mu_j^i(\hat{y}(t-k(l)))}{\partial a_i^j} = \begin{cases} 0 & \text{if } \hat{y}(t-k(l)) < a_i^j \\ 6\frac{(\hat{y}(t-k(l)) - b_i^j)(\hat{y}(t-k(l)) - a_i^j)}{(a_i^j - b_i^j)^3}\left[\frac{\partial \hat{y}(t-k(l))}{\partial a_i^j} - \frac{(\hat{y}(t-k(l)) - b_i^j)}{a_i^j - b_i^j}\right] & \text{if } a_i^j < \hat{y}(t-k(l)) < b_i^j \\ 0 & \text{if } \hat{y}(t-k(l)) > b_i^j \end{cases} \tag{C.24}$$

$$\frac{\partial \mu_j^i(\varepsilon(t-m(l)))}{\partial a_i^j} = \begin{cases} 0 & \text{if } \varepsilon(t-m(l)) < a_i^j \\ 6\frac{(\varepsilon(t-m(l)) - b_i^j)(\varepsilon(t-m(l)) - a_i^j)}{(a_i^j - b_i^j)^3}\left[-\frac{\partial \hat{y}(t-m(l))}{\partial a_i^j} - \frac{(\hat{y}(t-k(l)) - b_i^j)}{a_i^j - b_i^j}\right] & \text{if } a_i^j < \varepsilon(t-m(l)) < b_i^j \\ 0 & \text{if } \varepsilon(t-m(l)) > b_i^j \end{cases} \tag{C.25}$$

$$\frac{\partial \mu_j^i(\hat{y}(t-k(l)))}{\partial b_i^j} = \begin{cases} 0 & \text{if } \hat{y}(t-k(l)) < a_i^j \\ 6\frac{(\hat{y}(t-k(l))-b_i^j)(\hat{y}(t-k(l))-a_i^j)}{(a_i^j-b_i^j)^3}\left[\frac{\partial \hat{y}(t-k(l))}{\partial b_i^j}+\right. & \\ \left.\frac{(\hat{y}(t-k(l))-a_i^j)}{a_i^j-b_i^j}\right] & \text{if } a_i^j < \hat{y}(t-k(l)) < b_i^j \\ 0 & \text{if } \hat{y}(t-k(l)) > b_i^j \end{cases} \tag{C.26}$$

$$\frac{\partial \mu_j^i(\varepsilon(t-m(l)))}{\partial b_i^j} = \begin{cases} 0 & \text{if } \varepsilon(t-m(l)) < a_i^j \\ 6\frac{(\varepsilon(t-m(l))-b_i^j)(\varepsilon(t-m(l))-a_i^j)}{(a_i^j-b_i^j)^3}\left[-\frac{\partial \hat{y}(t-m(l))}{\partial b_i^j}+\right. & \\ \left.\frac{(\hat{y}(t-k(l))-a_i^j)}{a_i^j-b_i^j}\right] & \text{if } a_i^j < \varepsilon(t-m(l)) < b_i^j \\ 0 & \text{if } \varepsilon(t-m(l)) > b_i^j \end{cases} \tag{C.27}$$

$$\frac{\partial \mu_j^i(\hat{y}(t-k(l)))}{\partial c_i^j} = \begin{cases} 0 & \text{if } \hat{y}(t-k(l)) < c_i^j \\ 6\frac{(\hat{y}(t-k(l))-d_i^j)(\hat{y}(t-k(l))-c_i^j)}{(c_i^j-d_i^j)^3}\left[-\frac{\partial \hat{y}(t-k(l))}{\partial c_i^j}+\right. & \\ \left.\frac{(\hat{y}(t-k(l))-d_i^j)}{c_i^j-d_i^j}\right] & \text{if } c_i^j < \hat{y}(t-k(l)) < d_i^j \\ 0 & \text{if } \hat{y}(t-k(l)) > d_i^j \end{cases} \tag{C.28}$$

$$\frac{\partial \mu_j^i(\varepsilon(t-m(l)))}{\partial c_i^j} = \begin{cases} 0 & \text{if } \varepsilon(t-m(l)) < c_i^j \\ 6\frac{(\varepsilon(t-m(l))-d_i^j)(\varepsilon(t-m(l))-c_i^j)}{(c_i^j-d_i^j)^3}\left[\frac{\partial \hat{y}(t-m(l))}{\partial c_i^j}+\right. & \\ \left.\frac{(\hat{y}(t-k(l))-d_i^j)}{c_i^j-d_i^j}\right] & \text{if } c_i^j < \varepsilon(t-m(l)) < d_i^j \\ 0 & \text{if } \varepsilon(t-m(l)) > d_i^j \end{cases} \tag{C.29}$$

$$\frac{\partial \mu_j^i(\hat{y}(t-k(l)))}{\partial d_i^j} = \begin{cases} 0 & \text{if } \hat{y}(t-k(l)) < c_i^j \\ -6\frac{(\hat{y}(t-k(l))-d_i^j)(\hat{y}(t-k(l))-c_i^j)}{(c_i^j-d_i^j)^3}\left[\frac{\partial \hat{y}(t-k(l))}{\partial d_i^j}+\right. & \\ \left.\frac{(\hat{y}(t-k(l))-c_i^j)}{c_i^j-d_i^j}\right] & \text{if } c_i^j < \hat{y}(t-k(l)) < d_i^j \\ 0 & \text{if } \hat{y}(t-k(l)) > d_i^j \end{cases} \tag{C.30}$$

$$\frac{\partial \mu_j^i(\varepsilon(t-m(l)))}{\partial d_i^j} = \begin{cases} 0 & \text{if } \varepsilon(t-m(l)) < c_i^j \\ -6\frac{(\varepsilon(t-m(l))-d_i^j)(\varepsilon(t-m(l))-c_i^j)}{(c_i^j-d_i^j)^3}\left[-\frac{\partial \hat{y}(t-m(l))}{\partial d_i^j}+\frac{(\hat{y}(t-k(l))-c_i^j)}{c_i^j-d_i^j}\right] & \text{if } c_i^j < \varepsilon(t-m(l)) < d_i^j \\ 0 & \text{if } \varepsilon(t-m(l)) > d_i^j \end{cases} \quad \text{(C.31)}$$

C.2.3 With Gaussian Membership Functions

With *Gaussian membership functions* α_n could be $\bar{x}_j^i, \sigma_j^i$ and the gradients will be

$$\frac{\partial \mu_j^i(\hat{y}(t-k(l)))}{\partial \bar{x}_j^i} = -2\frac{(\hat{y}(t-k(l))-\bar{x}_j^i)}{{\sigma_j^i}^2}\mu_j^i(\hat{y}(t-k(l)))\left[\frac{\partial \hat{y}(t-k(l))}{\partial \bar{x}_j^i}-1\right] \quad \text{(C.32)}$$

$$\frac{\partial \mu_j^i(\varepsilon(t-m(l)))}{\partial \bar{x}_j^i} = -2\frac{(\varepsilon(t-m(l)))-\bar{x}_j^i)}{{\sigma_j^i}^2}\mu_j^i(\varepsilon(t-m(l)))\left[-\frac{\partial \hat{y}(t-m(l))}{\partial \bar{x}_j^i}-1\right] \quad \text{(C.33)}$$

$$\frac{\partial \mu_j^i(\hat{y}(t-k(l)))}{\partial \sigma_j^i} = -2\frac{(\hat{y}(t-k(l))-\bar{x}_j^i)}{{\sigma_j^i}^3}\mu_j^i(\hat{y}(t-k(l)))\left[\sigma_j^i\frac{\partial \hat{y}(t-k(l))}{\partial \sigma_j^i}-(\hat{y}(t-k(l))-\bar{x}_j^i)\right] \quad \text{(C.34)}$$

$$\frac{\partial \mu_j^i(\varepsilon(t-m(l)))}{\partial \sigma_j^i} = -2\frac{(\varepsilon(t-m(l)))-\bar{x}_j^i)}{{\sigma_j^i}^3}\mu_j^i(\varepsilon(t-m(l)))\left[-\sigma_j^i\frac{\partial \hat{y}(t-m(l))}{\partial \sigma_j^i}-(\varepsilon(t-m(l))-\bar{x}_j^i)\right] \quad \text{(C.35)}$$

C.2.4 With Triangular Membership Functions with 0.5 Overlap

For *triangular membership functions with 0.5 overlap* the α_n parameters could be m_{j-1}, m_j, m_{j+1} and the derivatives will be given by

$$\frac{\partial \mu_j^i(\hat{y}(t-k(l)))}{\partial m_j^i} = \begin{cases} 0 & \hat{y}(t-k(l)) < m_{j-1}^i \\ \frac{\frac{\partial \hat{y}(t-k(l))}{\partial m_j^i}(m_j^i - m_{j-1}^i) - (\hat{y}(t-k(l)) - m_{j-1}^i)}{(m_j^i - m_{j-1}^i)^2} & m_{j-1}^i < \hat{y}(t-k(l)) < m_j^i \\ \frac{\frac{\partial \hat{y}(t-k(l))}{\partial m_j^i}(m_{j+1}^i - m_j^i) - (\hat{y}(t-k(l)) - m_{j+1}^i)}{(m_{j+1}^i - m_j^i)^2} & m_j^i < \hat{y}(t-k(l)) < m_{j+1}^i \\ 0 & \hat{y}(t-k(l)) > m_{j+1}^i \end{cases} \tag{C.36}$$

$$\frac{\partial \mu_j^i(\varepsilon(t-m(l)))}{\partial m_j^i} = \begin{cases} 0 & \varepsilon(t-m(l)) < m_{j-1}^i \\ \frac{-\frac{\partial \hat{y}(t-m(l))}{\partial m_j^i}(m_j^i - m_{j-1}^i) - (\varepsilon(t-m(l)) - m_{j-1}^i)}{(m_j^i - m_{j-1}^i)^2} & m_{j-1}^i < \varepsilon(t-m(l)) < m_j^i \\ \frac{-\frac{\partial \hat{y}(t-m(l))}{\partial m_j^i}(m_{j+1}^i - m_j^i) - (\varepsilon(t-m(l)) - m_{j+1}^i)}{(m_{j+1}^i - m_j^i)^2} & m_j^i < \varepsilon(t-m(l)) < m_{j+1}^i \\ 0 & \varepsilon(t-m(l)) > m_{j+1}^i \end{cases} \tag{C.37}$$

$$\frac{\partial \mu_{j+1}^i(\hat{y}(t-k(l)))}{\partial m_j^i} = \begin{cases} 0 & \hat{y}(t-k(l)) < m_j^i \\ \frac{\frac{\partial \hat{y}(t-k(l))}{\partial m_j^i}(m_{j+1}^i - m_j^i) + (\hat{y}(t-k(l)) - m_{j+1}^i)}{(m_{j+1}^i - m_j^i)^2} & m_j^i < \hat{y}(t-k(l)) < m_{j+1}^i \\ 0 & \hat{y}(t-k(l)) > m_{j+1}^i \end{cases} \tag{C.38}$$

$$\frac{\partial \mu_{j+1}^i(\varepsilon(t-m(l)))}{\partial m_j^i} = \begin{cases} 0 & \varepsilon(t-m(l)) < m_j^i \\ \frac{-\frac{\partial \hat{y}(t-m(l))}{\partial m_j^i}(m_{j+1}^i - m_j^i) + (\varepsilon(t-m(l)) - m_{j+1}^i)}{(m_{j+1}^i - m_j^i)^2} & m_j^i < \varepsilon(t-m(l)) < m_{j+1}^i \\ 0 & \varepsilon(t-m(l)) > m_{j+1}^i \end{cases} \tag{C.39}$$

$$\frac{\partial \mu^i_{j-1}(\hat{y}(t-k(l)))}{\partial m^i_j} =$$

$$\begin{cases} 0 & \hat{y}(t-k(l)) < m^i_{j-1} \\ \dfrac{\frac{\partial \hat{y}(t-k(l))}{\partial m^i_j}(m^i_j - m^i_{j-1}) + (\hat{y}(t-k(l)) - m^i_{j-1})}{(m^i_j - m^i_{j-1})^2} & m^i_{j-1} < \hat{y}(t-k(l)) < m^i_j \\ 0 & \hat{y}(t-k(l)) > m^i_j \end{cases} \tag{C.40}$$

$$\frac{\partial \mu^i_{j-1}(\varepsilon(t-m(l)))}{\partial m^i_j} =$$

$$\begin{cases} 0 & \varepsilon(t-m(l)) < m^i_{j-1} \\ \dfrac{-\frac{\partial \hat{y}(t-m(l))}{\partial m^i_j}(m^i_j - m^i_{j-1}) + (\varepsilon(t-m(l)) - m^i_{j-1})}{(m^i_j - m^i_{j-1})^2} & m^i_{j-1} < \varepsilon(t-m(l)) < m^i_j \\ 0 & \varepsilon(t-m(l)) > m^i_j \end{cases} \tag{C.41}$$

C.2.5 With Polynomial Membership Functions with 0.5 Overlap

For *polynomial membership functions with 0.5 overlap* the α_n parameters could be m_{j-1}, m_j, m_{j+1} and the derivatives will be given by

$$\frac{\partial \mu^i_j(\hat{y}(t-k(l)))}{\partial m^i_j} =$$

$$\begin{cases} 0 & \text{if } \hat{y}(t-k(l)) < m^i_{j-1} \\ 6\frac{(\hat{y}(t-k(l)) - m^i_{j-1})(\hat{y}(t-k(l)) - m^i_j)}{(m^i_{j-1} - m^i_j)^3}\left[\frac{\partial \hat{y}(t-k(l))}{\partial m^i_j} + \frac{(\hat{y}(t-k(l)) - m^i_{j-1})}{m^i_{j-1} - m^i_j}\right] & \text{if } m^i_{j-1} < \hat{y}(t-k(l)) < m^i_j \\ 6\frac{(\hat{y}(t-k(l)) - m^i_{j+1})(\hat{y}(t-k(l)) - m^i_j)}{(m^i_j - m^i_{j+1})^3}\left[-\frac{\partial \hat{y}(t-k(l))}{\partial m^i_j} + \frac{(\hat{y}(t-k(l)) - m^i_{j+1})}{m^i_j - m^i_{j+1}}\right] & \text{if } m^i_j < \hat{y}(t-k(l)) < m^i_{j+1} \\ 0 & \text{if } \hat{y}(t-k(l)) > m^i_{j+1} \end{cases} \tag{C.42}$$

$$\frac{\partial \mu^i_j(\varepsilon(t-m(l)))}{\partial m^i_j} =$$

$$\begin{cases} 0 & \text{if } \varepsilon(t-m(l)) < m^i_{j-1} \\ 6\frac{(\varepsilon(t-m(l))-m^i_{j-1})(\varepsilon(t-m(l))-m^i_j)}{(m^i_{j-1}-m^i_j)^3}\left[-\frac{\partial \hat{y}(t-m(l))}{\partial m^i_j}+\right. & \\ \left.\frac{(\hat{y}(t-k(l))-m^i_{j-1})}{m^i_{j-1}-m^i_j}\right] & \text{if } m^i_{j-1} < \varepsilon(t-m(l)) < m^j_j \\ 6\frac{(\varepsilon(t-m(l))-m^i_{j+1})(\varepsilon(t-m(l))-m^i_j)}{(m^i_j-m^i_{j+1})^3}\left[\frac{\partial \hat{y}(t-m(l))}{\partial m^i_j}+\right. & \\ \left.\frac{(\hat{y}(t-k(l))-m^i_{j+1})}{m^i_j-m^i_{j+1}}\right] & \text{if } m^i_j < \varepsilon(t-m(l)) < m^j_{j+1} \\ 0 & \text{if } \varepsilon(t-m(l)) > m^i_{j+1} \end{cases} \tag{C.43}$$

$$\frac{\partial \mu^i_{j+1}(\hat{y}(t-k(l)))}{\partial m^i_j} =$$

$$\begin{cases} 0 & \text{if } \hat{y}(t-k(l)) < m^i_j \\ 6\frac{(\hat{y}(t-k(l))-m^i_{j+1})(\hat{y}(t-k(l))-m^i_j)}{(m^i_j-m^i_{j+1})^3}\left[\frac{\partial \hat{y}(t-k(l))}{\partial m^i_j}-\right. & \\ \left.\frac{(\hat{y}(t-k(l))-m^i_{j+1})}{m^i_j-m^i_{j+1}}\right] & \text{if } m^i_j < \hat{y}(t-k(l)) < m^i_{j+1} \\ 0 & \text{if } \hat{y}(t-k(l)) > m^i_{j+1} \end{cases} \tag{C.44}$$

$$\frac{\partial \mu^i_{j+1}(\varepsilon(t-m(l)))}{\partial m^i_j} =$$

$$\begin{cases} 0 & \text{if } \varepsilon(t-m(l)) < m^i_j \\ 6\frac{(\varepsilon(t-m(l))-m^i_{j+1})(\varepsilon(t-m(l))-m^i_j)}{(m^i_j-m^i_{j+1})^3}\left[-\frac{\partial \hat{y}(t-m(l))}{\partial m^i_j}-\right. & \\ \left.\frac{(\hat{y}(t-k(l))-m^i_{j+1})}{m^i_j-m^i_{j+1}}\right] & \text{if } m^i_j < \varepsilon(t-m(l)) < m^j_{j+1} \\ 0 & \text{if } \varepsilon(t-m(l)) > m^i_{j+1} \end{cases} \tag{C.45}$$

$$\frac{\partial \mu^i_{j-1}(\hat{y}(t-k(l)))}{\partial m^i_j} =$$

$$\begin{cases} 0 & \text{if } \hat{y}(t-k(l)) < m^i_{j-1} \\ -6\frac{(\hat{y}(t-k(l))-m^i_{j-1})(\hat{y}(t-k(l))-m^i_j)}{(m^i_{j-1}-m^i_j)^3}\left[\frac{\partial \hat{y}(t-k(l))}{\partial m^i_j}+\right. & \\ \left.\frac{(\hat{y}(t-k(l))-m^i_{j-1})}{m^i_{j-1}-m^i_j}\right] & \text{if } m^i_{j-1} < \hat{y}(t-k(l)) < m^i_j \\ 0 & \text{if } \hat{y}(t-k(l)) > m^i_j \end{cases} \tag{C.46}$$

$$\frac{\partial \mu_{j-1}^{i}(\varepsilon(t-m(l)))}{\partial m_{j}^{i}} =$$
$$\begin{cases} 0 & \text{if } \varepsilon(t-m(l)) < m_{j-1}^{i} \\ -6\frac{(\varepsilon(t-m(l))-m_{j-1}^{i})(\varepsilon(t-m(l))-m_{j}^{i})}{(m_{j-1}^{i}-m_{j}^{i})^{3}}\left[-\frac{\partial \hat{y}(t-m(l))}{\partial m_{j}^{i}} + \right. & \\ \left. \frac{(\hat{y}(t-k(l))-m_{j-1}^{i})}{m_{j-1}^{i}-m_{j}^{i}}\right] & \text{if } m_{j-1}^{i} < \varepsilon(t-m(l)) < m_{j}^{j} \\ 0 & \text{if } \varepsilon(t-m(l)) > m_{j}^{i} \end{cases} \tag{C.47}$$

Observe once more that when there is no feedback in the model $\partial \hat{y}(t - m(l))/\partial \alpha_n = 0$ the gradient expressions are reduced to the ones used for static functions.

The presence of the terms $\partial \hat{y}(t - m(l))/\partial \alpha_n$ in the gradients demands for computation of the gradients the calculation of a numerical solution of a discrete dynamic system, which must be updated each time a new data point is presented to the model.

D

Discrete Linear Dynamical System Approximation Theorem

Theorem D.1. *Any stable single-input–single-output discrete linear system with transfer function $f(z)$ and bounded input can be represented by a fuzzy system with normal triangular membership functions with overlap $\frac{1}{2}$.*

Proof: Suppose the transfer function $f(z)$ is given by

$$f(z) = \frac{y(z)}{u(z)} = \frac{b_0 + b_1 z^{-1} + \ldots + b_n z^{-n}}{1 + a_1 z^{-1} + \ldots + a_n z^{-n}} \tag{D.1}$$

By applying the inverse Z transform to the transfer function, we can convert the system into the difference equation:
This formula can be represented in vector form as

$$y(k) = -a_1 y(k-1) - \ldots - a_n y(k-n) + b_0 u(k) + \ldots + b_n u(k-n) \tag{D.2}$$

$$y(k) = T^T X(k) \tag{D.3}$$

where $T = [-a_1, \ldots, -a_n, b_0, \ldots, b_n]^T$, $X(k) = [y(k-1), \ldots, y(k-n), u(k), \ldots, u(k-n)]^T$ and $X(k), T \in \Re^N$ with $N = 2n+1$. A fuzzy system to map the function $y(k) = \mathcal{F}(X(k))$ representing the linear dynamic system can be constructed by placing an arbitrary number of normal, triangular membership functions with overlap $\frac{1}{2}$ covering the universe of discourse of each of the N inputs. A rule base is constructed using only AND operations and covering all possible combinations of antecedents. Each of the rules is initialized as follows:

$$\text{IF } x_1 \text{ IS } \mathcal{A}^1_{j_1} \text{ AND } \ldots \text{ AND } x_N \text{ IS } \mathcal{A}^N_{j_N} \text{ THEN } y^j$$

where

$$y^j = t_1 m^1_{j_1} + \ldots + t_N m^N_{j_N}$$

where x_i is the ith entry of the vector X, $\mathcal{A}^i_{j_i}$ is the fuzzy set defined at the ith input used as antecedent of the jth rule. This fuzzy set is described by the membership function:

$$\mu^i_{j_i}(x_i) = \begin{cases} \frac{x_i - m^i_{j_i-1}}{m^i_{j_i} - m^i_{j_i-1}} & \text{if } x_i \in [m^i_{j_i-1}, m^i_{j_i}] \\ \frac{m^i_{j_i+1} - x_i}{m^i_{j_i+1} - m^i_{j_i}} & \text{if } x_i \in [m^i_{j_i}, m^i_{j_i+1}] \\ 0 & \text{otherwise} \end{cases} \tag{D.4}$$

$m^i_{j_i}$ is the modal value of the membership function $\mu^i_{j_i}(.)$ and t_i is the ith entry of the vector T.

With this description of the fuzzy model, the evaluation of an arbitrary vector $X(k) \in [m^1_{j_1}, m^1_{j_1+1}] \times \ldots \times [m^N_{j_N}, m^N_{j_N+1}]$ will activate 2^N rules. To analyze the result of this evaluation assume without lost of generality $N = 2$. The point described by the vector $X(k)$ will be in the interval $[m^1_{j_1}, m^1_{j_1+1}] \times [m^2_{j_2}, m^2_{j_2+1}]$ and the set of active rules will be described as

1. IF x_1 IS $\mathcal{A}^1_{j_1}$ AND x_2 IS $\mathcal{A}^2_{j_2}$ THEN y^j
2. IF x_1 IS $\mathcal{A}^1_{j_1+1}$ AND x_2 IS $\mathcal{A}^2_{j_2}$ THEN y^{j+1}
3. IF x_1 IS $\mathcal{A}^1_{j_1}$ AND x_2 IS $\mathcal{A}^2_{j_2+1}$ THEN y^{j+2}
4. IF x_1 IS $\mathcal{A}^1_{j_1+1}$ AND x_2 IS $\mathcal{A}^2_{j_2+1}$ THEN y^{j+3}

where

$$\begin{aligned} y^j &= t_1 m^1_{j_1} + t_2 m^2_{j_2} \\ y^{j+1} &= t_1 m^1_{j_1+1} + t_2 m^2_{j_2} \\ y^{j+2} &= t_1 m^1_{j_1} + t_2 m^2_{j_2+1} \\ y^{j+3} &= t_1 m^1_{j_1+1} + t_2 m^2_{j_2+1} \end{aligned}$$

Then, the value of fuzzy system evaluated at the point $X(k)$ is given by the expression

$$\begin{aligned} &\mathfrak{F}(X(k)) = \\ &\frac{y^j(\mu^1_{j_1}(x_1)\mu^2_{j_2}(x_2)) + y^{j+1}(\mu^1_{j_1+1}(x_1)\mu^2_{j_2}(x_2))}{\mu^1_{j_1}(x_1)\mu^2_{j_2}(x_2) + \mu^1_{j_1+1}(x_1)\mu^2_{j_2}(x_2) + \mu^1_{j_1}(x_1)\mu^2_{j_2+1}(x_2) + \mu^1_{j_1+1}(x_1)\mu^2_{j_2+1}(x_2)} \\ &+ \frac{y^{j+2}(\mu^1_{j_1}(x_1)\mu^2_{j_2+1}(x_2)) + y^{j+3}(\mu^1_{j_1+1}(x_1)\mu^2_{j_2+1}(x_2))}{\mu^1_{j_1}(x_1)\mu^2_{j_2}(x_2) + \mu^1_{j_1+1}(x_1)\mu^2_{j_2}(x_2) + \mu^1_{j_1}(x_1)\mu^2_{j_2+1}(x_2) + \mu^1_{j_1+1}(x_1)\mu^2_{j_2+1}(x_2)} \\ &= y^j(\mu^1_{j_1}(x_1)\mu^2_{j_2}(x_2)) + y^{j+1}(\mu^1_{j_1+1}(x_1)\mu^2_{j_2}(x_2)) + \\ &\qquad y^{j+2}(\mu^1_{j_1}(x_1)\mu^2_{j_2+1}(x_2)) + y^{j+3}(\mu^1_{j_1+1}(x_1)\mu^2_{j_2+1}(x_2)) \end{aligned} \tag{D.5}$$

at the interval $[m^i_j, m^i_{j+1}]$, $\mu^i_{j+1}(x) = 1 - \mu^i_j(x)$. Then, the expression (D.5) becomes

$$\begin{aligned} \mathfrak{F}(X(k)) &= y^j \mu^1_{j_1}(x_1)\mu^2_{j_2}(x_2) + y^{j+1}(1 - \mu^1_{j_1}(x_1)\mu^2_{j_2}(x_2)) \\ &\quad + y^{j+2}\mu^1_{j_1}(x_1)(1 - \mu^2_{j_2}(x_2)) + y^{j+3}(1 - \mu^1_{j_1}(x_1)(1 - \mu^2_{j_2}(x_2)) \\ &= y^j \mu^1_{j_1}(x_1)\mu^2_{j_2}(x_2) + y^j \mu^1_{j_1}(x_1)\mu^2_{j_2}(x_2) + y^{j+1}\mu^2_{j_2}(x_2) \\ &\quad - y^{j+1}\mu^1_{j_1}(x_1)\mu^2_{j_2}(x_2) + y^{j+2}\mu^1_{j_1}(x_1) - y^{j+2}\mu^1_{j_1}(x_1)\mu^2_{j_2}(x_2) \\ &\quad - y^{j+3}\mu^2_{j_2}(x_2) + y^{j+3}\mu^1_{j_1}(x_1)\mu^2_{j_2}(x_2) - y^{j+3}\mu^1_{j_1}(x_1) + y^{j+3} \end{aligned} \tag{D.6}$$

replacing the values of the consequences in the expression (D.6):

$$\begin{aligned}\mathcal{F}(X(k)) &= t_1 m^1_{j_1}\mu^1_{j_1}(x_1) + t_2 m^2_{j_2}\mu^2_{j_2}(x_2) - t_1 m^1_{j_1+1}\mu^1_{j_1}(x_1) - t_2 m^2_{j_2+1}\mu^2_{j_2}(x_2)\\ &= t_1 m^1_{j_1+1} + t_2 m^2_{j_2+1} \end{aligned} \tag{D.7}$$

Finally, replacing (D.4) in (D.7) the expression becomes

$$\mathcal{F}(X(k)) = t_1 x_1 + t_2 x_2 = T^T X(k) \tag{D.8}$$

Being $\mathcal{F}(X(k)) = T^T X(k)$, then the constructed fuzzy system will be equal to the linear discrete dynamical system of Equation (D.1).

E

Fuzzy Control for a Continuously Variable Transmission

This appendix presents the design of a nonlinear PI-like fuzzy controller. The task of the fuzzy system is to generate the parameters (K_p and T_i) for a proportional-integral (PI) controller integrated in the control system to govern a continuously variable transmission (CVT). Nonlinear compensations were also designed to improve the performance.

E.1 Introduction and Process Description

The continuously variable transmission (CVT) is a type of transmission used in cars with a combustion engine. It is different from the classical gearbox transmission where the rotation ratio between the engine and the wheels changes in discrete steps (first, second, third gear and so on). Instead CVT can give a continuous speed ratio between the motor and the wheels. This specification improves the exploitation of the power given by the motor, such that the system can be commanded to reduce the fuel consumption and pollution (economic mode) or maximize the power transmitted to the wheels (sport mode).

The system is constructed using two conic pulleys connected to each other with a belt. The radius of the pulleys is variable and in fact increases if the two cones of the pulley approximate to each other and reduces if they split apart. The distance between the cones is controlled by means of a hydraulic cylinder connected to a hydraulic circuit, which feeds not only the pulleys but also the clutch system. The schematic representation presented in Figure E.1 explains the working principle.

The pressure applied to the pulleys (P_p, P_s) must be accurately controlled in such a way that the belt works with an optimal tension. If the tension is too high the belt will not move and it can break apart because of the friction and the stress. If the tension is too low the motor does not transmit the movement to the wheels and the belt can also slip off the pulleys, destroying the complete system (the rotation speed of the motor (N_e is about 1000–4000 r.p.m.).

The pressures for the cylinders are generated by a valve body, which behaves as a hydraulic amplifier. There is a third pressure that drives the wet-plate clutch. Therefore, the system has three control inputs. The pressures are controlled using a servo pressure and the servo pressures are regulated by a servo pulse width modulated (PWM) valve driven by electric signals. In this way, the signals generated by the controller of the CVT system are only electrical signals.

The three reference values for the pressures of the three valves are generated by a master control system. This master control system generates the set points of the pressures according to the conditions of the road, desired speeds and throttle positions (see Figure E.2).

The control of CVTs is challenging due to the complexity generated by the large number of variables involved (speed, temperature, oil viscosity, valve construction, etc.) and their interactions. The system is highly nonlinear and difficult to control by means of classical linear control techniques.

This appendix presents the design of the control system that regulates the pressure applied to the pulleys (P_p and P_s in Figure E.1). This control system receives the set-point signals from the master control system and guarantees the exact positioning of the pulleys despite the disturbances.

E.2 Performance Specifications

The main objective of the control system is to follow the set-point pressures generated by the master control system, with a minimum overshoot and with

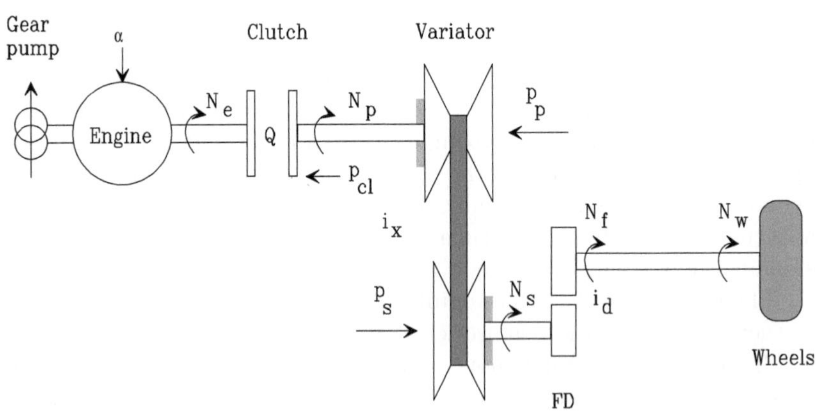

Figure E.1. Power train of a car with a CVT and wet-plate clutch. FD=Final reduction

a settling time of 60 to 70 msec. The condition of minimum overshoot is a consequence of the fact that overshoots can generate a release of the tension in the belt, generating a slip off of the pulleys and damage to the system. Hence, some overshoot is allowed when the pressures are increased (pulleys closing), but it has to be small when the pressures are decreased (pulleys opening).

E.3 A Physical Model for the CVT

A physical model of the system was obtained by Minten and Vanvuchelen [94]. This model is a full physical model oriented to functional simulation. The model is nonlinear and describes the full operating range of the CVT. This model is suitable for simulation of physical properties, but it is too complex for control design because of its long simulation time. Therefore, a simpler model for control was obtained. The system has one manipulated input, the voltage applied to the PWM servo valve V_{in}, and two measured disturbances, temperature T and engine speed N_{engine}. The only output considered in this study was the pressure applied to the pulley.

A simplified representation of the plant, for a given temperature T and on the engine speed N_{engine} is a nonlinearity in series with a second-order linear system. In other words, the system can be represented by a Hammerstein model where the coefficients are functions of the temperature T and the engine speed N_{engine}. Figure E.3 shows that the nonlinearity f depends mainly on the speed of the engine N_{engine} and that the dynamics of the second-order system $G(s)$ depends on the temperature T and on the engine speed N_{engine}.

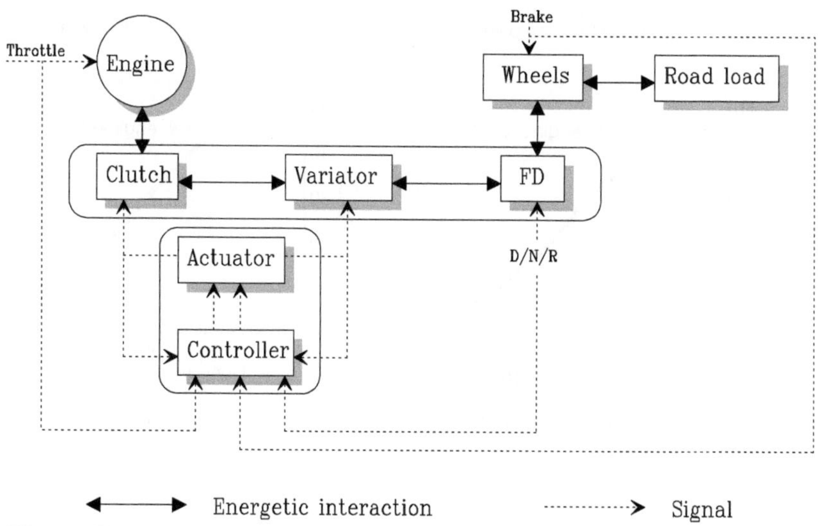

Figure E.2. Diagram of the system including the high-level controller

Figure E.4 shows the clear dependence of f on the engine speed. Using this assumption a set of nonlinearities and linear plant models are extracted for different operating points in temperature and engine speed.

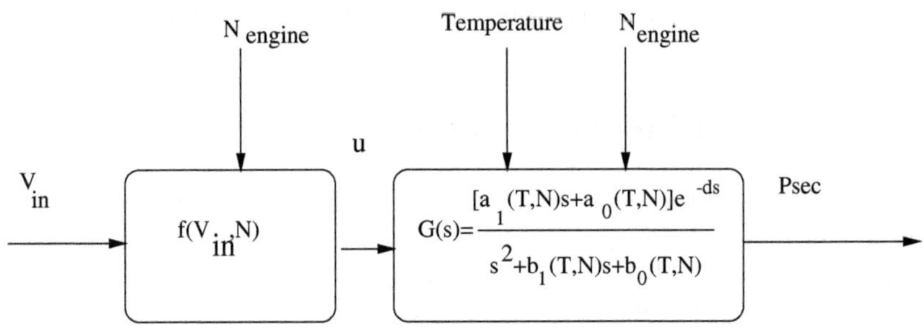

Figure E.3. Representation of the process by means of a static nonlinearity $f(V_{in}, N_{\text{engine}})$ and a dynamic linear system $G(s)$ with variable parameters

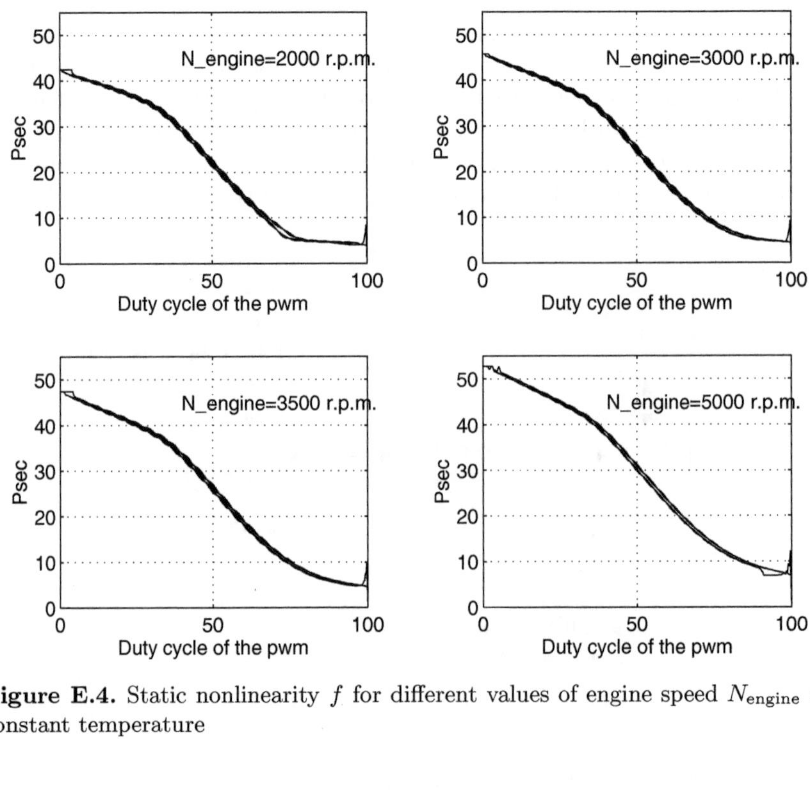

Figure E.4. Static nonlinearity f for different values of engine speed N_{engine} and constant temperature

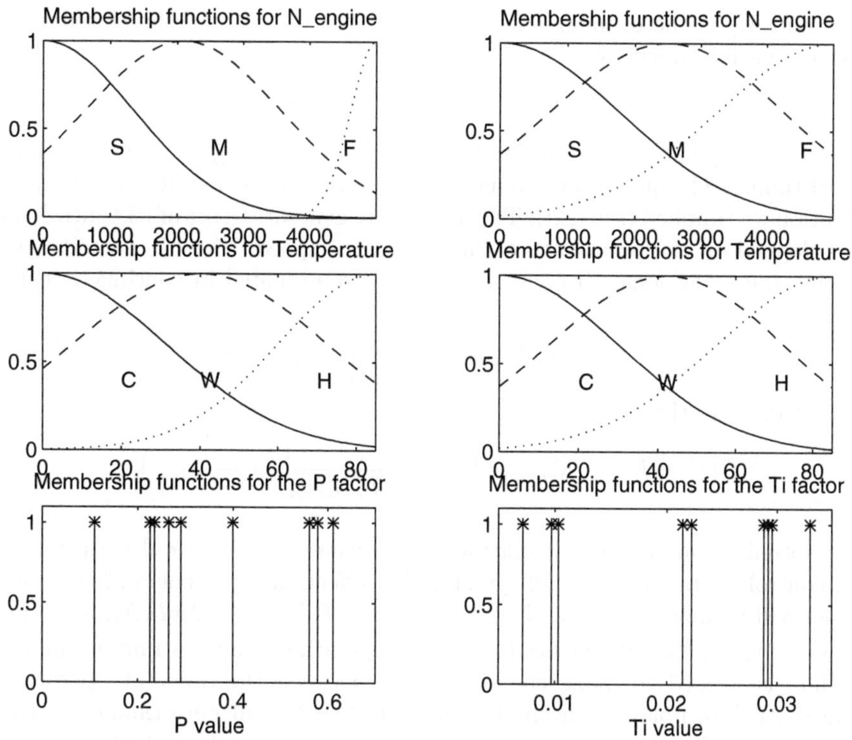

Figure E.5. Membership functions of the fuzzy inference systems that generate the parameters P and T_i for the PI controller.

E.4 Design of the Controller

The first step in the controller design is to compensate the nonlinearity $f(V_{in}, N_{\text{engine}})$ by means of $\hat{f}^{-1}(V_{in}, N_{\text{engine}})$. This function $\hat{f}^{-1}(.,.)$ is generated from measurement data using power series to approximate the function. Observe that this procedure can be applied due to the monotonicity of the function $f(.,.)$. Once the "local" linear models for different values of T and N_{engine} and the $\hat{f}^{-1}(.,.)$ function are obtained, an optimization procedure (see [95]) is applied to calculate a suboptimal proportional-integral (PI) controller for the operating point. The cost function of the optimization is defined as

$$J(K_p, T_i) = \lambda M_p + (1-\lambda) \int_0^\infty t(e^2(t)dt)^{\frac{1}{2}}$$

where K_p and T_i are the proportional gain and the integral time of the PI controller, $\lambda = 0.5$ is a weight that defines the importance of the overshoot $o.s.$ in the cost function, $e(t) = P_{\text{ref}}(t) - P_{\text{sec}}(t)$ is the tracking error, t is the time and M_p is the overshoot for a unitary step response of the closed loop system and is defined as:

$$M_p = \max\{\max(y(t) - 1), 0\}$$

The starting point for the optimization is obtained using the relay method for PI tuning. The quasi-Newton method for unconstrained multivariable optimization was applied (see [22]). This procedure generates a set of PI controllers that depend on the temperature and the engine speed. The PI controller including the set of local PI controllers can be represented in continuous time as

$$C(s) = K_p(T, N_{\text{engine}}) \left(1 + \frac{1}{T_i(T, N_{\text{engine}})s}\right)$$

and for discrete time as

$$C(z) = K_p(T, N_{\text{engine}}) \left(1 + \frac{1}{T_i(T, N_{\text{engine}})(1 - z^{-1})}\right)$$

The optimization of the controller was done using the discrete description of the controller. In this way, all quantization effects are present and the controller will be more accurate. The values of $K_p(T, N_{\text{engine}})$, $T_i(T, N_{\text{engine}})$ and $\hat{f}^{-1}(V_{in}, N_{\text{engine}})$ are defined only for some operating points. An interpolation method is needed to make the transitions between the different operating points. An interpolation method is needed. The overall performance of the controller is related to the number of operating points evaluated; this implies that a large number of parameters should be stored in a lookup table. We found that a good solution will be to approximate this lookup table with a fuzzy inference system (FIS). The main characteristics of this FIS are Gaussian membership functions, product-sum composition and defuzzification using center of gravity. The membership functions were uniformly distributed in the domain of the temperature and the engine speed. The model was tuned using a gradient descent algorithm. A picture of the membership functions can be seen in Figure E.5. The function $\hat{f}^{-1}(.,.)$ is also scheduled depending on the speed. The controller is shown in Figure E.6.

E.5 Stability Analysis

The closed-loop system can be described by means of a Takagi -Sugeno fuzzy model [assuming perfect cancellation of the nonlinearity $f(.,.)$] with rules like

$$\text{Rule}_i \text{ IF } \mathit{Temperature} \text{ is } \alpha_i \text{ AND } N_{engine} \text{ is } \beta_i \text{ THEN } \dot{x} = \tilde{A}_i x + \tilde{B}_i u$$

The condition for stability is the existence of a common P matrix such that

$$P > 0 \tag{E.1}$$

$$\tilde{A}_i^T P + P\tilde{A}_i < 0 \tag{E.2}$$

$$\forall i \quad i \in \{1, \ldots, N_{\text{rules}}\} \tag{E.3}$$

The search for this P matrix can be conducted by means of solving the feasibility LMI problem shown in Chapter 5. A feasible solution was found for this problem so that stability is guaranteed.

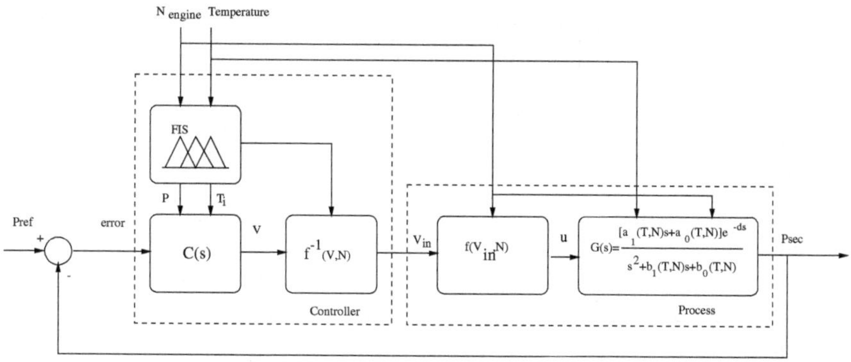

Figure E.6. Block diagram of the process with the nonlinear controller

Figure E.7 shows the improvements obtained with the fuzzy controller compared with the linear scheduled controller. It is important to observe that there is not only improvement in the performance, but also that any modification of the design will be easier to implement, due to the extra information provided by the rule base description. The rules of the controller have the following form:

IF T is A_j AND N_{engine} is B_j THEN P is O_j AND T_i is Q_j
IF N_{engine} is C_j THEN f^{-1} is R_j

E.6 Conclusions

This appendix has shown the design procedures of a control systems for a CVT. The control system is based on a fuzzy inference system that "schedules" the parameters of the controller according to some measured disturbances. In the procedure, several linear controllers were calculated by means of optimization for different operating points. In addition, different nonlinear compensators were obtained for different values of N_{engine}. Stability of the closed-loop system is guaranteed by means of the solution of an LMI. Improvements in the performance were observed when the new control system was compared with the scheduled linear version. Another advantage of the designed controller is the linguistic description of the scheduling action, which simplifies the retuning of the system in the industrial framework.

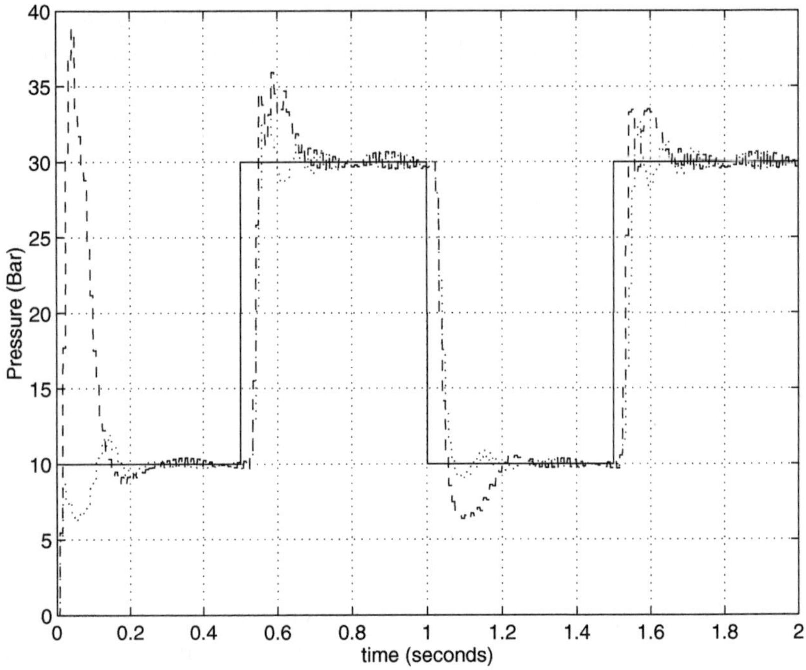

Figure E.7. Comparative response of the control systems. Continuous line (_) reference, dashed line (- -) linear controller with feed-forward action and dotted line (..) fuzzy controller. Observe that the overshoot and the settling time of the fuzzy with the fuzzy controller are significantly reduced.

References

1. G. Box, G. Jenkins: (1970), *Time Series Analysis, Forecasting and Control* (Holden Day, San Francisco, CA)
2. J. Bezdek: (1976), "A physical interpretation of fuzzy isodata" IEEE Trans. Syst., Man, Cybern. pp. 387–389
3. E. Gustafson, W. Kessel: (1979), "Fuzzy clustering with fuzzy covariance matrix," in *Proc. IEEE Control Decision Conference*pp. 761–766
4. R. Yager, D. Filev: (1994), *Essentials of fuzzy modeling and control* (John Wiley & Sons, Hoboken, NJ, U.S.A)
5. L. Zadeh: (1973), "Outline of a new approach to the analysis of complex systems and decision processes" IEEE Trans. Syst., Man, Cybern. Part B **3**(1), 28–44
6. D. Driankov, H. Hellendoorn, M. Reinfrank: (1993), *An Introduction to Fuzzy Control* (Springer-Verlag)
7. T. Takagi, M. Sugeno: (1985), "Fuzzy identification of systems and its application to modeling and control" IEEE Trans. Syst., Man, Cybern. **15**, 116–132
8. L.X. Wang: (1995), "Analysis and design of fuzzy identifiers of nonlinear dynamic systems" IEEE Trans. on Automatic Control **40**, 11–23
9. J.S.R. Jang, C.T. Sun: (1995), "Neuro-fuzzy modeling and control" Proceedings of the IEEE **83**(3), 378–406
10. R. Jager: (1995), "Fuzzy logic in control," Ph.D. thesis, T. U. Delft, The Netherlands
11. M. Sugeno, T. Yasukawa: (1993), "A fuzzy logic based approach to qualitative modeling" IEEE Trans. on Fuzzy Systems **1**, 7–31
12. N. Lori, P. Costa Branco: (1995), "Autonomous mountain-clustering method applied to fuzzy systems modeling," in *Intelligent Engineering Systems Through Artificial Neural Networks, Smart Engineering Systems: Fuzzy Logic and Evolutionary Programming*, ed. by C. Dagli, M. Akay, C. Philip, C. Chen, B. Fernández, J. Ghosh, Vol. 5 (ASME Press, New York, U.S.A.), pp. 311–316
13. J.S. Jang: (1994), "Structure determination in fuzzy modeling: A fuzzy cart approach," in *Proc. of IEEE International Conference on Fuzzy Systems*
14. S. Tan, J. Vandewalle: (1995), "An on-line structural and parametric scheme for fuzzy modelling," in *Proc. of the 6th International Fuzzy Systems Association World Congress, IFSA-95*pp. 189–192

15. W. Pedrycz, J. Valente de Oliveira: (1996), "Optimization of fuzzy models" IEEE Trans. Syst., Man, Cybern. Part B **26**, 627–636
16. W. Pedrycz: (1994), "Why triangular membership functions?" Fuzzy Sets and Systems **64**, 21–30
17. L.X. Wang: (1997), *A Course in Fuzzy Systems and Control* (Prentice Hall, Englewood Cliffs, NJ)
18. L.X. Wang: (1994), *Adaptive Fuzzy Systems and Control* (Prentice Hall, Englewood Cliffs, NJ)
19. J.S. Jang: (1992), "Neuro-fuzzy modeling: Architectures, analyses and applications," Ph.D. thesis, University of Berkeley-California
20. U. Bodenhofer: (1996), "Tuning of fuzzy systems using genetic algorithms," Master's thesis, Institut für Mathematik Johannes Kepler Universität, Linz, Austria
21. C. Peña: (2002), "Coevolutionary fuzzy modeling," Ph.D. thesis, École Polytechnique Fédérale de Lausanne
22. A. Grace: (1992), *Optimization Toolbox - For use with MATLAB*, Mathworks, Inc.
23. D.E. Goldberg: (1989), *Genetic Algorithms in Search, Optimization and Machine Learning* (Addison Wesley, USA)
24. Z. Michalewicz: (1996), *Genetic Algorithms + Data Structures = Evolution Programs* (Springer-Verlag, Berlin)
25. J. Espinosa, J. Vandewalle: (2003), "Extracting linguistic fuzzy models from numerical data-afreli algorithm," in *Interpretability issues in fuzzy modeling*, ed. by J. Casillas, O. Cordon, F. Herrera, L. Magdalena (Springer-Verlag, London)
26. D. Broadbent: (1975), "The magic number seven after fifteen years," Studies in Long Term Memory (Addison-Wesley)
27. J. Espinosa, J. Vandewalle: (2000), "Constructing fuzzy models with linguistic integrity-afreli algorithm" IEEE Trans. on Fuzzy Systems **8**(5), 591–600
28. J. Espinosa, J. Vandewalle: (1998), "Fuzzy modeling and identification, using afreli and fuzion algorithms," in *Proceedings of the 5th. International Conference on Soft Computing IIZUKA-98*Iizuka, Japan, pp. 535–540
29. J.S. Jang: (1998), *Fuzzy Logic Toolbox, User's Guide, V-2.0* (Mathworks, USA)
30. K. Nozaki, H. Ishibuchi, H. Tanaka: (1997), "A simple but powerful heuristic method for generating fuzzy rules from numerical data" Fuzzy Sets and Systems (86), 251–270
31. J. Espinosa, J. Vandewalle: (1998), "Fuzzy modeling with linguistic integrity," in *Proceedings of the International Workshop on Advanced Black Box Techniques for Nonlinear Modeling*Leuven, Belgium, pp. 197–202
32. L. Ljung: (1987), *System Identification Theory for the User* (Prentice Hall, Englewood Cliffs, NJ)
33. Y. Chikkula, J. Lee: (1997), "Input sequence design for parametric identification of nonlinear systems," in *Proceedings of the American Control Conference*Albuquerque, USA, pp. 3037–3041
34. H. Akaike: (1974), "A new look at the statistical model identification" IEEE Trans. on Automatic Control **19**, 716–723
35. X. He, H. Asada: (1993), "A new method for identifying orders of input–output models for nonlinear dynamic systems," in *Proceedings of the American Control Conference*San Francisco, CA, pp. 2520–2523

36. J. SJöberg, Q. Zhang, L. Ljung, A. Benveniste, B. Delyon, P. Glorennec, J. Hjalmarsson, A. Juditsky: (1995), "Nonlinear black-box modeling in system identification: A unified overview" Automatica **31**, 1691–1724
37. S. Chen, S. Billings: (1992), "Neural networks for nonlinear dynamic system modelling and identification" Intl. J. Control **56**, 319–346
38. G. Goodwin, K. Sin: (1984), *Adaptive Filtering Prediction and Control* (Prentice Hall, Englewood Cliffs, NJ)
39. S. Billings, W. Voon: (1986), "Correlation based model validity tests for nonlinear models" Intl. J. Control **44**, 235–244
40. T. Kailath: (1980), *Linear Systems* (Prentice Hall, Englewood Cliffs, NJ)
41. J.J. Østergaard: (1999), "High level process control in the cement industry by fuzzy logic," Fuzzy Logic in Control-Advances and Applications (World Scientific, Singapore)
42. M. Setnes: (1999), "Implementing fuzzy control in the manufacturing of washing powders," Fuzzy Logic in Control-Advances and Applications (World Scientific, Singapore)
43. C. von Altrock: (1995), "Fuzzy logic applications in europe," Industrial Applications of Fuzzy Logic and Intelligent System (IEEE-Press)
44. K. Astrom, T. Hagglund: (1995), *PID Controllers: Theory, Design and Tuning* (ISA)
45. J. Jantzen: (1991), *Fuzzy Control* (Technical University of Denmark)
46. J.S. Jang, C.T. Sun, E. Mizutani: (1997), *Neuro Fuzzy and Soft Computing* (Prentice Hall International, Englewood Cliffs, NJ)
47. J. Abonyi, H. Andersen, L. Nagy, F. Szeifert: "Inverse fuzzy-process-model based direct adaptive control"
URL `citeseer.nj.nec.com/abonyi99inverse.html`
48. J. Suykens: (1995), "Artificial neural networks for modeling and control of nonlinear systems," Ph.D. thesis, Faculty of Engineering, Katholieke Universiteit Leuven, Leuven, Belgium
49. A. Isidori: (1989), *Nonlinear Control Systems–An Introduction*, 2nd edn. (Springer-Verlag)
50. R. Marino, P. Tomei: (1995), *Nonlinear Control Design* (Prentice Hall, Englewood Cliffs, NJ)
51. V. Utkin: (1977), "Variable structure systems: A survey" IEEE Transactions on Automatic Control **22**(2), 212–222
52. R. Palm, D. Driankov, H. Hellendoorn: (1997), *Model Based Fuzzy Control* (Springer-Verlag)
53. H. Ying: (1998), "The takagi–sugeno fuzzy controllers using the simplified linear control rules are nonlinear variable gain controller" Automatica **34**(2)
54. K. Tanaka, M. Sugeno: (1992), "Stability analysis and design of fuzzy control systems" Fuzzy Sets and Systems **45**(2), 135–156
55. Mathworks, Inc.: (1997), *LMI Control Toolbox Version 1.0.4*
56. Ecole Nationale Supérieure de Techniques Avancées (ENSTA),Optimization and Control Group: (1998), *LMITOOL-2.0 package*
57. J. Zhao: (1995), "System modeling, identification and control using fuzzy logic," Ph.D. thesis, Université Catolique de Louvain, Belgium
58. O. Agudelo: (2003), "Control of a helicopter laboratory process using linear and fuzzy techniques," Master's thesis, Corporacion Universitaria de Ibague

59. M. Johansson, A. Rantzer: (1997), "Computation of piecewise quadratic lyapunov functions for hybrid system," in *Proceedings of the European Control Conference*Brussels, Belgium
60. E. Camacho, C. Bordons: (1995), *Model Predictive Control in the Process Industry* (Springer-Verlag, London)
61. D. Clarke, C. Mohtadi, P. Tuffs: (1987), "Generalized predictive control-parts i - ii" Automatica **23**, 137–160
62. J. Richalet: (1993), "Industrial applications of model based predictive control" Automatica **29**, 1251–1274
63. C. Garcia, D. Prett, M. Morari: (1989), "Model predictive control: Theory and practice-a survey" Automatica **25**, 335
64. R. Soeterboek: (1990), "Predictive control, a unified approach," Ph.D. thesis, Technische Universiteit Delft, The Netherlands
65. S. Loureiro de Oliveira: (1996), "Model predictive control for constrained nonlinear systems," Ph.D. thesis, California Institute of Technology, USA
66. L. Magni: (1998), "Nonlinear receding horizon control:theory and application," Ph.D. thesis, Università degli Studi di Pavia, Italy
67. F. Allgöwer, T. Badgwell, J. Qin, J. Rawlings, S. Wright: (1999), "Nonlinear predictive control and moving horizon estimation–an introductory overview," Advances in Control–Highlights of ECC'99 (Springer-Verlag)
68. J. Rawlings: (1999), "Tutorial: Model predictive control technology," in *Proceedings of the ACC*San Diego, CA
69. S. Yasunobu, S. Miyamoto: (1984), "A predictive fuzzy control for automatic train operation" *in Japanese* Systems and Control **28**(10), 605–613
70. J.A. Roubos, R. Babuška, P. Bruijn, H. Verbruggen: (1998), "Predictive control by local linearization of a takagi–sugeno fuzzy model" Proceedings of FUZZ-IEEE-98 pp. 37–42
71. R. Babuška: (1997), "Fuzzy modeling and identification," Ph.D. thesis, Delft University of Technology, The Netherlands
72. J.M. Da Costa Sousa: (1998), "A fuzzy approach to model-based control," Ph.D. thesis, Delft University of Technology, The Netherlands
73. J. Espinosa, M. Hadjili, V. Wertz, J. Vandewalle: (1999), "Predictive control using fuzzy models-comparative study" Proceedings of the European Control Conference-99
74. J. Espinosa, J. Vandewalle: (1998), "Predictive control using fuzzy models applied to a steam generating unit," in *Fuzzy Logic and Intelligent Technologies for Nuclear Science and Industry*, ed. by D. Ruan, H.A. Abderrahim, P. D'hondt, E. Kerre (World Scientific, Singapore)
75. J. Espinosa, J. Vandewalle: (1999), "Constrained predictive control using fuzzy models," in *Proc. of the Eight International Fuzzy Systems Association World Congress (IFSA-99)*Taiwan, pp. 649–654
76. J. Espinosa, J. Vandewalle: (1999), "Predictive control using fuzzy models," in *Advances in Soft Computing Engineering Design and Manufacturing*, ed. by R. Roy, T. Furuhashi, P. Chawdhry (Springer-Verlag, London)
77. M. Henson, D. Seborg: (1990), "Input-output linearization of general nonlinear processes" AIChE Journal **36**(11), 1753–1757
78. G. Lightbody, G. Irwin: (1997), "Nonlinear control structures based on embedded neural system models" IEEE Trans. on Neural Networks **8**, 553–567

79. B. De Moor: (1998), "Daisy: Database for the identification of systems" Department of Electrical Engineering -ESAT- K.U. Leuven, Belgium. http://www.esat.kuleuven.ac.be/sista/daisy/, Used data set: Continuous Stirred Tank Reactor, Section: Process Industry Systems, code: 98-002
80. E. Lawler, E. Wood: (1966), "Branch-and-bound methods: A survery" Journal of Operations Research **14**, 699–719
81. L. Mitten: (1970), "Branch-and-bound methods: General formulation and properties" Journal of Operations Research **18**, 24–34
82. G. Pellegrinetti, J. Bentsman: (1996), "Nonlinear control oriented boiler modeling-a benchmark problem for controller design" IEEE Trans. on Control Sys. Tech. **4**(1), 57–64
83. B. De Moor: (1998), "Daisy: Database for the identification of systems" Department of Electrical Engineering -ESAT- K.U. Leuven, Belgium. http://www.esat.kuleuven.ac.be/sista/daisy/, Used data set: Model of a steam generator at Abbott Power Plant in Champaign IL., Section: Process Industry Systems, code: 98-003
84. A. Gutierrez: (2000), "Fuzzy model-based predictive control for an hpde polymerization reactor," Master's thesis, Katholieke Universiteit Leuven
85. K. Choi, W. Ray: (1992), "The dynamic behavior of fluidized bed reactors for solid catalyzed gas phase olefin polymerization" Chemical Engineering **40**, 2261–2279
86. K. MacAuley: (1991), "Modeling, estimation and control of product properties in a gas phase polyethylene reactor," Ph.D. thesis, Mc Master University
87. W. Van Brempt, T. Backx, P. Van Overschee, B. De Moor, R. Tousain: (2001), "A high performance model predictive controller: application on a polyethylene gas phase reactor" Control Engineering Practice **9**, 829–835
88. M. Lobo, L. Vandenberghe, S. Boyd, H. Lebret: (1998), "Applications of second-order cone programming" Linear Algebra and Applications **284**, 193–228
89. A. Ben-Tal, A. Nemirovski: (1996), "Robust convex programming," Tech. rep., Minerva Optimization Center, Haifa, Israel
90. A. Ben-Tal, A. Nemirovski: (1997), "Robust solutions of uncertain linear programs via convex programming," Tech. rep., Minerva Optimization Center, Haifa, Israel
91. L. El Ghaoui, F. Outstry, H. Lebret: (1999), "Robust solutions to uncertain semidefinite programs" SIAM J. of Optim. **9**(1), 33–52
92. S. Boyd, L. Vandenberghe, M. Grant: (1994), "Efficient convex optimization for engineering design," in *IFAC Symposium on Robust Control Design*Rio de Janeiro, Brazil
93. S. Wright: (1997), "Applying new optimization algorithms to model predictive control" Chemical Process Control-V, CACHE, AIChe, Symposium Series **93**(316), 147–155
94. W. Minten, P. Vanvuchelen: (1991), "Modeling and simulation of an electronically controlled variable transmission," Master's thesis, K. U. LEUVEN, Belgium
95. P. Vanvuchelen, B. De Moor: (1993), "A multi-objective optimization approach for parameter setting in system and control design," in *Extended Abstracts of the IFIP Conference on System Modelling and Optimization*Compiegne, France

Index

α-cut, 62, 217

adaptive methods, 129, 131, 150
AFRELI algorithm, 60, 63, 89, 94
algorithm, 33, 38, 39, 50, 51
  AFRELI, 60, 63, 89, 94
  constrained nonlinear predictive control, 168, 173, 177, 178
  FuZion, 62, 65, 68, 86
  fuzzy c-means, 225
  genetic, 46
  Gustafson and Kessel, 226
  mountain clustering, 228
  predictive control, 152
  robust nonlinear predictive control, 202
algorithms
  genetic, 98, 100, 119

chaotic time series, 71, 80, 104
chemical reactor, 158, 160, 162
clustering, 21, 38–41, 60, 63, 65, 89, 94, 106, 210, 225
  fuzzy c-means, 39, 40, 63, 71, 77, 80, 85, 225
  GK, 226
  Gustafson and Kessel, 40, 226
  mountain, 39, 63, 64, 80, 85, 110, 228
consequences, 10, 15, 19, 21, 22, 25, 28, 29, 35, 39, 40, 60, 63, 68, 72, 76, 92, 94, 106, 107, 110, 111, 119, 134, 136, 177, 203, 223, 231, 234, 245
  estimation, 50, 51, 72
  initialization, 50, 106
  singleton-, 5, 66, 77, 231
constructing units, 8
core, 216
CSTR
  continuous stirred tank reactor, 158, 160, 162
CVT
  continuously variable transmission, 247, 249, 253

data mining, 4, 59
defuzzification, 3, 61, 93, 223, 252
direct adaptive methods, 131

error
  approximation, 6
estimation
  consequences, 50, 51, 72
evolutionary strategies, 21, 22, 46
experiment design, 91, 95, 119, 208

feedback linearization, 134, 139–141
FIS
  fuzzy inference system, 3, 5, 252
free response, 170, 172, 174–180, 200, 202
function
  approximation, 4
FuZion algorithm, 62, 65, 68, 86
fuzzification, 3, 61, 62, 94, 221, 222
fuzzy c-means, 225
fuzzy inference system, 3, 5, 6, 8, 20, 22, 29, 38, 50, 58, 61, 131, 136, 141, 221, 252, 253

fuzzy modeling, 3, 4, 59, 102, 191, 225
fuzzy models
  Mamdani, 3, 23, 38, 106, 173
fuzzy models
  Takagi–Sugeno, 18, 23–25, 35, 39, 40, 93, 119, 142, 178, 186, 202, 206

gas furnace, 109
genetic algorithms, 46, 98, 100, 119
GK, 40, 226
GMDH, 77, 102
GPC
  generalized predictive control, 157, 161, 162
gradient
  descent, 4, 21, 22, 26, 29, 35, 38–40, 60, 67, 90, 94, 106, 119, 133, 208, 252
  updating, 30–34, 231–235, 237, 239

HDPE, 187
  high-density polyethylene, 71, 85
height, 62, 217, 220, 223
helicopter, 145
heuristic search, 98, 109, 119, 160

least squares, 25, 28, 29, 35, 39, 50–52, 96, 101
  recursive, 25, 28, 39, 50, 52
linear matrix inequalities, 144, 145, 150, 253
linguistic integrity, 4, 21, 59, 63, 89
LMI
  linear matrix inequalities, 144, 145, 150, 253
LQR, 145
Lyapunov function, 137, 139, 150, 211

Mamdani, 3, 23, 25
  fuzzy models, 3, 23, 35, 38, 106, 173
  fuzzy systems, 23
membership function, 4–6, 8–10, 18, 20, 21, 23, 25, 26, 29, 35, 38–40, 46, 47, 51, 52, 61, 62, 65, 67, 71, 77, 80, 85, 89, 92–96, 101, 106, 110, 126, 128, 134, 142, 160, 177, 208, 215, 221, 225, 231, 234, 243, 244
  Gaussian, 14, 34, 233, 237
  polynomial, 12, 32, 33, 233, 235, 239
  trapezoidal, 5, 30, 65, 232, 234
  triangular, 5, 9, 31, 62, 64, 65, 68, 71, 237, 243, 252
mosaic, 22
mountain clustering, 39, 63, 64, 80, 85, 110, 228
MPC, 151, 186, 187, 201
  algorithm, 152

NLP
  nonlinear program, 199
NLQP
  nonlinear quadratic programming, 173
nonlinear predictive control, 154, 167, 168

PID, 55, 124, 128, 146, 160, 162, 183, 208, 211
power plant, XIII, 181, 183
predictive control, 151–153, 158, 167, 168, 177, 178, 181, 186, 196, 198, 201, 211
projection of fuzzy relations, 220
projection of fuzzy sets, 38–41, 60, 63, 64, 71, 74, 77, 80, 85, 89, 94, 106, 110, 210

QP
  quadratic programming, 171, 173, 175–178, 180, 181, 196, 199–201

recursive least squares, 25, 28, 39, 50, 52
Regularity criterion RC, 110
regularity criterion RC, 102
Regularity Criterion, RC, 109, 118, 160
RNLP
  robust nonlinear program, 199
robust nonlinear predictive control, 195
RQP
  robust quadratic programming, 196, 201

search methods, 98
semantic integrity, 60–62, 68
singleton
  consequences, 5, 66, 77, 231
sliding mode fuzzy control, 139, 140, 150

stability, XIII, 105, 118, 124, 136, 138, 139, 144, 145, 150, 153, 195, 206, 210, 211
steam generator, XIII, 181, 183
step response, 124, 152, 156, 157, 173, 174, 176, 186, 187, 193, 252
support, 62, 216
system identification, 91, 92, 95, 119, 195, 208

table lookup, 22
Takagi–Sugeno fuzzy models, 18, 23–25, 35, 39, 40, 93, 119, 142, 178, 186, 202, 206
trial-and-error, 129
trial-and-error design, 124

universal approximator, 8, 20

validation, 77, 80, 85, 89, 91, 102, 103, 108, 109, 118, 119, 160, 185, 210

MIX
Papier aus verantwortungsvollen Quellen
Paper from responsible sources
FSC® C105338

If you have any concerns about our products,
you can contact us on
ProductSafety@springernature.com

In case Publisher is established outside the EU,
the EU authorized representative is:
Springer Nature Customer Service Center GmbH
Europaplatz 3, 69115 Heidelberg, Germany

Printed by Libri Plureos GmbH
in Hamburg, Germany